# HOW TO GET PAID
# for Construction Changes

**Preparation Resolution Tools and Techniques**

**Steven S. Pinnell**

**McGraw-Hill**
New York   San Francisco   Washington, D.C.   Auckland   Bogotá
Caracas   Lisbon   London   Madrid   Mexico City   Milan
Montreal   New Delhi   San Juan   Singapore
Sydney   Tokyo   Toronto

**Library of Congress Cataloging-in-Publication Data**

Pinnell, Steven S.
   How to get paid for construction changes : preparation, resolution
tools, and techniques / by Steven S. Pinnell.
     p.  cm.
   Includes bibliographical references.
   ISBN 0-07-050229-3
   1. Construction contracts—United States.   2. Dispute resolution
(Law)—United States
KF902.Z9P56  1998
343.73' 078624—dc21                      98-10371
                                            CIP

## *McGraw-Hill*

*A Division of The McGraw-Hill Companies*

  2 3 4 5 6 7 8 9 0  DOC/DOC    0 3 2 1

ISBN 0-07-050229-3

*The sponsoring editor of this book was Larry Hager. The editing supervisor was
Scott Amerman, and the production supervisor was Pamela Pelton. This book
was set in New Century Schoolbook. It was composed in Hightstown, N.J.*

*Printed and bound by R.R. Donnelley & Sons Company.*

McGraw-Hill books are available at special quantity discounts to use as
premiums and sales promotions, or for use in corporate training programs.
For more information, please write to the Director of Special Sales,
McGraw-Hill, Professional Publishing, Two Penn Plaza, New York, NY
10121-2298. Or contact your local bookstore.

 This book is printed on recycled, acid-free paper containing
a minimum of 50% recycled de-inked fiber.

*To my brother, Tom.*

# Contents

# Preface

After working 25 years in construction and engineering, the need for a rational, nonadversarial, and cost-effective approach to resolving construction disputes was obvious. Too often I saw disputes between contractors and owners escalate into litigation, or contractors absorb a major loss to avoid drawn-out disputes and damaged business relations. Too often I found records inadequate for verifying the facts and quantifying costs. And too often I saw contractors and their attorneys being forced to rely on claim experts, who present plausible and reasonable testimony yet cannot adequately substantiate their professional opinions. The financial and emotional cost is excessive and creates an atmosphere of ill will and discord within the industry.

To address these issues, I wrote this book. *HOW TO GET PAID* provides the knowledge and skills to:

- Present an alternative, nonadversarial approach to getting paid for changes.

- Help contractor personnel prepare better change order requests and claims and then to negotiate fair and timely settlements.

- Provide the latest information on critical path scheduling and a number of innovative techniques and tools.

- Assemble cost and inefficiency data from a wide range of sources and present it with a rational approach to determining impact.

*HOW TO GET PAID* represents a crucially needed manual for anyone involved in construction disputes. It will aid in taking a professional approach to claim preparation and help resolve many of the contentious issues in claims resolution.

*Steven Pinnell*
*Portland, Oregon*

# Acknowledgments

My appreciation and gratitude to my good friend, Diane Fiest, who spent long hours editing chapters and gave me continued encouragement and guidance throughout the evolution of this book; to my editor, Catherine Glass, who provided one edit of this book and professional advice; and to Christine Jones, my fiancée, for her support and assistance.

I would also like to thank those who reviewed the manuscript and provided peer review. I am very grateful for their professional critique and counsel: Darrel Addington, Marsha Bailey, John Baker, John Bakkensen, Scott Bellows, William Bennett, Joe Boyd, John Bradach, Tom Brascher, Rob Braziel, Dave Brenneman, Roger Brown, Jeff Busch, David Buono, Mike Byer, Steve Campbell, Frederick Cann, William Cloran, James Confer, Gary Cummins, James Donaldson, David Douthwaite, Zoran Dimkic, Keith Dunlap, Ronald Eakin, Bill East, Perry Fowler, Richard Franzke, Walt Gamble, Walt Cauble, Ben Gerwick, James Grady, Ray Hansen, Nash Hasan, Roger Hennagin, Mel Hensey, Ralph Hochendoner, Paul Holma, Kenji Hoshino, Greg Howell, Roger Huntsinger, Don Irwin, Everett Jack, Roger Korvola, Sol Kutner, Theodore Kyle, Walt Lemon, Kevin Lybeck, Dennis Mandell, Kevin Marin, Gregg Martino, James Mason, Paul Mautner, Adele McKillop, Bill McManus, Ronald Messerly, Dragan Milosevic, Bill Moe, Roland Morris, Jim Nagle, Pete Oakander, Shane Osowski, Kathleen Parrish, Bill Peckham, Reg Perry, Frank Pfiffner, Kent Pothast, Dan Quartier, Rex Radford, James Rasmussen, Mark Rempel, Frank Riggs, George Ritz, Charles Schrader, Charlotte Schupert, Nicholas Scovill, Laurence Smith, Doug Stetler, Bill Striegal, Pamela Tittes, Douglas Van Dyk, Bruce Van Hein, Scott Warner, Ann White, Tom Woodworth, and the other friends and industry associates who provided insight and feedback regarding the content of this book.

I would also like to express my appreciation to Kim Wolthausen and the rest of the staff of Pinnell/Busch and to McGraw-Hill's fine staff.

# List of Abbreviations

| | |
|---|---|
| AAA | American Arbitration Association |
| ABC | Associated Builders and Contractors, a nonunion contractor's association |
| ACEC | American Consulting Engineers Council |
| AGC | Associated General Contractors |
| AIA | American Institute of Architects |
| APWA | American Public Works Association |
| ASCE | American Society of Civil Engineers |
| BOMA | Building Owners and Managers Association |
| CFMA | Construction Financial Managers Association |
| CM | Construction Management, a different contractual form of project delivery |
| CM, Agency | where the CM is an agent of the owner and doesn't guarantee the final price |
| CM/GC or CM/GMP | CM by a general contractor, with a guaranteed maximum price |
| COP | change order proposal, a different name for a change order request |
| COR | change order request, which is submitted before a claim is filed (*see* Chap. 5) |
| CSI | Construction Specifications Institute |
| ENR | *Engineering News Record* magazine, a construction trade journal |
| FAR | Federal Acquisition Regulations that govern U.S. Department of Defense (DoD), General Services Administration (GSA), and National Aeronautics and Space Administration (NASA) contracts |
| MCA | Mechanical Contractors Association |
| NECA | National Electrical Contractors Association |
| NSPE | National Society of Professional Engineers |
| PMI | Project Management Institute |
| RFI | Request for Information—submitted by the contractor for clarification or additional information from the owner or designer |
| TQM | total quality management |

# Introduction and Reader's Guide

## A. Introduction

*HOW TO GET PAID* is the only claims book you need. It describes the philosophy and procedures needed to get paid for extra work, delays, and impact—fairly, promptly, and without damaging business relationships. It is a primer for the novice, a workbook for the experienced contractor, and a reference with checklists and new techniques for the expert.

*HOW TO GET PAID* is written for contractors. Attorneys and other construction professionals will also find it to be invaluable. It describes how to prepare a change order request or claim, and then successfully negotiate a settlement with the owner. If you are a subcontractor or supplier it describes how to prepare a pass-through claim against an owner or a claim against the general contractor, and it describes how a general contractor should document backcharges. This is the first claims book to integrate a comprehensive, clearly defined analytical process with the "right brain" elements of human interaction and emotion. It is also the first claims book to embrace partnering—a formal process for building teamwork by the contractor, owner, designer, subcontractors, and others. Partnering improves attitudes, reduces conflict, and facilitates settlement. As explained in Chap. 2, it starts with securing a commitment from all parties, brings everyone together in a preconstruction workshop, defines a mission statement with common goals, improves communication, establishes conflict resolution procedures, and continues through to celebration of a successful project.

*HOW TO GET PAID* is not only compatible with partnering, it provides a strong foundation upon which to build a successful partnering relationship—while ensuring payment for extra work and impacts. It combines the latest concepts and technology with hands-on practical experience, real-life anecdotal examples, and sample forms that can be copied and modified for your specific needs. It is based on 20-plus years of experience in the U.S. design and construction industry and extensive research, supplemented by contributions from friends and associates throughout the country. It is intended for worldwide use anywhere a product or service is produced and delivered by an independent contractor.

Using *HOW TO GET PAID* will reduce the need for so-called expert opinion from highly paid consultants. They often have little choice but to make poorly supported assertions with little supporting documentation, and their opinions will be rebutted by the other party's expert, who is equally experienced, well qualified, and plausible. An experienced construction practitioner can use the techniques described in this book to develop a clear, convincing change order request or claim that leads to a fair and timely resolution.

Chapter 1 briefly describes the book's objectives and benefits, the contents of each succeeding chapter, and new and innovative techniques that will help you resolve disputes.

## B. Who Will Benefit from This Book and How

*HOW TO GET PAID* is written for general contractors, subcontractors, and suppliers. Where the book refers to the owner/general contractor, subcontractors can substitute general contractor/subcontractor.

### 1. *How To Get Paid* is for you if you are...

- The *claims preparer*—a consultant or contractor's employee preparing a change order request or claim. You may be a project manager, engineer, superintendent, estimator, or company executive.
- The *supervisor* of the claims preparer.
- A *construction executive* with ultimate responsibility for resolving disputes.
- An *attorney* responsible for settling, arbitrating, or litigating claims or advising a client on disputes.
- *Support staff* to the claims preparer—especially if hoping to assume greater responsibility.
- A *student or recent graduate* in construction or enrolled in a construction management program, who wishes to learn valuable skills.

### 2. Benefits to you

*HOW TO GET PAID* will benefit you whether you are a novice, expert, or experienced construction practitioner. Novices will learn what it took others decades to learn, the hard way. Recent engineering or construction management graduates will be able to prepare and negotiate change order requests and claims. They will be able to support a more experienced individual preparing a large claim, and be less likely to let a dispute get out of control, to fail to document change orders, or to forget timely notice of a change.

Construction practitioners with years of experience as a project engineer, project manager, operations manager, or superintendent, and having negotiated change orders before—will be able to prepare better change order requests and claims, resulting in quicker, better settlements with less effort.

Construction executives reading this book will be able to prepare change order requests and claims as well as most experts. They will be able to establish an effective companywide change order management program, train and manage in-house claims preparers, or select and manage a claims consultant.

Experts will benefit from *HOW TO GET PAID*'s comprehensive procedures, which serve as a checklist to ensure better results, and will find some new, innovative techniques for claims preparation. Experts can use *HOW TO GET PAID* as a reference book to obtain answers to specific problems.

Construction attorneys will increase their knowledge of the claims analysis techniques used by their experts, and can better select and manage claims consultants. They will be able to prepare claims themselves, with support from clients and paralegals. The chapter on contract law will help educate their clients in how to avoid claims. Attorneys in general practice will not need to refer construction disputes to a specialist but will be able to manage claims preparation for most cases with the information provided. The contract law chapter will provide a quick review of construction law issues with references to current sources with more detailed information.

After completing *HOW TO GET PAID,* you will know the best methods for resolving construction disputes and for preparing change order requests and claims, and you will have an invaluable reference for dealing with specific situations.

## 3. Types of projects and disputes covered

*HOW TO GET PAID* is based on years of experience in a wide variety of construction, including high-rise office buildings, high-tech and basic industry, utility construction, hydroelectric and coal-fired power plants, power transmission and substation construction, home building, steel fabrication and erection, mechanical construction, power distribution and control, painting, arctic and tropical construction environments, etc. The techniques apply to large, complex disputes and to small or midsize projects. They can also be applied to manufacturing contracts and can be adjusted for applicability to contracts worldwide.

These concepts and tools are applicable to a wide variety of contract types and project owners:

- Fixed-price, competitively bid public works contracts with cities, states, and the federal government.
- Negotiated contracts with private owners, including developers and industrial facility managers.
- Subcontracts and supply contracts, as well as prime contracts.
- Design/build, construction management (CM), and turnkey construction contracts.

*HOW TO GET PAID* describes how to prepare change order requests that are expected to be readily accepted or claims that undergo intense scrutiny and extended negotiations. *HOW TO GET PAID* explains how to negotiate a settlement and then enforce that settlement. It also tells how to prepare claims for mediation, arbitration, or litigation that will convince the fact-finder to award in your favor.

Although this book describes a dispute as a claim, you should focus on change order requests (requests for equitable adjustment). *A claim is made only after a change order request has been rejected.* Claims can be usually avoided by using the recommended techniques. A claim should be characterized as a change order request, unless you are in arbitration or litigation. The dispute resolution process and the sequence of submitting a change order request and preparing a claim are described in Fig. 1.1. Chapter references tell you where to find a detailed description of each step of the process.

## C.  Goals and Objectives

My primary goal is to help you prepare change order requests and claims that result in fair and timely payments. To do this, you must understand your specific needs and establish your own objectives. These objectives could include:

- Maximizing a *partnering approach* to claims management versus adversarial or "claims-conscious" approaches, and developing a philosophy and approach to best fit your personality and work environment.

- Understanding contracts and *contract law* well enough to apply the principles to typical situations encountered on the jobsite, while realizing when to seek expert advice.

- Understanding and using the best techniques for the specific circumstances *to prepare and successfully resolve a change order request or claim.*

- Being able to find information on a specific claim preparation or dispute resolution technique, understand it, and then apply that technique to *resolve current problems.*

- Preparing and settling change order requests and claims without bitterness, unnecessary conflict, or stress by using a *cooperative, positive approach.*

## D.  Unique Features of This Book

*HOW TO GET PAID* has a number of unique and relatively new techniques to help you resolve disputes.

### 1.  Getting paid without damaging business relationships

Some contractors are reluctant to submit change order requests or to file claims for fear of damaging ongoing business relationships. *HOW TO GET PAID*'s

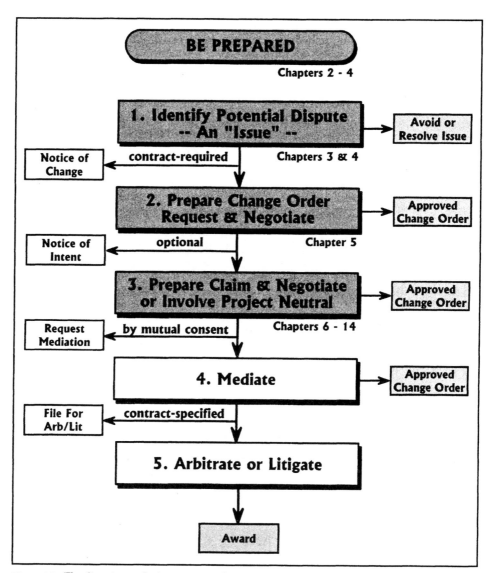

**Figure 1.1**  The dispute resolution process—An overview.

partnering approach enables you to insist on fair payment while preserving a working relationship.

## 2.  Compatibility with partnering

One of the most important, and unique, features of *HOW TO GET PAID* is that it supports partnering. Much of the literature on claims focuses on adversarial relationships and "claims-consciousness," which run counter to *the most significant trend sweeping through the industry—partnering. HOW TO GET PAID* encourages use of a partnering approach to change order requests and claims preparation, in order to achieve more reasonable settlements and enhance your partnering efforts and business relationships.

Prior approaches to claims focused on winning through confrontation, power positioning, and exploiting weaknesses in the contract documents or mistakes by the other party. Attitudes are changing and new, better methods of conflict avoidance and resolution are now available. U.S. culture and management styles have moved from command and control toward team building, consensus decision making, leadership, and win/win negotiation. Our work environment has changed. The old "sue the bastards" approach to conflicts and the resulting wasteful litigation have been replaced by new methods of negotiating and resolving disputes. These methods include avoiding disputes, resolving them, and going beyond disputes to collaborative problem solving and opportunity enhancement.

However, even with a partnering approach, you must be prepared for an adversarial response. Protect your contract rights, and be ready for arbitration or litigation, if necessary. Some owners will not negotiate a reasonable solution no matter how much you partner, because they have insufficient funds, are afraid of "political heat," or are just unwilling to pay more.

### 3. Innovative, new, and improved techniques and tools

*HOW TO GET PAID* describes techniques that most readers will find are new. You also will find improved versions of established techniques. These vary from commonsense tasks to sophisticated, multiphase programs as described in Fig 1.2.

### 4. Integration of people skills with analytical procedures

U.S. society and business have undergone a major transformation in attitudes due to sociological and political changes. There is an increased emphasis on the rights and capabilities of the individual empowerment of employees, consensus building, team building, and other developments. These have changed business practices.

Another source of change is international business competition, which has inspired some U.S. companies to adopt Total Quality Management (TQM) and other people-focused management concepts. These continuing changes have created a new environment for dispute resolution, and must be integrated with an effective Dispute Management Program.

Understanding behavioral styles, emotion, empathy and listening skills, instinct, and other right-brain aspects of human behavior enables us to work more productively with others. This book integrates that knowledge with the application of analytical processes for preparing and resolving claims.

### E.  Summary of Book Contents

*HOW TO GET PAID* is organized into 14 chapters, describing six phases and a step-by-step approach to preparing and resolving change order requests and claims.

| Common-sense Ideas and Improved Techniques | | |
|---|---|---|
| An Organized Methodology | Simple, powerful procedures and techniques for an experienced construction manager to prepare claims as an expert. | Chapters 4 - 12 |
| Change Order Management Program | Identify and resolve problems without impacting the work, ensure timely notice, and document the facts. | Chapter 4 |
| Loss Prevention Plan | Identify design problems and mitigate the risks of onerous, risk-shifting contract clauses. | Chapter 4 |
| Staff Training | Train an "in-house claims expert" and project staff in the fundamentals of contract rights and responsibilities. | Chapter 4 |
| Chronological Document Review | Understand the project not only as it was built by the project team, but from the viewpoint of all parties. | Chapter 9 |
| Highlight Key Text | Facilitate review and understanding of critical text in the source documents. | Chapter 9 |
| Summary Note Chronology | Summarize the facts affecting the disputed issues, to maximize analysis and document retrieval. | Chapter 9 |
| Narrative Text | Collect, organize, and analyze the facts; document your analyses; and communicate your findings and work plans. | Chapter 9 |
| Measured Mile Technique | Determine labor inefficiency from the cost records when needed information is lacking. | Chapter 12 |

| Innovative New Techniques | | |
|---|---|---|
| Timescale Arrow Diagrams | Graphically display a network diagram that is easier to understand than other methods. | Chapter 10 |
| Detailed As-Built Schedules | Merge facts from job documents onto one, timescaled drawing revealing vital information not otherwise discernible. | Chapter 10 |
| ELIPSE Schedules | Merge events, labor hours, impacts, productivity, scheduled and actual progress, and environmental conditions. | Chapter 10 |
| Banded Comparison Schedules | Graphically convey the magnitude of differences between the as-planned and the as-built schedules. | Chapter 10 |
| Rational Analysis of Inefficiency | A blend of expert opinion, modified total cost, industry studies and formulas, measured mile, and scientific methods. | Chapter 12 |

**Figure 1.2**  Innovative, new, and improved techniques and tools.

*Chapter 1* describes who should read this book, how it will help resolve construction change order requests and claims, its unique features, and how to use the book effectively.

*Chapter 2* describes the history and trends in construction dispute resolution. It explains partnering and the essential elements of a successful change order request or claim.

*Chapter 3* explains contract law and contracts. It also explains those contract clauses affecting change order requests and claims, and describes how to minimize their impact with a Loss Prevention Plan.

*Chapter 4* recommends a Change Order Management Program with effective recordkeeping, notice, and change order requests. It includes employee training and use of an in-house claims expert.

*Chapter 5* details preferred methods of preparing and negotiating change order requests.

*Chapter 6* describes how to select and manage a claims consultant or an in-house claims expert.

*Chapter 7 is a synopsis of the whole book.* It describes a preliminary analysis process for the phased analysis of claims that is more efficient and effective than rushing headlong into a claim.

*Chapter 8* describes the documents needed to prepare a claim and how to organize them. It also describes how to interview witnesses and how to make a site visit effective.

*Chapter 9 is the heart of the book.* It details a logical process of reviewing the documents chronologically, as the project unfolded, from the viewpoints of multiple participants. This enables you to determine the facts, to detect basic patterns hidden in the details, to understand the fundamental forces driving the dispute, and to retrieve key documents supporting your position. Chapter 9 describes how to analyze entitlement and present your findings in a clearly written document to convince the claim reviewer. It describes the contents and organization of a claim, a number of easy-to-use but powerful techniques to determine the facts and analyze entitlement, and how to deal with common problems. Chapter 9 also briefly discusses defending against owner counterclaims.

*Chapter 10* explains scheduling claims. It includes several unique tools, including a powerful scheduling technique, and tells you how to create a detailed as-built schedule from multiple source documents.

*Chapter 11* explains how to compute equitable adjustments. It describes the fundamentals of cost analysis and how to avoid common mistakes in estimating direct cost, impact, overhead, and profit.

*Chapter 12* explains how to compute inefficiency using a "Rational Analysis" approach. This combines the measured mile approach, learning curve inefficiencies, industry-accepted formulas and rules of thumb, modified total cost, expert opinion, and a work improvement approach based on time and motion studies.

*Chapter 13* covers how to assemble the claim document and prepare exhibits. It provides example exhibits, and explains communication principles that ensure the claim reviewer's comprehension and retention.

*Chapter 14* describes preparing for and negotiating a final settlement using win/win techniques.

## F.   How to Use This Book

This is a comprehensive reference book. How you use it will depend upon who you are and upon your experience and needs.

Many readers will review the book from start to finish. If you aren't currently preparing a claim, note the recommendations on how to avoid problems and how to be prepared for claims, and implement them on your current project. When involved in a new claim, reread the pertinent chapters, and refer to the appendix for example analyses and computations. If you want immediate answers to pressing problems, refer to Figs. 1.2 and 5.7 which describe the process of resolving disputes and where to find the information you need.

Every chapter starts with a list of section headings and has a "How To" flowchart that summarizes the procedures described in the chapter. The procedural steps addressed are shaded, with chapter references provided for the other steps.

If you encounter an unfamiliar word or phrase, refer to the index for a reference to where it may be used in context or with a description.

Contact Pinnell/Busch, Inc., in Portland, OR, or check the web page at www.pinnellbusch.com to obtain updated information or to comment on the book.

# Background and Overview of Change Orders, Claims, and Dispute Resolution

This chapter describes industry trends, partnering, the Dispute Management Program, and the key requirements for a successful change order request or claim. It includes the following sections:

A. Construction industry trends affecting disputes and claims

B. The Dispute Management Program—what it is and what it can do

C. A partnering approach to change order requests and claims

D. Key requirements for successful change order requests and claims

## A. Construction Industry Trends Affecting Disputes and Claims

In the past few decades, some industry trends have increased the number and severity of construction claims, and the difficulty of resolving them. Fortunately, new attitudes and management techniques are now reducing the number of claims and facilitating their resolution. The current focus is on partnering, dispute resolution, and risk management rather than on risk avoidance, claims-consciousness, and aggressive litigation. However, many in the industry still practice the old methods, so be prepared for an adversarial relationship, while maintaining a partnering approach. If you understand these trends, you will be better able to work with the participants to resolve disputes.

### 1. More competition, reduced margins, and tighter schedules— the 1970s and 1980s

The 1970s and 1980s were a time of change with more competition, reduced margins, and tighter schedules. Average pretax profits on a well-designed, well-managed project dropped from a range of 7 to 10 percent in 1970 to 2 to 4 percent in 1984.[1] From 1991 to 1995, net income as a percentage of revenues averaged from 1.1 to 1.5 percent.[2] Contractors faced stiffer competition and reduced profit margins to obtain work. Concurrently, projects grew more complex and schedules tightened as owners became aware of the cost savings from faster completion. Contractors no longer priced "contingencies" in their bids and could not absorb the costs of delays, extra work, and impact. They had to get paid, or lose money.

Many contractors shifted to new markets. Consequently, some estimators, managers, and supervisors were not versed in the new type of work or local conditions, and mistakes were more frequent. Geographical expansion required long-distance management, which made subcontractor selection and the hiring of reliable workers and supervisors more difficult.

## 2.  Contractor claims-consciousness and owner risk avoidance

The result of reduced margins and shifting markets was an increase in claims. A few contractors, notably on public works contracts, adopted a claims-conscious attitude, and claims became a profit center. These contractors submitted excessively low bids or unbalanced bids, taking advantage of design errors and making up the difference on change orders. Although most contractors were reasonable in requesting extra payment and time, disputes increased because owners and designers didn't understand the cost impact of their decisions or didn't recognize the contractor's contract rights. More change order proposals were submitted, but many were denied; more claims were filed, and many ended up in court.

Partly in reaction to the increase in claims (and partly a cause of their increase), owners began to transfer more risk to contractors. Onerous contract clauses became more frequent, which led to more claims and litigation. The resistance of some owner representatives' to contractors' legitimate requests for compensation further aggravated the situation.

The relationship between contractors and subcontractors deteriorated. Some general contractors incorporated onerous clauses in their subcontracts, bid-shopped for lower prices after contract award, delayed payment for the slightest excuse, and indiscriminately levied backcharges. Some subcontractors bid more work than they could handle, failed to adequately staff their jobs, delayed projects or installed defective work, and filed unwarranted claims and liens.

Business litigation in the United States increased to the point that the courts became overcrowded, and construction dispute settlements were delayed. The costs of litigation became enormous as legal tactics became more adversarial. In response, arbitration became a widely accepted, cheaper, and faster alternative. Construction arbitration cases submitted to the American Arbitration Association (AAA) increased from 1300 in 1973 to 3150 in 1984, a 140 percent increase.[3] Cases filed with the AAA increased 100 percent to 5440 from 1983 through 1990, but then declined for the next 3 years as alternative methods of resolution were employed.[4]

Contractors became more conscious of their contract rights and responsibilities, and implemented better management practices. They realized the value of critical path scheduling and better job cost accounting. At the same time, the makeup of the industry changed. The original owners, many of whom came up through the trades, began to retire. Their children, graduates of engineering and business schools, took over. Graduates from university construction management programs filled the ranks of superintendents, project engineers, and project managers and eventually started their own companies. The Associated General Contractors (AGC), the Associated Builders & Constructors (ABC), and other trade associations implemented supervisory training programs for thousands of field personnel to enhance their supervision and management skills.

## 3.  New contract forms

In reaction to the litigiousness of traditional competitive bidding, and owing to the time and potential cost savings, many private owners used negotiated construction. Two new contract delivery methods, construction management and design/build, became commonplace in privately funded construction and then migrated to public works. The contractor's involvement in the design process; the need for teamwork by the owner, designer, and contractor; and the need to maintain good relations to ensure new business—all contributed to fewer claims. As this practice spreads, partly replacing competitive bidding on public works, contractors are becoming more aware of the importance of marketing and focusing more on service—which reduces claims (but not necessarily change orders).

## 4.  Changes in the quality of design and designer responsibility

Owner overemphasis on cost-consciousness can lead to inadequate design fees, which results in lower quality and less detailed design. Architects and engineers are sometimes forced to do the bare minimum in order to maintain a profit margin. As a result, plans and specifications are at times incomplete and poorly coordinated. There has been an increase in performance specifications, which reduce design fees—with both good and bad results. In addition, contract administration fees are often inadequate.

Paralleling the sometimes decreased detail and quality of design was an abdication of responsibility by designers during construction—owing to the risk of litigation and the demands of insurers. Architects no longer assumed the role of "Master Builder." Together with the engineers, they withdrew from inspection and other construction involvement, resulting in even less knowledge of the construction process. In public works agencies, reduced budgets due to taxpayer mistrust led to cuts in design and review time.

Countering these trends is the increase in value engineering and constructibility reviews which provide contractor input to the design process. This is one of the major advantages of design/build and CM/GC contracts. Another compensating trend is the increase in construction management and owner representative contracts, involving a knowledgeable construction expert on the side of the owner. However, some of these experts contribute to the problems with a negative approach, rather than cooperating to achieve a successful project.

The advantages of computer-aided drafting and design, although significant, are not yet fully realized. Except in special circumstances, contractors often lack access to electronic design that could aid in quantity takeoffs, layout, construction simulation, as-built preparation, etc. Groupware and the Internet also have great potential for increased communication and reduced conflicts.

Fortunately, there is some evidence of more risk sharing by innovative public agencies. They have found it beneficial to share some of the risks of piledriving, tunneling, and other uncertain operations that contractors have traditionally been forced to accept. These agencies have also adopted and promoted partnering, dispute review boards, and a positive approach to dispute resolution.

## 5. Dispute avoidance and resolution trends—the 1990s

Current trends are reducing the number, severity, and difficulty of resolving disputes. This began with the rise in arbitration and other Alternative Dispute Resolution (ADR) concepts. ADR was a reaction to the overly claims-conscious attitude of some contractors and the risk-shifting approach of some owners and designers. ADR includes dispute review boards, neutral experts, mediation, and other methods to diffuse conflict and resolve disputes.

Arbitration, instead of litigation, aids in resolving disputes. Although sometimes expensive and time-consuming, it is still better than litigation and can be made more effective, less expensive, and faster.

Mediation is very successful in reducing litigation and arbitration. Settlement rates of 75 to 90 percent are common, costs are much less than with arbitration or litigation, and the parties are more likely to continue business relationships, a significant benefit.

Partnering, however, is having the largest impact. Partnering improves attitudes—from confrontation, claims-consciousness, and win/lose negotiation toward team building, dispute resolution, and collaborative problem solving. It integrates the goals of all parties (the owner's need for a quality, on-time, under-budget project and the contractor's need to make a profit) with a common mission statement and a methodology to avoid and resolve disputes.

Win/win negotiation is another tool for dispute resolution. It allows the parties to seek solutions benefiting both sides, rather than one side winning and the other losing. On larger projects, Dispute Review Boards help reduce conflicts and promote early settlement.

## 6. Practices outside the United States

Based on a limited review of the literature, other countries appear to have similar but less severe problems with claims. In the United Kingdom, for example, contractors are concerned about claims and how to better manage the change order process. Vidogah and Ndekugri's report on a survey of contractors indicated that U.K. contractors are concerned with the costs of claim preparation, the negative image of appearing claims-conscious, and the need to deal with disputes as part of a managed process throughout the course of construction rather than having to "lodge a heavy document at the end of a project...."[5] Therefore, it appears that the methods recommended in this book are applicable in other countries.

## B.   The Dispute Management Program—What It Is and What It Can Do

A Dispute Management Program (DMP) is a philosophy and a comprehensive set of techniques for avoiding and resolving disputes. It is not a new body of knowledge, and a number of its individual elements are already used by some organizations. A DMP is unique because it marries partnering concepts with modern techniques for dispute avoidance and resolution within an integrated philosophy and set of procedures.

A Dispute Management Program provides a broad spectrum of alternative dispute resolution techniques, progressing from proactive to reactive. It begins with good project management, continues with a collaborative and partnering approach to problem solving, transitions (if necessary) to a cooperative approach, and regresses to an adversarial relationship only if disputes cannot be resolved. It includes:

1. *Project management policies and procedures* that ensure projects are better managed, while minimizing errors and other sources of conflict.

2. *Training* the project team in interpersonal skills, for more productive interpersonal dynamics, with less tension and conflict.

3. *Partnering* to promote a more successful project environment, in which all parties work together and claims are avoided or readily resolved.

4. *Dispute avoidance and collaborative problem-solving techniques* to reduce costs, improve quality, and facilitate resolution. (Dispute review boards or neutral experts can help this effort.)

5. *Win/win negotiation techniques* to foster prompt, fair resolution of conflicts.

6. *A Change Order Management program* that supports dispute resolution, without adversely affecting partnering, and enables winning in court if necessary. This includes thorough documentation with timely notice of changes, without posturing or blame, and ensuring that the facts are known so that everyone can participate in problem solving.

7. *Alternative Dispute Resolution* (ADR) to mediate or adjudicate disputes not resolved by the project team, in order to avoid the delay, cost, and negative impact of litigation. This includes the use of neutral experts, dispute review boards, mediation, and binding arbitration.

8. *Fair but firm legal strategies and tactics,* with an emphasis on winning without legal gamesmanship that would delay resolution or increase costs.

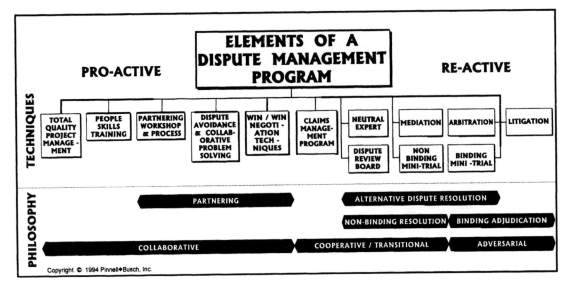

Figure 2.1    The elements of a dispute management program.

## C.   A Partnering Approach to Change Order Requests and Claims

A partnering approach provides the best assurance of a successful change order request or claim. Partnering is crucial to the recommended philosophy and procedures in this book. It builds the relationships and trust needed for building the project and negotiating contract changes. It starts at the preconstruction conference and continues throughout the project.

### 1.   What partnering is and what it can do

Partnering is a relatively new concept and methodology that improves teamwork and minimizes disputes and claims. It is not a contractual agreement nor does it change legally enforceable rights or duties. Although described in the contract documents, its execution is outside the contract. In many ways, it is a return to the way that construction used to be done.

*Partnering is fundamentally a change in attitude,* from an adversarial relationship to a cooperative attitude and approach to problem solving based on mutual trust and respect. It requires a change in the culture of the project team. All parties—the owner, designer, contractor, subcontractors, suppliers, and affected members of the public—form an informal partnership to ensure a more successful project for all. Changing attitudes is difficult, however. Formal procedures with considerable continuing effort and better use of people skills are necessary to make partnering work.

The benefits of partnering are immense. For example, one industrial contractor in a long-term, strategic partnership with an owner found productivity savings of 16 to 17 percent on 18 projects surveyed. A government agency experienced better cost control, reduced paperwork, attainment of value engineering objectives, and no litigation on the projects partnered.[6]

Another positive trend is partnering among the project owner (especially public works agencies), regulatory agencies, and project neighbors. Partnering can also help relationships between different divisions of an agency and during design among the project owner, the designers, and other stakeholders. This results in better designs and contracts, with fewer design errors and changes and a subsequent reduction in disputes and claims.

## 2.  How to partner a project

The steps essential for successful project partnering are[7]

1. *Partner all projects.* If a partnering clause isn't in the contract, ask that it be added and offer to split the cost with the owner. It will pay for itself many times over.

2. *Secure top management commitment.* Conduct a preworkshop partnering awareness seminar for top management and key project personnel, if they aren't familiar with partnering.

3. *Identify a strong "champion"* on the project team from each organization who takes primary responsibility for the success of the partnering effort. The champion ensures partnering success.

4. *Invite all stakeholders to participate in the partnering workshop,* including the designer and key subcontractors. Include the executive level of each organization, unless they have participated in several partnering efforts.

5. *Select the best available partnering facilitator*—a neutral "people" person highly skilled in personal relations, communication, conflict resolution, team building, and goal setting. The facilitator also needs construction experience.

6. *Conduct a preworkshop analysis and planning effort* to familiarize the facilitator with the basic project elements. Review critical dates and tasks, the personalities and their history of working together, the basic expectations and concerns of each stakeholder, and critical issues to be resolved. Customize the workshop for the project and participants, and provide everyone with background material prior to the workshop.

7. *Conduct the partnering workshop at a neutral facility* away from the jobsite. On small projects or when the participants have partnered before, this may be an informal half-day session. On most projects, schedule one day or one and one half days; very large projects may require two or three days.

8. *Accomplish the following tasks* at the workshop:
   - Introduce everyone and establish a relaxed atmosphere.
   - Set communication guidelines and workshop ground rules.
   - Explain general partnering concepts.
   - Briefly examine personality characteristics and behavioral style assessments.

- Discuss principles of communication, problem solving, and conflict resolution.
- Use workshop exercises to build personal relationships, teamwork, and awareness.
- Work on team communications and establish reporting procedures.
- Discuss mutual interests, expressed positions, hidden agendas, and project needs.
- Determine each party's expectations and needs.
- Develop a mission statement (project charter) and post it throughout the jobsite.
- Identify potential problems, briefly analyze, and plan for avoidance of specific problems.
- Develop quality indicators and methodology.
- Discuss the joint cost of project delay and agree how to minimize delays and total job cost.
- Discuss onerous contract clauses and how to avoid their applicability and minimize impact.
- Develop a responsibility matrix for partnering action.
- Define a conflict resolution process.
- Prepare an access list of all participants—phone, fax, e-mail, etc.
- Set stages of team evolution (e.g., when additional subcontractors come on board).
- Develop follow-up tasks for partnership.

9. *Encourage team members to establish personal relationships* during the workshop through one-on-one conversations, team building, and moving beyond business issues to discussing personal interests.

10. *Declare common goals and measurable objectives* in the joint project mission statement. Each party's objectives, once accepted, must be shared by all. Project safety, quality, schedule, and budget are the first priorities. Also include the contractor's profit and other objectives, and establish procedures to verify accomplishment of the objectives.

11. *Document the conflict resolution process.* Develop procedures for collaborative problem solving at the lowest possible level and raising unresolved disputes to the next level of management for both organizations (i.e., field, jobsite management, home office, chief executive). This almost guarantees settling disputes at the jobsite level, as each management level for both organizations want to resolve their own problems and their supervisors dislike being unnecessarily involved.

12. *Document the workshop's achievements* with framed mission statements, team photographs, and other symbols for distribution to workshop participants at an occasion such as the ground-breaking ceremony. A joint/project logo, team coffee cups, and other simple promotions can have a significant positive effect on creating a team spirit.

13. *Establish a proactive risk identification and collaborative problem-solving effort.* Identify risks 120 days out and develop plans to avoid or minimize

impacts. Update the list quarterly at the monthly progress meeting. When problems occur, involve everyone in finding a solution.

14. *Schedule a half-day follow-up workshop as more subcontractors start work.* Review the champion's roles and responsibilities and the team's progress toward meeting the mission statement objectives. Use the workshop to coach the champion in leading subsequent partnering sessions.

15. *Conduct periodic telephone checkups* to identify unresolved problems or a slackening of partnering efforts and to prevent conflicts from progressing too far for easy resolution. A follow-up workshop can revitalize the partnering effort and resolve lingering disputes.

16. *Celebrate your success* upon achieving major milestones, accomplishing the objectives in the charter/mission statement, and completing the project.

## 3. Relationship with the owner's representative

When preparing a change order request or claim, avoid an adversarial tone. Use a positive, partnering approach, but be ready to adopt an assertive position on a specific issue and proceed to arbitration or litigation if the owner's representative fails to reciprocate. Do not, however, allow a dispute over an issue to disrupt the overall partnering effort.

If a dispute threatens to damage the working relationship between jobsite personnel and the owner's representative, *bring in someone else* to present claims and negotiate change orders. This can be someone from the home office or a consultant.

Assume that you will be working with the project owner and designer in the future, and maintain good relations, whether on competitively bid public works construction or negotiated private work.

## 4. Relationships between general contractors and subcontractors or vendors

Partnering between the general contractor and subcontractors can lead to even greater savings than partnering with the owner, and is crucial when pursuing a major claim against owners involving those subcontractors. Owners may seek to exploit conflicts between the general contractor and the subcontractors and will question any claim in which the subcontractors don't participate. The general contractor and subcontractors should resolve disputes between themselves (at least tentatively) before pursuing subcontractor pass-through claims against the owner.

Partnering with the general contractor is essential for a subcontractor negotiating a claim against a general contractor. Subcontractors must convince the general contractor that a reasonable settlement is in the general contractor's best interest and preferable to arbitration or litigation—without damaging the prospect for future business.

### 5.  Use partnering to help resolve disputes

Use the partnering process and commitments to build relationships and resolve disputes. Comply with the partnering charter and don't inflate claims. When submitting a change order request or claim, evoke the partnering agreement along with equity arguments to help settle the issues. If necessary, call for a midproject partnering session to defuse any conflicts that have developed and restore a cooperative atmosphere. Do not label your request a "claim." Refer to it as a change order request or request for equitable adjustment.

For major, potentially disruptive changes use partnering techniques to define the scope of work and develop a solution. *Jointly* develop with the owner, designer, and subcontractors a work plan, schedule, and cost agreement that meet the needs of all parties.

### 6.  Use partnering to implement collaborative, problem-solving contract modifications

Use project partnering to introduce innovative, problem-avoiding, and opportunity-enhancing modifications to the contract. For example, propose a value engineering change proposal clause, which allows the contractor to propose money-saving changes to the contract and share in those savings.

### 7.  Protect your contract rights, but do not abandon partnering because of a dispute

While focusing on partnering, *preserve your contractual rights*. A partnering approach will not always work. Partnering must be reinforced by a firm position, supported by the contract, timely notice, and extensive documentation. You must still give notice of changes, obtain written directives before performing extra work, and maintain adequate records of extra work.

Do not, however, set aside partnering efforts because of an adversarial relationship over a single issue. Try to segregate actions on that one issue from your relations on other issues. If a dispute does affect the partnering relationship, resume partnering once the conflict is resolved or mitigated.

## D.  Key Elements for Successful Change Order Requests and Claims

The requirements for a successful change order request or claim reflect your goal—*to convince* the claim reviewer of the validity of the change order request or claim and the need to make adequate compensation. To do this, the request must be:

- *Understandable.* Clarify the issues. Be clear and concise. Most people will not risk agreeing to something they don't understand and few have the confidence to admit they don't understand something.

- *Factual.* Reference and attach supporting documents, and avoid unsupported allegations or incorrect statements.

- *Contract-compliant.* Comply with the contract to prevent denial based on exculpatory clauses or disclaimers.

- *Realistic.* Be tactful and minimize the importance of errors by the reviewer. Or, clearly define responsibility and state that a prolonged dispute will focus on those issues and their performance.

- *Compelling.* Convince the reviewers of the benefits of resolving the issue now, and help them to justify it to their superiors.

Achieving these objectives will be easier if you do the following:

---

**Seven Steps To Successful Changes and Claims**

1. Implement a Change Order Management Program.

2. Do a Better Job of Settling Change Order Requests.

3. Understand the Claim Reviewer and Other Interests.

4. Address All Elements of Settling the Request or Claim.

5. Be Timely.

6. Conduct an Efficient, Effective Claim Preparation Effort.

7. Satisfy Your Client.

---

## 1.  Implement a Change Order Management Program

Adequate documentation is needed for preparing change order requests or claims, but is often lacking. Many contractors need to maintain better records in order to prepare more successful claims. Timely notice, as required by the contract, is also needed but also may be deficient. Fortunately, both problems can be avoided in the future by implementing a Change Order Management Program.

The Change Order Management Program, as described in Chap. 4, should include the following features for recordkeeping, giving notice, and ensuring a partnering approach:

- A *partnering approach* to dispute avoidance and resolution.

- *Prebid review* to identify potential problems and include a contingency in the bid.

- *Loss Prevention Plan* to be implemented when onerous contact clauses are encountered.

- *Preaward constructability review* to identify problems before contract signing and start of work.

- *Effective job cost accounting* system and procedures to track and report costs.

- *Recordkeeping procedures* that are simple, effective, and implemented.

- *Timely notice of extra work* in a nonadversarial manner.
- *Personnel training and motivation* to understand contract law and your procedures.
- *Communication* with subcontractors and suppliers regarding changes and delays.
- *Prompt resolution of disputes* when they occur—in a fair, prompt, and firm manner.

### 2. Do a better job of settling change order requests

There are only five ways to get paid for extra work:

- A negotiated change order resulting from a change order request.
- A negotiated change order resulting from a dispute review board hearing and recommendation.
- A negotiated or mediated change order resulting from a claim.
- A unilateral change order issued by the owner over the contractor's protest.
- An award by an arbitrator, judge, or jury.

It is easier, faster, less expensive, and far more pleasant to negotiate rather than to arbitrate or litigate. Partnering and win/win negotiation will be more successful than a combative attitude and style. A partnering approach allows claims to be presented and resolved in a professional, businesslike manner, without damaging ongoing business relationships. Disputes can be characterized as contested change order requests rather than claims. Positive language and labels promote positive responses.

Contractors can improve their preparation and negotiation of change order requests, and avoid the delay, expense, and stress of claims.

### 3. Understand and address the claim reviewer and consider other interests

The change order request or claim must address the reviewer(s) of the claim—usually the owner's representative and possibly a decision maker within the owner's management structure. Consider their personalities, biases, needs, expectations, and understanding of the project and issues in dispute, as well as the availability of funds. Characterize the dispute in the most acceptable terms to avoid unnecessary resistance. Observe, for example, how they receive and process information. You *must* address their expectations if they are to understand what you're trying to communicate.

**Achieve understanding.**  Clearly communicate the facts so that the reviewer can understand your change order request or claim. This may require explaining issues the reviewer should already know, and presenting them so that they can be understood. Use concrete-specific, sensory-based words which have the same connotations to everyone. Otherwise, your claim will be rejected.

**Overcome adverse interests.**  Sometimes, the reviewer must be convinced of a mistake—before agreeing to your change order request or claim. For example, if the reviewer is the designer, you may have to convince them of a design error. If the reviewer is the owner's on-site representative, you may have to show that their response to a request for information was untimely. Then, you must prove the cost of their mistakes, when they may be held responsible. Needless to say, this is difficult. Even if the reviewer is not responsible for an issue, his or her performance may be measured by how little you accept.

**Address multiple levels of review.**  There can be multiple levels of review which must be satisfied. One person or different individuals may conduct the review at each of three different levels: factual/technical, legal, and business.

**Know the players.**  In most cases, others besides the claim reviewer can hinder or help with the claim preparation and resolution. They include:

- *Claims preparer.* The individual (or team) responsible for preparing the change order request or claim may be a project team member or an employee or consultant expert.

- *Contractor's project team.* The project manager, project engineer, project superintendent, and other jobsite management, administrative, and supervisory personnel who built the project normally prepare change order requests and claims.

- *Contractor's claim support team.* Contractor personnel and others responsible for working with and assisting the claim preparer may have authority to approve the preparer's work plan and budget, recovery strategies, negotiation efforts, and settlement amounts. An attorney is included if legal issues are unclear or if the claim is likely to be mediated, arbitrated, or litigated.

- *Contractor's decision maker.* Usually a construction company executive, but sometimes the project manager or attorney, makes final decisions on preparation, presentation, and settlement of the change order request or claim.

- *Designer.* The architect or engineer who prepared the plans and specifications.

- *Owner's representative.* The Contracting Officer's Representative (COR) on U.S. government contracts. The individual(s) representing the owner are responsible for inspection, monitoring progress, approval of pay requests, negotiation and approval of change orders, etc. The representative may be the designer, an employee of the owner, or an independent construction manager.

- *Owner's claim reviewer.* The individual who reviews the change order request or claim for the owner and decides whether to accept it or whether to recommend acceptance to the actual decision maker. The reviewer may be the owner's representative.

- *Owner's decision maker.* The individual in the owner's organization who makes the final decision to settle a change order request or claim. In some cases, this may be a combination of people, with none having clear-cut authority. On U.S. government contracts, the decision maker is the contracting officer.

**4.  Pay attention to all elements of the change order request or claim**

Many assume a successful change order request or claim results from simply preparing a well-written document. Achieving success is more encompassing, for it includes additional tasks and intangibles such as attitudes and expectations. In addition to the physical document it includes:

- *The relationship* between the contractor and the reviewer/approver of a change order request or claim, the reviewer's assessment of the contractor's honesty and competence, and empathy toward the contractor's project team.

- *Communications* between the change order or claim preparer and the reviewer. Communication should create a sense of teamwork—involving the reviewer in preparation of the change order request or claim (primarily by indicating how they want it presented), and committing them to an open minded review of the claim and a fair decision.

- *The written document* summarizing the change order request or claim, with a narrative of the project, justification for payment, and references to supporting documentation.

- *Exhibits* included with the written document, and displayed at the presentation meeting.

- *The presentation meeting* at which the change order request or claim is presented along with an oral overview, a question-and-answer period, and some negotiations, if warranted.

- *Subsequent meetings* or other communications responding to the reviewer's questions and providing additional information as requested.

- A *series of well-organized negotiations* that are conducted in an atmosphere of frank communication and cooperation, and that lead to a mutually acceptable settlement.

- *The reviewer's awareness that you will arbitrate or litigate,* or take other actions if negotiations fail. This requires clearly communicating your resolve, and a history of follow-through on prior disputes.

**5.  Be timely**

Timing is vital to success—especially when giving notice, requesting clarification, submitting claims, or attempting resolution. Establish patterns and expectations early, before the reviewer(s) expects too much compromise. Attempt to resolve anticipated problems at the beginning of a project, while everyone is in a positive, cooperative mode. Submit change order requests as soon as possible after requested or when discovering a problem. If the request is rejected, file a claim while construction is ongoing. When owners need contract performance, they are more likely to cooperate to maintain good working relations. If the project team is too overworked to handle a claim, bring in help.

One example of good timing is a manufacturing subcontractor's submittal of a large claim against the prime contractor only weeks before the prime, a subsidiary of a large national defense contractor, was sold to another firm. The claim was sent directly to the prime's corporate counsel, who had to reveal it to the purchasing company, bypassing the mid-level managers who might have buried it until after the sale. Funds were presumably set aside for payment, as negotiations for the $4-million-plus contract change required only two short days. In a second example, a large claim against a public agency was filed just months before an important bond election. Settlement was swift, reasonable, and unpublicized.

## 6.   Conduct an efficient, effective change order request or claims preparation effort

Better managing the change order process saves time and money, guarantees a better work product, and achieves your goal. The two basic requirements for good management of change order and claims resolution efforts are:

1. *Follow the plan-action-feedback loop* in Fig. 2.2. Good project management consists of three basic steps:

   - *Plan.* Set clearly defined objectives and prepare a work plan to achieve the objectives.
   - *Action.* Take action based on the plan.
   - *Feedback.* Measure progress toward the objectives and compare actual versus planned.

   Repeat the cycle: Develop a corrective action plan, take action based on the revised plan, obtain more feedback, and continue.

2. *Work step-by-step in phases.* For efficiency and effectiveness, break the work into phases and follow a step-by-step process. As each phase is completed, compare the interim work product to planned, compare cost and schedule performance, and take corrective action, if needed. More importantly, examine the current findings and recommendations to see if any revisions in strategy or work product are necessary. Claim preparation often leads to unexpected results, both positive and negative. Be open to change and respond accordingly.

   Repeat the step-by-step process during each analysis phase. The steps are not separate, sequential processes but usually overlap and often run

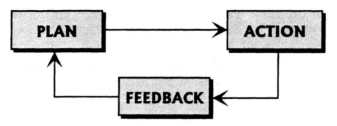

**Figure 2.2**   Plan-action-feedback loop.

concurrently or in a different order, depending upon the specifics of the claim. Special tasks may be required for some claims, while standard tasks may be omitted for others.

## 7.  Satisfy your client

As either an in-house/employee expert or a consultant, remember that a good job and the best possible settlement for your company or client isn't enough. *Everyone* must know what you are doing and what you did. Keep your boss or client *fully* informed, obtain written authorization for increases in your budget. If you surprise them with bad news, you may have an unhappy supervisor or client and a dispute over costs. I once helped a contractor recover $1.1 million in arbitration on a very difficult impact claim, only to be partially stiffed on fees as I was over budget and the contractor's president expected $1.25 million. Had I communicated better, this would have been avoided.

# 3

# Overview and Practical Understanding of Contracts and Contract Law

Before preparing a claim, you need to understand proof of entitlement and the theories of recovery. For example, why is the owner (or general contractor, if a subcontractor is making a claim against the general) responsible for the extra time and cost? What will make them pay? The answers are dictated by the contract, state or federal law (statute) and applicable regulations, common law, industry practice, and the actions of the parties. On U.S. government contracts, the Federal Acquisition Regulations (the FARs) also apply.

This chapter provides a comprehensive introduction and working explanation of contracts and construction law, a discussion of contract interpretation procedures, and a brief discussion of key contract clauses affecting change orders and claims. Contractors will find this chapter to be an invaluable aid in understanding contracts and handling day-to-day contract interpretation issues. Experienced construction managers will better understand and apply legal principles in their day-to-day activities or when preparing a claim. Construction attorneys can use this chapter to educate their clients in contract law.

The following topics are covered:

A. Requirements for recovery

B. Elements of the contract

C. Fundamentals of contracts and contract interpretation

D. Other legal concepts and terms

E. Contract clauses affecting change order and claims resolution

F. Types of extra costs and damages—based on entitlement

G. How to proceed if unable to negotiate a settlement

*Do not rely on this book for legal advice.* Although this chapter provides general information on contracts and contract interpretation, confirm all statements relating to the law with a construction attorney before applying the information to a specific dispute. The law varies substantially between states and between countries, and some general statements may not be applicable in your jurisdiction.

Contractors need to understand the basics of contracts and contract law—to negotiate contracts, to protect contract rights and fulfill contract responsibilities, to prepare change order requests or claims, and to negotiate changes. More importantly, contractors must know when to seek legal advice. They should establish a working relationship with a construction attorney who becomes familiar with their business and helps them avoid arbitration or litigation.

This book assumes that a request for equitable adjustment (a change order request or claim) is being prepared without legal advice. However, if there are questions about legal issues, or if the dispute is likely to lead to arbitration or litigation, consult with an attorney.

**A partnering approach.**  This book is founded on a partnering approach to dispute resolution. The tone of Chap. 3 may sound adversarial at times, because *you must be prepared for conflict* and must know your contract and legal rights and responsibilities. Contractors and their attorneys are encouraged to apply the information provided in a nonadversarial manner.

**Basic recommendation.**   The most important steps to protect yourself contractually are:

- Deal only with people you trust.
- Read and understand your contract. If you have questions, have your attorney review it before signing.

## A.  Requirements for Recovery

Recovery refers to the receipt of money, the ultimate objective of a claim. The process begins when a contractor recognizes a loss and believes it may be compensable. To continue, you must persevere and implement the following steps.

### 1.  Determine if assets exist that can be collected

First, determine if the other party has funds or assets that can be attached (collected) to pay your claim. If in doubt, use a credit reporting service such as Dun & Bradstreet—before entering into a contract. A construction lien on improved, privately owned property is one alternative, a claim on a contractor's payment bond is another. Miller Act claims can be made on federal projects. The states have similar laws, for "Little Miller Act" claims on publicly funded projects. An insurance claim is yet another alternative. However, a

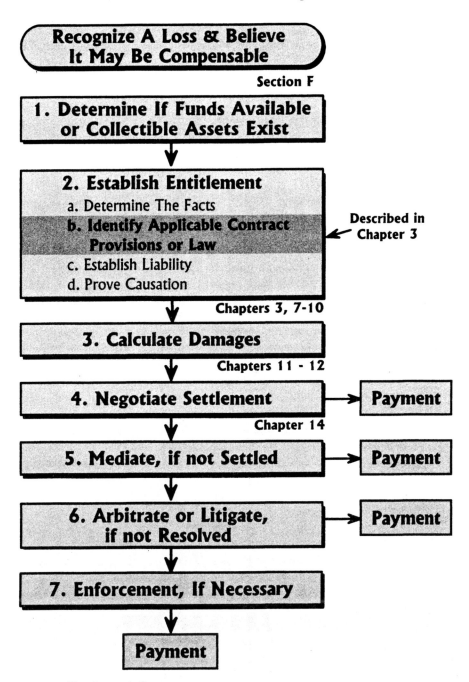

**Figure 3.1**  Requirements for recovery.

claim against an insolvent, unbonded, defaulting subcontractor with the repair cost of defective work exceeding their retainage may be worthless. If recovery is unlikely, do not waste your time and money preparing a claim.

## 2. Establish entitlement

Entitlement is a legal term meaning your legal or equitable right (i.e., you are *entitled*) to additional time or money, based on the rights and responsibilities defined by the contract and applicable law. *You must establish entitlement with reasonable certainty, in order to recover damages.*

**a. Determine the facts.**   First, determine the facts. Review the project source documents and analyze what actually happened, as described in Chaps. 7 through 9.

**b. Identify the applicable contract provisions or contract law.**   Next, develop a theory of recovery for each issue in dispute. This requires understanding the basics of contract law, knowing the contract, identifying the facts that govern the dispute, establishing a contract right (or quasi-contract approach) and determining one or more theories of recovery. If involved, an attorney will determine the theories of recovery based on the facts gathered by the claim preparer. Retain legal counsel if a claim is large and complex, involves unfamiliar legal issues, or is likely to lead to mediation, arbitration, or litigation.

If possible, develop multiple theories of recovery for each issue, in case one or more theories are rejected. For example, suppose you assume that the contract provision for changed work is the primary theory of recovery. An alternate theory may be that the owner breached the contract by requiring performance of work not specified in the contract (extra work). A third possible theory may be quasi-contract, requiring payment of the reasonable value of the work (sometimes called *quantum meruit*) contract.

**c. Establish liability or the right to a contract adjustment or equitable adjustment.**   Apply the applicable law and the terms of the contract to the facts to determine if you are entitled to relief, which may include payment of money, a time extension, a scope reduction, or performance of specific tasks by the other party. Analyze the facts and the applicable contract provisions or contract law. Consult an attorney if appropriate.

**d. Prove causation.**   Next, trace the linkage from the initiating cause (event, action, omission, or condition), for which the owner (or other adverse party) is responsible to the resulting costs suffered by the contractor. Although causation is normally a term used in tort law, it is used in this book to describe constructing a logical relationship between cause and damages. Chapters 9 and 10 describe how to establish causation.

## 3.   Calculate damages (or the amount of contract adjustment or equitable adjustment)

Although this chapter refers to the "recovery of damages," attorneys use the term *damages* only if there is a breach of contract (e.g., the owner refuses to pay for a legitimate contract change). Most change orders are paid as either contract adjustments (based on procedures specified in the contract such as a variation in quantities clause) or as equitable adjustments (based on negotiation of an equitable settlement within the contract terms). Only if a dispute proceeds to arbitration or litigation are damages awarded. However, contractors frequently use the word damages for all change order costs, and this book will also.

*To recover damages, you must prove a specific amount of cost.* General assertions of extra costs, without quantification, are usually insufficient. Guesstimates or opinions, unless by a qualified expert, are not adequate, and the amount recovered will usually be far less than the costs actually incurred. Proving a specific amount requires the good recordkeeping and cost accounting systems described in Chap. 4 and the rational methods of calculating damages covered in Chaps. 11 and 12.

Recovery of damages is limited to actual, provable loss (plus a reasonable profit). These costs may include consequential damages, lost profits, and attorney fees if allowed by the contract or applicable statute. You cannot receive more than your actual losses (punitive damages are almost unheard of in construction disputes).

## 4.   Present the request for equitable adjustment and negotiate settlement or win an award

Next, negotiate settlement, as discussed in Chap. 14. If unsuccessful, request mediation. If refused, or if mediation is unsuccessful, demand arbitration, if it is specified in the contract. If it isn't, try to negotiate an agreement for binding arbitration. If unsuccessful, commence litigation.

## 5.   Enforcement

You need to consider enforcement when preparing and negotiating claims. Some claim reviewers will negotiate only if they believe: (1) you can and will litigate or arbitrate to protect your rights; (2) you would eventually win an award and obtain payment; and (3) it will be cheaper (in both money and their personal loss of reputation) for them to settle now. Also, if a settlement is negotiated or an award is made in arbitration or litigation, and the party is unwilling or unable to pay, you must know how to *force* them to pay. Without the ability to force payment, you have little leverage to negotiate and must rely on the other party's goodwill and sense of fairness.

Ensuring the ability to enforce recovery precedes signing a contract. For example, a contract with a corporate shell (consisting of an assumed business name and few assets) can result in disaster if the improved property has insufficient value to cover your lien (or is otherwise encumbered).

The best method of enforcement is a construction lien on improved private property or a Miller Act or "Little Miller Act" bond claim on public works projects. Preserving lien rights or promptly filing bond claims is therefore essential; the requirements should be confirmed with an attorney. Filing a lien or bond claim may elicit an angry response from the owner (or general contractor), but often forces a return to serious negotiations. Your recourse against subcontractors may be their performance bond or to withhold their progress payments. The subcontractor's bond, their financial status, and the retainage on their work are therefore critical. However, the cost of repairing defective work or of delay to the critical path may far exceed their contract amount.

In arbitration, the arbitrators' decision may need to be entered as a judgment in the courts. The judgment will be executed by garnishing (unilaterally seizing through legal action) accounts or assets in satisfaction of the judgment.

## B.  Elements of the Contract

The terms of the contract govern entitlement and recovery of damages, unless conflicting with state or federal statute. The courts tend to strictly scrutinize unambiguous but onerous clauses (e.g., no damages for delay clauses) and may employ creative interpretation to prevent their enforcement. You should therefore understand the pertinent terms and conditions of the contract before signing and seek legal advice when appropriate. Ambiguous clauses (i.e., those having more than one reasonable meaning) are interpreted in accordance with the rules listed in Sec. C.3 below.

The following is a much-simplified explanation of contracts, and should not be relied upon when involved in a legal dispute. Obtaining competent legal counsel is essential in such cases.

## 1.  Types of design and construction contracts

There are eight types of contracts in the design and construction industry that may affect a claim:

1.  Design contracts between the owner and designer.

2.  Design subcontracts between the lead design consultant and subconsultants.

3.  Construction contracts between the owner and general contractor.

4.  Subcontracts between the general contractor and subcontractors, and between subcontractors and sub-subcontractors.

5.  Purchase orders (which are contracts for the purchase of goods) between contractors and their suppliers.

6.  Consultant contracts between a professional construction manager or owner's representative and the owner.

7. Design/build contracts between the owner and design/build contractor.

8. Design contracts between the contractor and a design professional preparing plans for temporary construction facilities (e.g., shoring), or elements of the completed work.

This chapter focuses on the contract between the owner and the general contractor, with some discussion of subcontract issues (which also apply to subcontractor/sub-subcontractor relationships).

## 2.  Elements of construction contract documents

Construction contract documents are generally grouped into at least five parts. Each element should be reviewed when preparing a change order request or claim.

**Part I—Bidding requirements.**  The bidding requirements are:

- Invitation to bid—the text of the advertisement alerting bidders to the contract letting.
- Instructions to bidders—information for submitting a bid on the contract.
- Proposal form—a sample proposal form, containing alternatives or unit quantities and space for insertion of unit prices or a lump sum price.
- Bid bond form—normally supplied by the bonding company on their own form.
- State and federal requirements—as required on publicly funded projects.

When preparing a change order request or claim, refer to the bidding requirements for a general description of the project (in the invitation to bid) and for information such as the bid date and the time from bid to award.

**Part 2—Contract forms.**  The contract is usually on a standard form, prepared by a design professional association or by an owner who frequently constructs projects. If not a standard industry form, read the contract carefully for unusual provisions, and check standard contract forms for additions or deletions. Performance and payment bond forms are usually included.

**Part 3—Conditions of the contract.**  The conditions of the contract include the general and special conditions and supplementary or special conditions. Carefully check the general and special conditions, as they contain the most important clauses affecting a claim.

Many project owners and designers use standard general conditions, applicable to all of their construction contracts. These standard general conditions are integrated with a standard owner-contractor agreement, and together form the contract. They eliminate the need to review every detail of the general conditions for every contract.

American Institute of Architects Document A201 is the most widely used set of contract general conditions in the United States for building construction. It is often used in conjunction with A101, the standard form of agreement between an owner and the contractor on lump sum, fixed price contracts. Another widely used set of contract documents, often applied to civil works, is that prepared by the Engineers' Joint Contract Documents Committee (EJCDC). These standard agreements and general conditions (No. 1910-A-1 and 1910-8-A-2, 1990 editions) are published jointly by the National Society of Professional Engineers (NSPE), the American Consulting Engineers Council (ACEC), the American Society of Civil Engineers (ASCE), and the Construction Specifications Institute (CSI). They are endorsed by the Associated General Contractors of America (AGC).

In addition to the U.S. government, each state department of transportation, numerous cities and counties, several state chapters of the American Public Works Association (APWA), and other organizations have developed standard technical specifications for civil works.

General conditions are modified by supplementary conditions that change and add conditions applicable to the specific project. The modified clauses usually include insurance requirements, contract duration and liquidated damages, and wage rates (on federally funded projects).

**Part 4—Technical specifications.**   The technical specifications describe the materials required, the quality of workmanship, required procedures or techniques, and the completed project. Technical specifications are normally organized on two levels, as follows:

**1.   Level 1—Organization into divisions.**   The technical specifications on most projects are organized into 16 divisions in accordance with the CSI (Construction Specifications Institute) Index. The most important is Division One, but also check those governing the work in dispute. The divisions are:

1. General requirements: meetings, submittals, scheduling, quality control, close-out, etc.
2. Site work: demolition, earthwork, foundations, utilities, and landscaping.
3. Concrete.
4. Masonry.
5. Metals.
6. Wood and plastics.
7. Thermal and moisture protection.
8. Doors and windows.
9. Finishes: paint, carpet, flooring, sheetrock, etc.
10. Specialties: lockers, bulletin boards, toilet and bath accessories, etc.
11. Equipment.
12. Furnishings: furniture, cabinets, rugs, etc.

13. Special construction: integrated systems.

14. Conveying systems: elevators, etc.

15. Mechanical: HVAC, plumbing, fire protection, pumps, etc.

16. Electrical: distribution, motor control, lighting, telephones, special circuitry, and controls.

**2. Level 2—Organization of divisions into sections.** The index divides each of the 16 divisions into sections, with the standard sections being:

- Scope—defines the work covered.

- Materials—standards, brands, types, strengths, etc.

- Workmanship—means and methods, testing.

- Measurement and payment, on unit price contracts.

**Part 5—Drawings.** To obtain an overall picture of the project scope and specifics, begin your review of the contract documents by examining the drawings. Look for details on the drawings affecting the work in dispute.

The drawings are organized by the subconsultant responsible for their design. Unfortunately, work for some trades may be indicated on the plans for another trade and not on the plans where the work is normally defined. There also may be physical or system conflicts between the work of different design disciplines. In addition, work may be required in the technical specifications but not shown on the plans. For building construction, the drawings are generally organized as follows:

- A project location plan and general information

- Civil work including excavation, utilities, and site work

- Architectural

- Structural

- Mechanical

- Electrical

Drawings normally contain a revision number and date, the initials of the drafter and checker, and notes of changes. When a change is made, a triangle with a revision number is normally placed on the drawing next to the change and the added work is shown in a cloud. Verify that you have the current applicable version of the drawings with all addenda changes posted.

Field sketches are issued by the designer during construction to clarify ambiguous details or to resolve actual site conditions. Collect and organize the sketches and reference them to the appropriate drawings so as not to overlook changes. Examine the sketches to determine whether the design was modified and costs consequently increased. Also, determine if late release of drawings impacted the work.

**Other elements of the contract documents.**   If earthwork is in dispute, check the soils (geotechnical) report and any special studies or tests included or referenced in the contract documents.

**Addenda and contract modifications (approved change orders).**   Addenda, or prebid changes to the contract, are issued after the bid documents are released to bidders but before bids are received. Ensure that all addenda are posted to the plans and specifications. This should have already been done by the project team.

Record approved changes on the plans, specifications, and bid schedule (or schedule of values) to avoid omissions when preparing a claim. Also, verify that the cost accounting budget has been revised to reflect the change order costs and work quantities.

### 3.   Federal Acquisition Regulations (FARs)

Contracts with the federal government are governed by the Federal Acquisition Regulations, known as the FARs. If preparing a claim against a federal agency, check the applicable FAR and consult with an attorney experienced in federal contracts. The FARs are extraordinarily complex and very procedurally orientated.

### 4.   Modification and clarification of the contract—outside the contract

Contract provisions can be supplemented or superseded by state or federal statute and common law. If the contract language is ambiguous or does not address an issue, the contract will be interpreted using the principles listed in Sec. C.3 below.

**Statutes.**   State statutes (laws) and common law may supersede certain contract provisions. "No damages for delay" clauses, for example, have been declared void on public works contracts by the legislature in some states. State prompt payment acts governing retainage and the time of payment may apply, and pay when/if paid clauses may be declared unenforceable. In addition, state statutes will dictate licensing or registration requirements and construction lien proceedings. Contractors and claim preparers should consult with a construction attorney to determine what statutes may modify the contract or otherwise affect recovery.

**Common law.**   Legal precedent, also called common law or case law, consists of previously decided cases or court opinions. Common law is used to interpret contracts and provides guidance for conditions not contemplated by the contract. For example, "industry practice" is a common law principle. Also, for example, the actions of the parties (on this or prior projects) may be used as evidence of how a contract should be interpreted. The parties' actions may modify or waive a contract provision or be used to interpret the intent of the parties. For example, an owner's history of paying for verbally directed extra work can supersede a clause requiring a signed change order before the contractor performs extra work. Contractors cannot rely on this principle, however, and should always obtain written confirmation of contract changes.

**Equity.** In laymen's terms, equity means fairness. Equitable principles and rules have been developed over many years in courts of equity. In the United States, the equity courts and common law courts have merged and the earlier equitable doctrines survive. For example, quantum meruit is one equitable theory of recovery. Although seldom used as a legal theory, equity can be invoked as a theory of recovery when the contract and case law are unclear or an issue is not addressed by the contract.

## 5. Subcontracts

Agreements between the general contractor and subcontractors, and between subcontractors and sub-subcontractors, are subcontracts. The AGC and several subcontractor associations have published a standard subcontract form, but it is not widely used. The AIA subcontract form is more widely used. Most contractors have their own subcontract forms, which favor the general contractor more than the AGC or AIA forms and require careful examination.

**Flow-down provisions from the prime contract.** Nearly every subcontract provides that applicable provisions of the prime contract between the owner and the general contractor "flow down" to the subcontract. Subcontractors must therefore understand the terms of the prime contract and should obtain a copy from the general contractor. Subcontractors should verify that contract provisions in the prime contract protecting the general contractor flow down to protect the subcontractor.

**Privity of contract.** Privity of contract means having a direct contractual relationship with another party. Although most subcontracts bind the subcontractor to "applicable terms" of the prime contract, this provision does not create a direct contractual relationship between the subcontractor (or supplier) and the project owner. Therefore, *subcontractors cannot pursue claims directly against the owner but must "pass through" their claim to the general contractor,* who submits the claim to the owner.

If unable to obtain a pass-through agreement with the general contractor, subcontractors can file a lien on private projects which forces the owner to respond through the general contractor. In addition, they can sometimes claim one of the following theories to require the owner to use the disputes clause in the prime contract:

1. Express contract (applicable terms). Requirement in the prime contract that subcontracts be bound to "applicable terms" of the contract.

2. Express contract (incorporation). Incorporation of the prime contract by reference in the subcontract.

3. Implied or quasi-contract. Based on third party beneficiary, implied warranty, or misrepresentation theories.

**The *Severin* Doctrine.**   The owner is not liable for damages to a subcontractor, unless the general contractor is obligated to pay the subcontractor (for damages caused by the owner).

**Subcontractor claims against the general contractor.**   Claims by subcontractors against a general contractor, or against other subcontractors, are becoming more common, especially for scheduling delays. Preparation of delay claims against a general contractor is similar to a delay claim by a general contractor against the owner. The negotiation process differs slightly, and obtaining a fair settlement is more difficult, since subcontractors depend on the goodwill of the general contractor for fairly administrating the current contract and awarding future contracts.

The relationship between a general contractor and subcontractor is intertwined and less well defined by the subcontract than the relationship between a general contractor and the project owner. Many issues are governed by industry practice or the standard procedures of the general contractor. More importantly, most subcontracts offer little protection to the subcontractor and allow the general contractor a great deal of latitude in job management and contract administration.

**General contractor claims against subcontractors and suppliers.**   General contractors collect damages from subcontractors or suppliers by backcharging the estimated or actual cost and withholding this amount from the next progress payment. Some general contractors abuse their power and unfairly assess backcharges. Subcontractors must be prepared to press for payment, with a performance and payment bond claim or lien against the project being their best weapons.

**Liens.**   Even without privity of contract, subcontractors and suppliers have lien rights or bond claim rights to obtain payment. Liens are more than just a legal process to obtain payment. They apply tremendous pressure on owners and general contractors. Construction loan agreements typically require that the project be free of liens, leaving the owner the choice of paying or "bonding around" the lien. Otherwise, the bank may call (terminate) the construction loan or increase the interest rate. Owners therefore become very upset if liens are filed on their property and often pressure the general contractor to settle the dispute with the subcontractor.

A lien may be declared invalid if it fails to conform to the exact language of the lien law. In addition, legal sanctions may be imposed for unwarranted or excessive lien filings. Therefore, be very careful to comply with *all* requirements when filing a lien. Legal counsel is advised when filing liens.

**AGC standard subcontracts.** The latest AGC standard subcontract form (No 600) contains two clauses on pass-through claims—6.2 and 6.6. These clauses require the subcontractor to submit claims promptly to the general contractor and limit the damages to a share of the general contractor's recovery.

This new subcontract was developed by a joint AGC/subcontractor committee and adopted by the national AGC and subcontractor associations. However, it has not been adopted by most local chapters and is not often used due to controversy over several of its provisions, especially the payment clause.

## 6.  Purchase orders

Purchase orders are used for agreements with vendors or suppliers of goods and should generally not be used for services or subcontract work. If problems arise between contractors and suppliers, interpretation of purchase orders for the sale of goods is governed in most states by the Uniform Commercial Code (UCC), which covers:

- Merchantable title, free and clear of liens and encumbrances.

- Express warranties by affirmation, promise, description, or sample.

- Implied warranty of merchantability or fitness for a particular purpose.

- Loss or damage in transit: F.O.B. (Free On Board) point of origin (contractor responsibility) or F.O.B. destination (supplier responsibility).

- Defective or nonconforming goods.

- How industry practices and past actions of the parties apply.

A common problem encountered when analyzing a dispute over materials obtained by a purchase order is that the terms and conditions were never fully defined by the contractor and supplier. If so, the conduct of the parties on this and other projects will be used to help interpret the contract. Or, the UCC (Uniform Commercial Code), which has been enacted into statute in some form in most states, may define terms not specified in the contract.

Read the seller's standard purchase order terms. If unacceptable, promptly reject the designated terms in writing. When a contractor and a vendor each submit conflicting written terms of sale, they should refer the "battle of the forms" to an attorney.

## 7.  Shop drawings and other submittals

Items that are not sufficiently detailed to be fabricated from the plans, or items whose installation varies depending upon the equipment manufacturer, require shop drawings detailing the work. Shop drawings are common for structural and reinforcing steel, most equipment, and many other items. Other submittals include catalog cuts, manufacturers' certifications, operation and maintenance manuals, safety plans, traffic control plans, quality assurance/control plans, etc.

Late materials delivery, project delay, and extra costs arising out of shop drawings and submittals are frequent causes of conflict between the owner, contractor, subcontractors, and suppliers. Designers often make changes or add requirements which increase costs without recognizing the change as a contract change. Designers and owner's representatives may fail to return submittals on time. If the project is delayed by late return of submittals, the contractor is due a time extension. Even return within the time specified in the contract can be cause for a time extension, if: (1) the contractor notified the designer of the need for early review, (2) the designer could have returned the submittal earlier, and (3) the late response delays the project.

Shop drawing preparation and review often reveal problems due to vague design details, wrong dimensions, and plans that do not fit together. When examining a claim involving shop drawing review and approval, obtain a copy of the submittal stamped by the designer and study the transmittal forms and submittal log. To determine if a change has been made in the approval process, compare the shop drawing and notes to the original drawings. If necessary, ask for a review by the subcontractor/supplier involved or by an independent expert.

Some problems with shop drawing review are caused by the general contractor not reviewing shop drawings before forwarding them to the owner's representative. Many contracts require such a review, but some general contractors spend insufficient time reviewing subcontractor's and vendor's shop drawings.

## 8.  Methods of specifying

There are five basic methods by which designers specify construction performance. The contractor's responsibilities and ability to recover damages vary depending upon how the work is specified. Contractors need to understand the implications of each method and how to minimize their risks.

**1.  Proprietary.**   One of the most common methods of specifying manufactured materials or equipment is by stipulating one or more (often three) manufacturers' products or an " approved equal." Sole source specifications can greatly increase costs and put the contractor at the mercy of a greedy or unresponsive supplier.

The problem with the "or equal" clause is the risk that a less expensive material proposed by a supplier may be rejected. If the cost difference is significant, contractors risk losing the bid if they do not use the less expensive material, or losing money if they win the contract and the alternative isn't accepted as an equal. Unreasonable rejection of an essentially equal product is cause for a claim.

**2.  Prescriptive.**   The second most common type of specification is prescriptive. It details the materials and methods to be used. If the contractor uses the materials and follows the means and methods specified, and the result is unsatisfactory or cannot be accomplished, the end result is the owner's responsibility (based on the Spearin doctrine).

**3. Performance.** Performance specifications describe the end result, not the means. The CSI definition of a performance specification is "a statement of required results, verifiable as meeting stipulated criteria, and free of unnecessary process limitations." A true performance specification states only the desired end result and places no restriction on the methods and materials used.

Mechanical and electrical systems are often specified under a performance spec, also called a design/build specification. When specifying performance, it is essential that the specifications be well written, comprehensive, and lead to a satisfactory end product.

One inherent risk in performance specifications is they usually require that the contractor's design conform to all laws, regulations, and building codes. A combination of factors, or a building official's overly strict code interpretation, may result in a more stringent code requirement than anticipated and increased costs.

**4. Reference standard.** Sprinklers and mechanical and electrical systems are often specified by reference standards, such as those of the American Society for Testing and Materials (ASTM), the American Concrete Institute (ACI), and others. In some cases, designers will require that the work meet any and all applicable codes and ordinances. Although suitable for a performance specification, as for sprinkler and HVAC systems, reference standards create problems when applied to trades that do not normally design their work.

**5. Cash Allowance.** Electrical and plumbing fixtures, carpet, and owner furnishings to be selected later by the owner are sometimes specified as a cash allowance stated in the specifications. The contract amount is adjusted for the actual cost of the selected material.

## 9. Contract price

Contracts can be priced in three ways, as follows:

**1. Firm fixed price.** A firm fixed price is either lump sum (for buildings) or unit price (for civil works). Lump sum contracts frequently have one or more additive or deductive bid alternates. These alternatives may be used with the base bid to determine the low bidder, depending on how the bids compare with the budget. Unit price contracts often use a combination of unit prices and lump sum prices for individual bid items. The bid items are listed in a bid schedule with the estimated quantity of work and space for the contractor to write in a unit bid price. Unit price contracts are useful when the quantities may vary from plan or cannot be determined in advance.

Contractors spread their overhead and profit among the various bid items on unit price contracts, since there isn't a bid item for overhead. Excessive variations in the quantities of unit price contracts can be cause for a claim. For example, a federal agency listed 800 sewer liner cutouts in the bid schedule, but the actual quantity (and the estimated quantity from their in-line television survey) was about 120. The contractor claimed the underrun, even though there was no variation in quantities clause, because of the extreme variation and the concealment of a material fact. The contractor was paid most of the costs, but only after extended negotiation.

**2. Cost plus or time and materials.** Cost plus contracts are based on cost reimbursement for actual direct costs and jobsite overhead plus a fee. The fee can be either a fixed amount or a percentage of final cost, with or without a guaranteed maximum cost. Force account work is paid as cost plus.

Disputes over a cost plus contract may include the unit rate of contractor-furnished equipment; whether small tools are included in the fee; and whether some items are job expenses or part of home office overhead and therefore included in the fee instead of being reimbursable.

**3. Guaranteed maximum price (CM/GMP, also called CM/GC).** A construction management (CM) contract with a guaranteed maximum price (GMP) is cost plus with a limit on the total cost. In some cases, the contractor shares a portion of the savings. CM/GMP contracts are used more in building construction than in civil works. They are often fast-tracked, with the contractor giving input to the design process and providing the GMP at an early stage of design (often at 30 percent of design completion). From then to design completion, the contractor provides "value engineering" input to keep the cost of the project within the GMP while maintaining the scope as defined when the GMP was prepared.

The most frequent types of dispute on a CM/GMP contract are the scope of work included in the GMP and delays by the designer. The completeness and thoroughness of the GMP plans and the contractor's GMP estimate working papers are crucial to proving what was included in the scope. CM/GMP contractors need to prepare a detailed estimate with clearly defined quantities and assumptions, and to provide their complete budget computations to the owner. They need to be assertive in pushing for adoption of their value engineering proposals to keep budgets from being exceeded. If overruled, CM/GMP contractors need to insist that the GMP be increased for the increased scope.

Avoid design delays on CM/GMP contracts by (1) including a drawing release schedule in the contract, (2) tracking design to ensure on-time completion, and (3) giving prompt notice if the design is late.

## C. Fundamentals of Contracts and Contract Interpretation

Simply stated, a construction contract is an agreement between two parties for performance by one party and payment by the other. The law recognizes performance and payment as duties and provides a remedy if one party fails to perform (i.e., breaches) the terms of the agreement.

### 1. Fundamentals of contracts

The contract should state the work to be performed, the price, and when the contractor will be paid. The contract determines what has to be done and how to compute time and compensation for work beyond the contract. The contract terms are critical in determining the rights and obligations of the parties in the event of a dispute. If the issues are not spelled out, you must rely on provisions of applicable common law or statues.

## 2. Elements of a contract

There are six major elements to a contract:

**1. An offer.** There must be an offer, not just an expressed desire. The offer must be clear, complete, and communicated. It is valid for a "reasonable" duration and can be revoked prior to acceptance unless otherwise specified. A bid, for example, is an offer, but a request for bids is not. A contractor's bid to build a house is probably valid a month later, absent some limiting condition in the proposal or a notice canceling the proposal. However, it is not reasonable to hold a builder to a price quoted 6 months earlier.

**2. Acceptance.** Acceptance may be explicit or implied. Acceptance must be absolute and unambiguous. Silence is not acceptance, but action based on an offer may be. If acceptance is qualified or in any way differs from the offer, it becomes a counteroffer which then must be accepted by the other party before it becomes binding. Acceptance may also be made by commencing performance.

**3. Agreement.** There must be a meeting of the minds. A mutual mistake can allow either party to void a contract, but a unilateral error does not. Clerical errors can be set aside to avoid injustice. Ambiguities are usually construed against the drafter of the document. Fraud or misrepresentation may enable the victim to void the contract.

**4. Valid consideration.** Valid consideration is the actual or promised exchange of money, goods, services, or some action or inaction.

**5. Between competent parties.** The parties must have valid authority to contract for their organization. For example, if a public body exceeds its legal charter, it cannot be forced to comply with its own contract. Another example is contract signatures; often only the governing body can sign a contract. On public works contracting, a contractor must know whether the person representing the agency can sign contracts to bind the agency. A contract signed by either a minor or someone who is mentally impaired (e.g., from excessive alcoholic consumption) can be voided. In addition, unlicensed contractors usually cannot enforce payment even if they have fully performed their obligations.

**6. Legal form and content.** Contracts must conform to legal requirements. For example, contracts for the sale of land or contracts extending more than one year must be in writing. Contracts for actions that are illegal or against public policy are unenforceable.

Oral contracts and oral modifications to contracts are generally enforceable, but difficult to prove. Most construction contracts contain provisions requiring written authorization to perform extra work, which would generally negate an oral agreement or contract modification. Written agreements are highly recommended. Legal advice should be sought if trying to enforce an oral contract.

*The most important factor ensuring a good contract is the reputation and good faith of the other party. No contract, no matter how well written and carefully enforced, will adequately protect you if dealing with someone not acting in good faith.* If you have a legal dispute, the cost and potential loss from the legal action will be excessive even if you win.

All contracts, even brief letter contracts, must contain all six of the elements described above. Initially, some of the elements may not be fully defined, such as the contract amount, schedule, or other requirements, but the parties must agree to try to fill in the blanks later. For example, a letter contract for a fast-track, privately financed CM project may require the contractor to start work immediately while design is being completed, with the guaranteed maximum price to be determined later. Or, the parties' history, or course of dealing with these issues in the past, may be implicitly relied upon.

## 3. Rules of contract interpretation

Contract interpretation begins with a careful reading of the contract. *Only if ambiguous will a contract be interpreted outside the contract* (i.e., from other documents, legal precedents, or the conduct of the parties). The rules of contract interpretation include:

1. The expressed intent of the parties in the written agreement is the most important factor. If this is unclear, the intent of the parties may be inferred from the documents or from their conduct. For example, a history of paying for a certain type of changed work on other projects may be used to require payment for that same work on this contract.

2. The interpretation of ambiguities must be reasonable, as made by an experienced, prudent contractor.

3. The ordinary meaning of words is used, with technical or industry-specific terms given their normal interpretation, unless the intent is clearly otherwise.

4. The implied terms necessary to comply with the explicit terms become part of the contract (e.g., forming concrete is implicitly required by a contract to place concrete).

5. Words and phrases must be interpreted within the context, with all parts given equal weight.

6. Special conditions prevail over general conditions, handwritten terms prevail over printed terms, and words prevail over numbers.

7. Absent any other reasonable interpretation, ambiguities are construed against the drafter of the document. However, the other party must demonstrate reliance on the contrary interpretation and prove the interpretation was reasonable.

8. Clerical errors are discounted or ignored, unless reasonably relied upon. A mutual mistake (i.e., both parties misunderstood the true facts of an issue) can be reason for reforming a contract to reflect both parties' intent.

9. Oral or superseded written agreements may be used to interpret an ambiguous written contract. Prior negotiations are assumed to be merged into the document and cannot modify unambiguous clauses. They can be used to clarify ambiguous terms.

10. The conduct of the parties helps to interpret a contract or to imply a contract, and can modify or supersede portions of an agreement.

11. Separately negotiated, or added, terms and conditions supersede originally included material or materials that weren't separately negotiated. Therefore, all handwritten changes should be initialed and dated by both parties.

Contract interpretation is a question of law that is ultimately decided by judges or arbitrators, not by juries. Disputed questions of fact (including the parties' intent) are decided by fact finders—a jury, a judge acting without a jury, or an arbitrator.

**An example of contract interpretation.**   A contract that requires excavation to a "reasonable depth" before backfilling is ambiguous. In resolving the issue, a judge or arbitrator first reads the contract as a whole to see if there is any explanation within the contract for "reasonable depth." If another part of the contract requires excavation for a similar condition to at least 10 feet, then 10 feet may be interpreted as reasonable. If the contract is silent, the judge or arbitrator then goes outside the contract, i.e., to extrinsic evidence. First, the negotiations and communications prior to contract award are reviewed. Second, the prior dealings between the parties are examined. For example, a comment during a prebid conference that poor soil conditions may extend down 6 feet would indicate that 6 feet was reasonable. Or, an 8-foot depth required on a previous contract might be considered reasonable. If negotiations and prior communication shed no light on the issue, the judge or arbitrator looks at local or national practice, which may, for example, be 4 to 5 feet.

**Patent versus latent ambiguity.**   If an ambiguity is patent (obvious), a contractor has a duty to alert the owner. Failure to alert the owner may make the contractor responsible for the increased cost. In most cases, it is difficult to see how an error can be patently obvious to a contractor during the 2- to 4-week bid period, when it wasn't obvious to the designer and owner during the 6- to 12-month design process.

**Industry practice.**   Industry practice can be crucial to interpreting a contract. If the contract is unclear or does not address an issue, industry practice is examined to determine what is standard and how that standard is applied. For example, if the specifications do not provide tolerances, industry practice may be used to determine whether sheetrock is so irregular as to require replacement. The UCC definition of "usage of trade" (U.C.C. Sec. 1-205(2)) is

> ...any practice or method of dealing having such regularity of observance in a place, vocation, or trade as to justify an expectation that it will be observed with respect to the transaction in question.

## 4.   Privity of contract vis-à-vis the designer

Damages to a contractor due to actions or inaction by the designer ordinarily must be claimed against the owner. If held liable, the owner may seek recovery from the designer in a separate action.

Under the American Institute of Architects' (AIA) document A201 (general conditions of the contract) and case law, the designer assumes certain legally enforceable contractual duties to the contractor. These include noninterference, reasonable contract administration, impartiality in the resolution of disputes, etc.

The owner implicitly warrants the accuracy and sufficiency of the contract documents and is therefore liable for the contractor's costs resulting from inadequate design. In most states, the Economic Loss Rule, which prevents the recovery of purely economic losses from a party absent a contract, precludes a tort action by a contractor against the designer. In some states, however, contractors may pursue a tort action against the designer. This litigation is outside the contract and is based on the common law principle that the designer has a duty of reasonable care to third parties who either build or use the project. Third parties injured on the project after completion of construction, or subsequent owners of the project who discover construction deficiencies can also pursue a tort action against the designer or contractor for alleged design or construction errors (within a specified period as determined by the applicable statute of limitations).

## 5.   Contract performance

Performance is fulfilling the requirements of the contract. Compliance is defined in a number of ways.

**Substantial performance.**   Although strict performance is expected, contract obligations can be satisfied by substantial performance, if all of the following conditions are met:

- All important work is performed, with *unintentional* variations only in unimportant details.
- The overall project is unimpaired and will function as desired.

- Defects or variations can be readily corrected by the owner at little expense.
- The owner is adequately compensated by a reduction in the contract price or continued performance.

**Strict versus substantial compliance.**  Strict compliance is required of work that can be precisely measured (e.g., cubic yards of concrete, number of tables, height of doors, etc.). Substantial compliance is allowed for work that cannot be precisely defined (e.g., workmanlike manner, smooth, etc.).

**Substantial completion.**  Substantial completion varies according to the laws of the particular jurisdiction and any definitions in the contract. It generally requires a high degree of completion *and* beneficial occupancy of the project (i.e., when it can be used for its intended purpose). For example, a building that is 99 percent complete but without code-required fire alarms is not substantially complete; nor is a school that is occupied for classes but lacks the specified gymnasium and cafeteria.

**Time of performance.**  Failure to perform strictly within a specified time is a material breach if the contract states that "time is of the essence" or provides a specific completion date.

**Approval of performance.**  When approving performance, the owner's representative must act in good faith and cannot be arbitrary. In the case of utility or fitness, performance that would satisfy a reasonable person, rather than the individual making the decision, usually suffices if challenged in court. Rejections on the basis of aesthetics are hard to contest.

Approval does not excuse the performing party from damages for latent defects, but approval (or even failure to reject) of obviously defective work (after a reasonable time) excuses the contractor from having to correct it.

**Excuses for nonperformance.**  Nonperformance is a breach of contract. The following exceptions excuse nonperformance:

- *Impossibility of performance*—for example, the building to be painted burns down. Impossibility does not refer to extreme difficulty, inconvenience, beyond the capabilities of the party responsible for performance, or even to being economically ruinous.
- *Economic waste*—is a legal theory that may be used to avoid replacing inadvertently noncompliant (but largely conforming) work that will perform as well as the work specified when replacement would be a waste of economic resources.
- *Intervening illegality*—when the actions agreed to in the contract become illegal.
- *Act of God*—natural disasters beyond the control or contemplation of the parties.

- *Fraud*—by one party may excuse the other party from performance.

- *Material breach or prevention by the other party*—excuses the first party from performance and permits legal action for damages.

**Remedies for nonperformance.**  Failure to perform as specified in a contract, when nonperformance materially affects the outcome, is a breach of contract. The remedies for breach of contract are as follows.

- *Damages*—Compensatory damages are to compensate the injured party for losses and are not intended to materially better the condition of the injured party. Damages for delay can be either liquidated or actual. When damages occur, the damaged party must attempt to minimize costs, as they cannot recover for damages that could have been mitigated.

  Liquidated damages are specified in the contract as a dollar amount per day for failure to complete on time. To be enforced, liquidated damages must be (1) reasonable, (2) difficult to fix with certainty and estimated in good faith before the fact, and (3) approximately equal to anticipated damages. Liquidated damages need not equal actual damages to be enforced.

  Actual or compensatory damages are the actual costs incurred by the damaged party, as determined after the fact. They can be enormous but must be reasonably foreseeable by the parties when entering into the contract. For example, the repair of defective work would be recoverable but lost profits may not be, if they are beyond what would be reasonably foreseeable. Penalties are disallowed, although a bonus for specific performance (e.g., early completion) can be offset by an equal credit for failure to perform.

- *Restitution*—Restitution applies to contracts for goods. It provides for the goods to be returned to the damaged party.

- *Specific performance*—Courts may order specific performance, for example, contract completion, as when a seller improperly refusing to complete a contract to sell real estate is compelled to do so.

- *An injunction*—An injunction is a court order requiring the party at fault to take, or refrain from taking, specific actions (e.g., forcing a contractor to return to work). Injunctions are used infrequently, and only if monetary damages cannot compensate the injured party.

## 6. Contract termination

There are five ways to terminate a contract:

- *Performance*—The primary means of terminating a contract is performance by both parties as specified in the contract.

- *Pursuant to the terms of agreement*—The contract may have a provision for terminating a contract separately from performance.

- *Mutual agreement or unilateral contract right of the owner*—The parties can mutually agree at any time to terminate the contract in whole or in part, or to form a new contract. In addition, most public works contracts allow the owner to terminate the agreement under specified conditions for the owner's convenience.

- *Impossibility of performance, or intervening illegality*—A contract will be terminated if it becomes impossible or illegal to perform.

- *Breach of contract and judicial termination*—If one party materially breaches the contract (by not performing or by preventing the other party from performing), the other party may terminate the contract and seek damages by seeking judicial relief.

## 7. Duty to disclose

The owner has a duty to disclose superior knowledge of conditions materially affecting contract performance. If the owner withholds such information and construction is adversely affected, the contractor can claim the additional cost. The contractor has a duty to seek clarification of obvious errors, omissions, or ambiguities. Both parties must disclose vital information that may result in unsuccessful performance. For example, if aware that a contractor is installing work incorrectly, the owner has a duty to inform the contractor.

The most obvious example of failure to disclose concerns subsurface conditions, such as the presence of rock, overly wet material, a high water table, etc. Other examples include:

- Unusual soil conditions which in combination with another event can cause disastrous impacts (e.g., some clay soils become extremely unstable or swell when saturated).

- Unusual site conditions that result in extremely difficult working conditions not apparent from a normal site investigation during dry weather (e.g., unexpected flooding from a small creek after normal winter rainfall). An exception might exist if the flooding was well known by local contractors and discernible from asking local residents.

- Limited access not apparent from the drawings and a site visit.

- Existing building conditions for a remodel not apparent from a walk-through.

- Other contractors on the site or adjacent to it are likely to interfere with the contractor's work.

## 8. Preservation of contract rights

When signing change orders, contractors should include a time extension and the cost of impact and delay in the change order request, or reserve their rights to make a later claim.

When an owner gives a unilateral change order requiring the contractor to proceed immediately with costs to be negotiated later, contractors should immediately notify the owner that they (1) will proceed, (2) are establishing a job cost code for the extra work, (3) are preparing a time impact analysis to determine the schedule impact, and (4) will submit a change order request and a time extension request (if there is delay) as soon as possible. Likewise, if an owner's interpretation of the contract causes a constructive change, the contractor should notify the owner that they are proceeding under protest and will be recording the costs and filing a change order request (or claim) when the work is complete.

### 9. Additional information

For more information on contract law, refer to Dunham and Young's book, *Contracts, Specifications, and Law for Engineers*[49]; Ralph C. Nash, Jr.'s *Government Contract Changes,* published by Federal Publications, Inc.[88]; Justin Sweet's *Sweet on Contract Law,* published by the American Bar Association Forum on the Construction Industry, 1997; or one of the other popular references on contract law.

## D.   Other Legal Concepts and Terms

### 1. Lien and bond rights and the Miller Act

Contractors and material suppliers should consult an attorney regarding notice requirements under state lien and bond statutes. Some states require predelivery notices be sent to the owner and mortgagees to establish priority or to preserve lien rights or rights to attorney fees. Many states have short notice and filing time requirements from the completion of work to when a lien must be filed.

Liens are not possible on contracts with public entities, but payment can be pursued under the federal Miller Act for U.S. government contracts, or under little Miller Acts for state and municipal contracts. The Miller Acts require contractors to provide performance bonds and payment bonds. Performance bonds protect the owner in case of default by the contractor, and payment bonds protect against the general contractor's failure to pay workers, subcontractors, or suppliers. Payment bonds also make it easier for a subcontractor to get paid for disputed work, although notice must be given within a specified time.

Be aware that public agencies occasionally fail to obtain the required bonds and that sureties sometimes become insolvent.

### 2. Discovery, attorney-client privilege, and attorney work product

In litigation, the general rule is that all relevant information, including documents and testimony, may be required to be disclosed to the adverse party. Exceptions are called privileges. Certain communications between attorney and client are privileged and are exempt from discovery or testimony. Attorney work product prepared in anticipation of litigation, such as an issue analysis prepared by an independent expert for an attorney, is normally privileged. Discovery begins after the complaint is submitted in litigation or a demand has been made for arbitration. Discovery is the process by which the parties exchange documents, identify witnesses, and depose witnesses (question under oath by opposing counsel). Discovery in arbitration is quite limited, unless the parties have agreed otherwise.

Claims preparers should prepare their work product for the attorney, if one is involved, to prevent or at least delay discovery. Assume that any work prepared for a claim that is not prepared under the request or authority of an attorney is discoverable. See Sec. B.13 of Chap. 9 for comments on the risks of discovery of an expert's work product and Sec. B of Chap. 8 for recommendations on how to assist with discovery.

### 3. Hearsay and the rules of evidence

Contractors should know that if a dispute is litigated, they may not be able to tell their story as they would in a normal conversation. The rules of evidence, which regulate how evidence (witness testimony and documents) is presented to the fact finder, will apply. For example, documents must be authenticated, i.e., proved to be business records that are kept contemporaneously and customarily in the course of business. Records maintained in the course of business are those documents that a contractor normally and regularly maintains; they are not special documents related to a claim. Hearsay will not be allowed. This means that the project manager normally will not be able to testify as to what the supervisor said. Only the supervisor can testify to that.

Arbitration is considerably less strict. Arbitrators, mediators, and dispute review boards seldom decline to hear testimony because it is hearsay. They do, however, consider the strength of the evidence and often discount such testimony.

### 4. Witnesses

The two types of witnesses are fact witnesses and expert witnesses. Fact witnesses may testify to what they saw and heard and are not generally allowed to give their opinions in testimony. Expert witnesses are hired to give expert opinion based on past experience and knowledge of similar projects.

## E. Contract Clauses Affecting Change Order and Claims Resolution

Claim preparers and construction personnel involved in contract issues should know how common contract clauses affect change order requests and claims. Contractors' estimators should check these clauses when bidding and notify management of problems so an appropriate markup for risk is added and a risk prevention plan is implemented. If a clause appears overly onerous, do not bid the job. Project managers, project engineers, superintendents, and foremen should be aware of contract provisions posing special risks and should know how to respond if the provisions become applicable.

---

**Clauses to Review When Preparing a Claim**

1. Permits and Fees
2. Submittals and Substitutions
3. Scheduling
4. Coordination and Coordination with Others
5. Interpretation, Intent, and Design Errors
6. Reference Standards
7. Performance Specifications
8. Exculpatory Disclaimers
9. Interpretation of the Contract
10. Differing Site Conditions
11. Inspection and Administration
12. Notice Requirements
13. Changes
14. Disputes
15. Value Engineering Change Proposals
16. Suspension and Termination
17. Delays and Time Extensions
18. No Damages for Delay
19. Final Inspection, Close-Out, Final Payment, and Warranties
20. Hold Harmless, Defense, and Indemnity
21. Progress Payments
22. Pay When Paid and Pay If Paid
23. Variation in Quantities
24. Liquidated Damages
25. Geotechnical Reports, and Environmental Clauses
26. Certification

---

**Risk allocation and onerous contract clauses.**  Risk allocation is the process of assigning the financial risk of unexpected conditions, events, and problems between project participants. Risk allocation is traditionally based on who has control of the problem, who can bear the financial cost, and who has superior bargaining power when forming the contract agreement. Contracts generally favor the author, usually the owner and designer for prime contracts and the general contractor for subcontracts.

Recent trends are changing the traditional allocation of risk and pose special problems in some construction markets. The weaker party to the agreement seldom adds sufficient contingency for the additional risk, has little or no control over the risk, may experience financial ruin if the event or condition occurs, but must accept the risk to compete in that market.

Contractors should identify the contract clauses that allocate risk and determine if they differ from industry practice. If the risk allocations are onerous, they should either not bid the project or deal with the risks proactively. Don't hope they will not be invoked. The recommended approach is a Loss Prevention Plan as described in Sec. D.2 of Chap. 4.

Some clauses are so grossly unfair that the courts are reluctant to enforce them as they are believed to be against the public interest. A definition of an onerous contract clause is one that (1) is written by a party with significantly superior bargaining power; (2) is not in accordance with industry practice; (3) reallocates a risk that the first party has more control over to the second party; and (4) could result in serious financial loss to the second party. Unfair risk allocation is generally encountered in contracts between parties with unequal bargaining power or knowledge. Although they *may* serve the narrow short-term interests of the party preparing them, they prevent a partnering relationship between the parties, are detrimental to the health of the industry, and may be negated or circumscribed by an effective Loss Prevention Plan and a well-argued presentation before a sympathetic arbitration panel, judge, or jury.

## 1. Permits and fees

Unusual, or newly imposed, fees might not be anticipated or the amount readily determined. Late issuance of permits (particularly railroad permits) may delay the project. If notice to proceed is tied to permit issuance, the contractor has little control over the start date and may be unable to collect damages for a late start that forces work into winter. Late issuance of permits can also result in uncompensated losses on design/build contracts, as the contractor is responsible for design.

## 2. Submittals and substitutions

Submittal review time can be a problem, especially if the time allowed for review is lengthy and the project schedule is tight. A number of ways to minimize submittal problems are described below.

- *Implement good scheduling practices* that identify material delivery dates, thereby determining the dates that submittals are needed and avoiding construction delays from late delivery.

- *Submit requests for substitution* as soon as possible, to allow time to negotiate or to find alternatives in case the substitution is denied.

- *Implement a timely procurement effort* to ensure purchase orders are issued early enough for timely delivery and to enforce specified submittal and delivery dates.

- *Use a submittal log* that tracks scheduled and actual receipt of submittals from subcontractors and suppliers, their transmittal to the designer, and return.

- *Implement a quality assurance program* that ensures suppliers provide complete, contract-compliant submittals in a timely manner, and checks submittals before forwarding them to the designer.

- *Phase submittals and prioritize them for review* when large numbers of submittals are required.

- *Identify when a response is needed* on all submittals to ensure timely owner response.

- *Inform the designer if a submittal is needed earlier than specified* and warn the designer that failure to make reasonable efforts to respond may be a cause for damages.

- *Expedite* when delays occur.

Substitutions under the "or equal" clause are a frequent source of conflict between the contractor and the designer. A contractor is never certain that the less expensive material is truly "equal" and that the designer will accept it. Contractors should ask the owner to list the most important characteristics of the equipment or material, in order to understand what requirements must be met. Failure to make prebid inquiries does not preclude a later claim on the issue, unless the issue is obvious and should have been brought to the owner's attention prior to bid. If the contractor makes a prebid inquiry and the owner fails to respond, then the contractor can submit a claim on that issue as long as the interpretation is reasonable.

## 3. Scheduling.

Critical path method (CPM) scheduling is the most effective scheduling tool. Although it appears to be complex, CPM is actually rather simple. For a detailed discussion, see Chap. 10.

**Ownership of the float.**    Conventional practice and many CPM scheduling clauses state that float belongs to the project and the first party needing float. Some even state that float is for the benefit of the owner. However, float is needed by contractors as it allows a contractor to shift crews from one noncritical activity to another activity if problems are encountered. It also protects a contractor from scheduling errors or overly optimistic estimates of activity durations. A schedule with little float and multiple critical paths is more expensive to build than one with only one critical path and reasonable float on other activities.

The contractor prepares the schedule, and the owner has little recourse but to accept it unless it appears grossly inadequate. Therefore, contractors are free to prepare a schedule with multiple critical paths and little float. In the extreme, this approach is unsatisfactory as it degrades the value of the schedule for managing the project. See Sec. F.5 of Chap. 10 for a detailed discussion of float and how contractors can protect it from owner abuse.

**Recommended allocation of float.**   The following allocation of float would better serve the industry than conventional practice and should be pursued by contractors:

   1.  **Contemporaneous resolution.**   If the owner delays a noncritical activity, the float goes to the owner—unless the contractor protests based on the issues described under 2 below.

If the contractor later delays the same noncritical path, exceeding the float and delaying the project, the contractor absorbs the delay costs and the delay is noncompensable. However, under those circumstances it seems unfair to assess liquidated damages, although that is the conventional practice. It would be more just if the contractor's delay (up to the amount of the owner's delay) were excusable but noncompensable. There are other strong incentives for the contractor to finish as quickly as possible to save extended overhead, without being punished for inadvertent delays.

   2.  **Unresolved delay.**   If, instead of contemporaneous resolution (when the delay occurs), (1) both parties delay a noncritical path over a period of time; (2) no contemporaneous time impact analyses are made and no resolution is reached between the parties regarding the delay; (3) when delaying the project the owner did not rely upon the work being noncritical; and (4) the responsibility for delay is later determined by a thorough schedule analysis, then the allocation of float *should* be as follows:

- Weather delays (and other excusable but noncompensable delays) and contractor delays in the order that they occur. If exceeding the float, subsequent weather delays are excusable but noncompensable, and subsequent contractor delays are nonexcusable.

- Owner delays, compensable to the contractor if exceeding the remaining float, unless the owner relied on the activities being noncritical when delaying the work, in which case the delay should be excusable but noncompensable.

**Early completion.**   Many projects can be completed in less time than specified by the contract. Astute contractors prepare a prebid schedule determining the duration actually required and include jobsite overhead for that duration only. These savings are then passed on to the owner in the form of a lower bid, and the contractor may win the project because of better scheduling.

Some owners refuse to approve schedules with an early completion date, in spite of the contractor's clear right to complete early, because the owner fears a delay claim. Contractors should preserve their right to complete early by changing the schedule only under protest, and by showing punch list, demobilization, or other activity prior to substantial completion. The owner must be put on notice that any owner-caused delay, even if it does not extend the project past the contract-specified completion date, will damage the contractor and that compensation will be due. The notice should declare that the owner has a legal obligation that supersedes the terms of the contract to not interfere with the contractor's right to finish early.

**Proof of delay claims.**   To be compensated for a delay, contractors must prove (1) the delay was owner-caused, (2) the entire project was delayed, and (3) the amount of delay. Contractors will also need to disprove concurrent contractor-caused delay and prove that every reasonable effort was made to mitigate the delay and its impacts.

Delay can be direct (if work is suspended or stretched out by owner actions) or consequential (if work not directly affected by the owner's action is also delayed).

The only practical way to prove delay is with a critical path schedule analysis, which examines both the as-planned and the as-built schedules and determines why they are different. Contractors should prepare a detailed schedule before starting work and revise it when a delay or other schedule change occurs. Too many contractors simply record the actual start and finish dates of the original as-planned activities and fail to record revised logic, deleted or added activities, etc.

To prove damages, you must show a direct, causal link between the delay and the alleged damages.

**Concurrent delay.**   Concurrent delay exists when both the contractor and the owner simultaneously delay the project, or when one party delays the project at the same time as an excusable delay (e.g., abnormal weather). The contractor is not assessed liquidated damages and is not compensated for the delay costs for concurrent delay. Each party normally bears its own costs. In the past, when unable to separate the responsibilities with an accurate as-built schedule, the courts sometimes apportioned the costs between the parties. This should not be necessary if you create a detailed as-built schedule.

Contractors should not slow down during an owner delay without notifying the owner's representative of their efforts to mitigate damages. Otherwise the owner's claims analyst may claim concurrent delay.

**No damages for delay clauses.**   See Sec. D.18 below.

## 4.  Cooperation and coordination with others

The coordination clause requires a contractor to cooperate with others, including other contractors on the site, with whom there is no privity of contract. The owner is the umbilical cord connecting all of the contractors working onsite and others whose performance is necessary to the work being performed. Thus, if one contractor interferes with another's work, the second contractor can claim constructive change by the owner.

**Multiple prime contract coordination clauses.** These clauses require several independent contractors on a project to coordinate among themselves. The owner's representative takes little or no responsibility for resolving disputes. Some contracts give a construction manager authority to adjudicate conflicts, while others provide a master schedule with milestone interfaces and coordination points. In any case, problems frequently occur and resolution is difficult.

Luckily there are ways to minimize the problems:

1. Prepare a detailed prebid CPM schedule. If appropriate and approved by legal counsel, submit it to the owner prior to the bid as unsolicited information.
2. Upon notice of award, prepare a detailed schedule and submit it to the owner. Provide notice that the bid price (including jobsite overhead, supervision, equipment, etc.) is based on the schedule, the start and finish dates, and accomplishment of key milestones, and does not include overtime or other acceleration costs.
3. Update the schedule monthly, revise it when required, and take appropriate action, if necessary, including giving notice to the owner.

**Award of later onsite contracts by owner or other interference.** The owner should honor the contractor's schedule and not award a later contract that interferes with the work (e.g., untimely carpet installation). The later contract should conform to the contractor's first schedule, unless that schedule can be easily adjusted to accommodate the second contract.

**Action against other contractors.** Coordination clauses may result in the contractor, instead of the owner, having to take action against another contractor for delaying the project. Such clauses make resolution difficult and expensive.

## 5. Intent and design errors

Intent clauses require the contractor to complete the project as the designer "intended," even if the design is not clearly specified. The wording in most intent clauses requires minor items, customarily performed and obviously necessary, to be included in the price bid. Some owner's representatives try to interpret these clauses to require contractors to correct design errors.

A typical clause reads as follows:

> *Omissions.* Omission from the contract documents of details of work which are manifestly necessary to comply with the intent of the drawings and specifications, or which are customarily performed, shall not relieve the Contractor from performing such work.

Some clauses, however, contain subtle wording that assigns responsibility for the designer's errors and omissions to the contractor, regardless of the scope. Some require the contractor to notify the owner of all discrepancies prior to bidding. Such a requirement is impractical and patently unfair.

To challenge a designer's assertion that an item is required under the intent clause, explain that the disputed work is one of the following:

- Not essential for completing the work that is detailed.
- Not usually performed by the craft doing the work that is detailed.
- Not the only way to accomplish the desired result.

**Defective plans and specifications.**    As established by the U.S. Supreme Court in the Spearin case, the project owner implicitly warrants the suitability of the plans and specifications. The contractor is not responsible if the completed project does not perform as expected, or if it cannot be constructed as designed. To protect themselves from delays and impact, contractors should perform constructibility reviews before starting construction, as described in Chap. 4. Otherwise problems will be encountered as the work is being installed, with much more disruption and delay.

The owner's designer is responsible for specifying the materials that are available, but the contractor also has a responsibility to identify problems during bidding.

## 6.  Reference standards

Reference standards (e.g., ASTM, ACI, UBC, etc.) are often used as a catchall to force contractors to comply with code requirements that should have been incorporated in the plans but were missed by the designer. This can be successful if the work is not detailed and is labeled a performance specification. If the work is detailed and the item is shown incorrectly or is omitted, the contractor should be compensated.

I once asked a large, international design firm to see one of the references quoted in their specifications. Even their main office library didn't have a copy, which suggests a defense when some obscure reference standard is quoted to avoid paying a contractor for extra work. Ask the designer if they have read the reference, if it has been enforced on other contracts, and whether it was their intent that such a provision be applicable to this project.

## 7.  Performance specifications

Performance specifications are sometimes called design/build but are not the same as the design/build of an entire project. Performance specifications are most common for mechanical, sprinkler, and electrical systems. Problems develop if means and methods are specified for one part of the system and performance standards for other parts, and the two are incompatible. Other potential risks to the contractor include professional liability for design errors, possible failure to meet building codes, and performance standards that are not reasonably achievable.

## 8. Exculpatory disclaimers

The most common exculpatory disclaimers are those for geotechnical data. The data are provided for the contractor's use, while the owner disclaims responsibility for the data's accuracy or completeness. Exculpatory clauses attempt to shift the risk of unknown conditions to the contractor. In many jurisdictions, overreaching exculpatory clauses are not enforced, especially if the contractor is encouraged to rely on the data, or if the clauses conflict with a differing site conditions clause.

Nonstandard exculpatory disclaimers often foreshadow design problems and an uncooperative owner. Use caution when bidding these contracts.

## 9. Interpretation of the contract

Most contracts designate the designer as the interpreter of the contract documents. This seems logical, since the designer prepared the documents. Unfortunately, designers sometimes confuse what they intended the documents to say with the way a reasonable bidder interprets them. Another problem is that a designer's interpretation may result in a constructive change to the contract, although the authority to make changes is retained by the owner. In any case, contractors should request clarification of ambiguous clauses in writing as soon as possible.

## 10. Differing site conditions

Differing site conditions are categorized in one of two ways: (1) differing materially from what was indicated on the bid documents or (2) differing materially from conditions ordinarily encountered or inherent in this type of work. Differing site condition clauses generally refer to unknown subsurface conditions such as the presence or absence of rock, unfavorable soil, groundwater, or some other natural condition. However, manmade conditions from earlier or ongoing construction may also qualify as differing site conditions. Weather is not a differing site condition but in combination with unusual soil (extreme sensitivity to moisture) can result in a differing site condition.

Many contracts require the contractor to have made a reasonable investigation of the site during bidding, before allowing a differing site condition claim. In federal cases, a contractor can claim differing site conditions without having investigated the site but will be held to the results of a reasonable site inspection. A contractor need not conduct independent testing or use an expert but generally must visit the site, review the soil reports and boring logs or samples, and verify local weather and soil conditions from records, local residents, or past experience. Missing a hidden Indian burial ground does not preclude a differing site condition claim, but failure to anticipate high groundwater in gravel next to a river would.

The presence of a differing site condition clause or a disclaimer determines whether there is entitlement for increased time or cost. Some contracts include geotechnical baseline reports. Federal contracts include a differing site condition clause, as do most state and local public works contracts, but many private contracts do not. However, a theory of warranty of plans and specifications may apply. Failure to disclose actual knowledge (e.g., superior knowledge) generally establishes a contractor's rights for damages, regardless of the existence of disclaimers or the absence of a differing site conditions clause.

## 11.  Inspection and administration

The owner's representative exercises control of the contractor's performance by administering and inspecting the work. Administration and inspection must be made in good faith, as specified in the contract.

**Overinspection.** If the owner's representatives (1) are overly strict on inspection, (2) exceed the intent or defined requirements, or (3) require work not specified in the contract, the contractor is due compensation and additional time for a constructive change or interference. The distinction between reasonable and overly strict inspection is difficult to prove. It is far better to resolve the problem through partnering and on-the-job negotiation.

## 12.  Notice requirements

Untimely notice of a change may preclude cost recovery for extra work and always makes negotiation more difficult. *Contractors must give timely notice of a change, as specified in the contract.*

The purpose of notice is to give the owner an opportunity to avoid or minimize damages. The owner may not know the magnitude of the cost and delay from a change, or even be aware of a change. Once informed, the owner may reduce or eliminate the costs and delays. Failure to give timely notice prevents mitigation. In theory, and generally in practice, the notice clause is a reasonable requirement.

Unfortunately, some owners (and their designers and attorneys) make the notice requirements so stringent that it is difficult to comply without starting a paper war. Some owners try to avoid responsibility for the extra cost by ignoring effective notice and by relying on the onerous contract provision. Meanwhile, most contractors' jobsite personnel are busy and often forget to give notice in the form and timeframe required.

Always check the contract notice provisions for giving notice. If time remains, give notice immediately. If the deadline has expired, give written notice and note that it confirms earlier communications (if true) and that the owner was aware of the problem (if true).

In giving notice, two questions arise: (1) when did the contractor discover the condition and (2) when did the contractor realize it would affect the project.

Contractors usually use the latter date in giving notice, since the former results in an unreasonable number of unnecessary notices. Owners, however, frequently assert that notice should have been given when the condition was discovered.

The courts and arbitrators may broadly construe the notice requirements, especially in federal court, if informal or constructive notice was made and if the owner's rights are not prejudiced. However, failure to comply with the notice requirements may result in procedural disputes and complicate or prevent a negotiated settlement. Full notice compliance, however, can start a paperwork war.

To avoid a paper war while protecting yourself from an overly stringent notice requirement, agree with the owner's representative (in writing) to simplify the notice requirements (after confirming the representative's authority to modify the contract). Propose the plan in the partnering workshop and obtain written agreement as soon as possible.

## 13. Changes

All contracts need a changes clause giving the owner the right to make changes to the work or to require extra work. These clauses describe how changes and extra work are authorized and priced.

**Three ways to price changes.**   Change clauses normally provide for work to be priced one of three ways: by agreement before doing the work, by force account (while maintaining detailed records), or by negotiation after-the-fact based on whatever records were maintained.

**Written change order only.**   Unfortunately, owners' representatives often fail to give written directives, and contractors' jobsite personnel often fail to obtain it before proceeding with changes to the work. It takes little effort, however, to issue multipart, carbon-copy speed memos to field personnel and to instruct them to confirm verbal directives and to request written confirmation on the return part of the speed memo. Although low-tech, speed memos are effective.

If extra work is complete or ongoing without a written directive, request the owner's representative to confirm verbal directives in writing. If necessary, use a request for information, which generates a written response (but make sure the respondent is authorized to make the change). Even without a written directive, however, owners cannot hide behind a requirement for written change orders if other verbally directed changes have been paid.

A contractor is often uncertain whether to proceed without a written directive, to avoid impact. Sometimes, it is better to proceed without the specified notice but always notify the owner in writing that you are being forced to proceed without a written directive in order to mitigate damages.

**Authority to direct changes.**   Most contracts specify who is authorized to direct extra work. Review the contract and notify the authorized representative if others are directing additional work.

**Changes outside the scope of the original contract.** Contractors are normally pleased to perform extra work. However, if there are unresolved disputes over other work, concerns about the owner's ability to pay, or if the added work is beyond their ability to perform, contractors may refuse additional work. Consult with an attorney first.

**Inadequate markup.** Specified markups for change order work are usually too low. Although the normal 15 percent markup does not cost a contractor money out of pocket (for most changes), it often doesn't compensate for the actual overhead plus a reasonable profit. The markup provision is grossly unfair if there are many changes which require substantial time to administer and implement, if the changes disrupt other work, or if they delay the project. Specified markups are often used in public contracts or contracts with uneven bargaining power between the contractor and owner. They are routinely upheld on the theory that the contractor did not have to sign the contract and therefore assumes any risks not adequately accounted for in the bid.

The markup provision is usually disregarded when filing a major claim or arbitrating/litigating a dispute, because the magnitude of the changes is beyond the contemplation of the parties when the contract was executed. One can also argue that the specified markup conflicts with the general statement assuring adequate payment for extra work. You also can note that AIA document A.201, Art. 7.3.6.5, 1987 version, allows cost recovery for all field staff working on changed work.

Owners may argue that the markups are adequate since the contractor did not need to add staff. This argument overlooks the fact that the contractor's staff are there for the contractor's benefit, to control costs and maintain progress. Their diversion to price and to manage change order work interferes with their designated duties and leads to increased costs on bid work. For all these reasons, owners should pay at least full markup. Since change order work is often discovered when work is ongoing or imminent, it is disruptive and consumes more jobsite supervision than bid work. Markup for extra work should therefore be higher than for bid work.

On small direct cost changes, and especially on combination add and deduct changes, the contractor's costs of estimating, coordinating the change in work, procuring materials, and negotiating and processing the order often exceeds the value of the change order. Thus, the contract-specified markup will be insufficient. In this scenario, contractors must insist on compensation for change order preparation costs and the additional management, administrative, and supervisory time.

**Deductive change orders.** Owners often ask for credit for the overhead and profit on deductive change orders. Full credit for overhead is certainly unfair as overhead often increases for a deductive change and some overhead costs always exist. The contractor's overhead is usually fixed and does not decrease for deleting many work items. Likewise, the contractor has already incurred considerable risk and effort to earn the profit markup.

**Including time extensions in change orders.**  Contractors seldom request extended overhead on a change order that extends the contract time. This practice is partly due to the difficulty in negotiating both the change order amount and the additional costs of jobsite and home office overhead. However, these costs are real and are absorbed in overhead or taken from the profit if not included in the change order. Therefore, if applicable, every change order involving additional time should include extended overhead and escalation.

**Including impact in change orders.**  Contractors may charge more than appears fair on a change order, only to discover that they lost money due to impact on other work. Contractors must include the "expected value" of impacts or expressly reserve the right to file for impact costs later. Either course leads to resistance from most owners.

**Proceeding under protest.**  If forced to perform additional work while dissatisfied with payment or other issues, contractors should proceed under protest. Place the owner on written notice that work proceeds under protest before proceeding.

**Constructive change.**  One type of constructive change is forced on a contractor by situations for which the owner is responsible. These constructive changes require contractor performance outside the contract requirements. The most frequent example is constructive acceleration, caused by an owner failing to promptly grant a time extension for excusable delay and the contractor accelerating to avoid liquidated damages. Another type of constructive change is caused by the owner misinterpreting the contract and forcing the contractor to perform work outside the contract requirements. For example, if the owner's representative requires that the work be performed a certain way, the contractor risks having it rejected unless it is.

The constructive change concept can also apply to overinspection, design errors and ambiguities, revised owner requirements, and other requirements not mentioned in the contract.

**Cardinal change.**  A cardinal change occurs when a project is so altered that it no longer resembles the project described in the original contract. The change can be in either the finished product or the manner in which the work is performed. Damages are computed as cost plus.

**Quantum Meruit.**  *Quantum meruit* is a legal theory allowing payment to a contractor for the reasonable value of work performed, i.e., cost plus. It is used for a cardinal change or for additional work outside of contract requirements for performance and payment.

**Total cost and modified total cost claims.**  Total cost claims require adding up the actual costs, subtracting the estimated cost (plus change order costs), and claiming the difference plus markup. Total cost claims are easy to prepare, provide full compensation with profit, and are therefore popular with contractors. However, they are generally not accepted by project owners or the courts, except in certain limited situations. Total cost claims must meet the following criteria in order to be accepted:

1. It is impossible to accurately compute damages from specific causes.
2. The contractor's bid was reasonable.
3. The contractor's costs are reasonable.
4. The contractor is not responsible for any of the added costs.

Because of the difficulty in proving the above conditions, the courts allow contractors to modify the total cost claim approach by deducting bid errors and construction errors for which the contractor is responsible. This is called a modified total cost claim.

A modified total cost analysis is also a good check on the reasonableness of your claim damages and helps convince an owner's representative of its validity.

**Cost plus claims.** A cost plus claim is similar to a total cost claim but doesn't rely on the estimate. It consists of totaling up the actual costs, adding markup, subtracting payments to date, and claiming the balance. These claims are unpopular with owners but are used if the change in scope is so extensive as to constitute a breach of contract and a cardinal change.

**Trading.** An owner's representative will sometimes agree to overlook a provision of the specifications if the contractor's field personnel perform some minor extra work without charge. These agreements are convenient but pose risks for both parties and must be considered when preparing a claim or defending an owner's counterclaim.

## 14. Disputes

The disputes clause describes the process for resolving disputes not settled under the changes clause. It may appear as a provision in the changes clause or constitute a separate clause.

**Claim submittal requirements.** Some dispute clauses contain detailed requirements for substantiating entitlement and computing damages within impossible time frames. Comply, using the best available information, and note that additional necessary information will be forthcoming.

**Dispute resolution process.** Most disputes clauses specify a progressive process for review and decision before proceeding to arbitration or litigation. Usually the process includes submission of a change order request to the designer or owner's representative for a decision. In some cases, mediation may be suggested or required before the dispute is adjudicated. More frequently, contracts specify arbitration in lieu of litigation, although some limit arbitration to disputes under a specified amount.

**Duty to proceed.** The contract usually requires contractors to proceed with construction during a dispute. Failure to continue working is normally a material breach of contract, which can result in severe damages. However, stopping work is an option when directions are absent or unclear or when the owner fails to pay for work performed as provided in the contract. In addition, there are degrees of compliance and variations in how quickly a contractor performs additional work under protest. In any case, consult with an attorney before stopping work or refusing to proceed.

**Venue and choice of law.** Out-of-state project owners may specify that any arbitration or litigation must occur in the jurisdiction of their home office and under the law of that state. Such clauses put the contractor at a substantial disadvantage in attempting to resolve disputes and should be renegotiated if possible.

15. **Value engineering change proposals**

    Some contracts encourage the contractor to propose changes to contract-specified work to reduce costs while performing the specified function, and call for sharing the savings in a specified formula. These value engineering (VE) clauses benefit both parties and should be used on all projects. If your contract doesn't have such a clause, suggest it during the partnering workshop. Be aware, however, that your proposals may be rejected, or accepted late and delay the project (if you don't condition the proposal on a response date). If the work can be done less expensively within the contract, there is, of course, no need for a contract change or sharing of the savings.

16. **Suspension and termination**

    Suspension clauses allow the owner to suspend work, if, for example, unexpected problems must be resolved before work can proceed. Termination clauses allow for termination of the contract for the convenience of the owner when the project must be abandoned or the delay is extensive, or for cause when the contractor's performance is unsatisfactory.

    **Suspension.** Suspension clauses allow the owner to suspend the work, and require payment of the contractor's costs including a proportional share (or all) of the profit. Such clauses may be titled suspension of work, stop work order, or owner delay of work. All such clauses must be read very carefully. They may allow the owner to delay the work for a "reasonable" but undefined period of time, and grant the contractor only a time extension. Others may allow the contractor both time and money, if the delay is unreasonable, but may also impose fairly stringent notice requirements and otherwise limit recovery.

    **Termination for cause.** Typical clauses reserve the owner's right to terminate a contractor for default, and enumerate the grounds for default. For example, failure to complete on time or failure to make reasonable progress which endangers timely completion are normally grounds for termination. Failure to perform important elements of the work, such as using apprentice electricians instead of licensed electricians, is also cause for termination. If threatened with termination, contractors must review the termination clause to determine what constitutes cause for termination and whether the owner must send a cure notice or show cause notice beforehand. *When threatened with termination for cause, contractors must consult an attorney immediately.*

**Termination for convenience.**  Both public works and private contracts may allow for termination for the owner's convenience. If, for example, excavation encounters an Indian burial ground precluding further work or promising several years' delay, the owner may wish to abandon the project and terminate the contract. Termination for convenience clauses are optional under the AIA documents, and typically specify that the contractor is entitled to costs incurred plus some profit.

**Wrongful termination.**  The consequences of wrongful termination depend upon the contract. Some termination clauses state that if the termination is wrongful, it is to be treated as a termination for convenience. Otherwise, or if the termination is issued in bad faith to harm the contractor, the contractor is entitled to common law damages, including all profits had the contract been completed.

## 17.  Delays and time extensions

The delay and time extension clause usually allows a time extension but no compensation for delays due to (1) natural disasters (acts of God), (2) acts of war or public disorder, (3) acts of the government, (4) abnormal weather conditions, and (5) labor strikes and slowdowns. Regardless of the terms of the contract, the contractor is due a time extension (plus compensation) for delays caused by the designer or owner, in accordance with contract law.

## 18.  No damages for delay

No damages for delay clauses limit or preclude recovery of extended overhead and impact costs due to owner-caused delays, and only grant a time extension. The 1997 edition of the AIA standard contract documents require the owner and contractor to waive claims for consequential damages against the other, which may affect recovery for some delay costs such as loss of use by the owner and extended home office overhead by the contractor.

Some states, including Oregon, Washington, and California, prohibit no damages for delay clauses on public work contracts. However, this does not protect contractors on private construction projects and may offer limited protection to subcontractors. A variation of the no damages for delay clause, which attempts to skirt the law, defines the costs included in extended overhead or otherwise limits the amount of recovery.

**Loss prevention.**   All parties to a contract are obligated to mitigate damages to the other party and may not hinder or delay the other party. This argument *should* apply to the no damages for delay clause and enable contractors to recover damages under the theory that the owner could have readily avoided hindering or delaying the contractor. To maximize the probability of recovery, proceed as outlined in the loss prevention plan described in Chap. 4.

When the contract contains a no damages for delay clause, contractors should *greatly* increase the effort devoted to scheduling and tracking progress. Extended overhead costs (home office plus jobsite) often exceed one thousand dollars a day, and the impact of pushing weather-sensitive work into winter conditions can exceed that by an order of magnitude. It is unwise to use standard scheduling methods.

If delays are major, argue that the delays were beyond the contemplation of the parties and therefore the clause does not apply. In some states, the courts may find ways to compensate a contractor sustaining actual damages. In one case, the courts treated the cause of delay as a cardinal change and awarded damages *quantum meruit.*

No damages for delay are unenforceable if any of the following conditions apply:

- The delay was not within the contemplation of the parties when entering into the contract.

- The delay constitutes abandonment of the contract by the owner.

- The delay is caused by fraud or bad faith (simple negligence does not negate the clause).

- The delay is caused by active interference, such as allowing the jobsite conditions to be different from those specified in the contract or issuing a notice to proceed when aware that work could not proceed.

- State law forbids their use.

## 19.  Final inspection, punch list completion, and final payment

Contracts usually specify a procedure for final inspection. The designer or inspector conducts the inspection and provides a "punch list" of uncompleted or defective items to be corrected by the contractor. The certificate of substantial completion is normally issued at the same time or after certain specified work is competed. Upon completion of all punch list items and required paperwork, the project is closed out and final completion is certified.

The requirements for final payment are normally:

- Correction of all punch list items, or a credit for those items that the owner will complete or that are nonconforming but acceptable for use.

- Release of claims and liens.

- Written consent of surety.

- Materials documentation (tests, certifications, warranties, etc.).
- Submission of operation and maintenance manuals, warranties, and as-builts as specified.

## 20.  Hold harmless, defense, and indemnity

These commonplace and seemingly innocuous clauses can cause a contractor financial disaster. Consider them very carefully.

Hold harmless clauses generally preclude suing the other party for its alleged actions and require you to defend and indemnify the other party from the claims of third parties. Defense clauses require defending the other party at your expense from suits by third parties. Indemnity agreements require paying the costs of defending the other party as well as for damages assessed by a court or arbitration panel.

Typical indemnification clauses vary from limited liability, primarily for one's own actions, to a broad form of contractual liability for actions by others beyond your control. Some make you liable to another party for their negligence. Carefully study these clauses and consult with an attorney if uncertain of their consequences. Extreme versions of hold harmless/defense/indemnity clauses are not in the public interest, may be unenforceable in many states, and should be challenged if recommended by your attorney.

Most liability insurance policies *exclude* coverage for losses incurred by these clauses. Contractors must obtain a "contractual liability" endorsement in their general liability insurance policy to cover the additional risk.

## 21.  Progress payments

Progress payment clauses specify the submittal and approval procedures for monthly progress payment requests. For unit price contracts, payment is based on the bid schedule. For lump sum contracts, the contractor prepares, and the owner's representative approves, a schedule of values for payment purposes. The contractor prepares monthly payment requests based on the schedule of values, and the owner's representative reviews and approves them. For cost plus work, the contractor submits detailed billings with labor hours, subcontractor and material invoices, etc., although in some cases interim payments are based on percent complete and only the final payment is supported by detailed invoices.

For lump sum work, some contracts require the contractor to cost load the schedule activities with a dollar value for each. If so, the monthly progress payment is automatically generated by the scheduling software by multiplying each activity's percent complete by its value.

**Retainage and offsets.** Retainage is withheld from progress payments to ensure that the work is satisfactorily completed. It is often 10 percent during the first half of the project and none thereafter, if progress is satisfactory. Some states limit retainage on public works or bonded projects to 5 percent. On many contracts, when the project is essentially complete, the retainage may be reduced (at the owner's option) to twice the estimated cost of the work remaining.

Many states require paying interest on retainage or placing it in an interest-bearing account. Some states also assess interest damages for late payment of retainage.

Some contracts allow the owner to withhold payments for other reasons, such as a contractor's failing to pay prevailing wage on a public works project or for the contractor's debts or obligations on other projects. The owner may also withhold payments when previously accepted work is found to be defective. In many states, the contractor can "bond off" such claims by posting a bond to cover the dispute and thereby free up the progress payment.

Offsets are debts the contractor allegedly owes the owner on other contracts or under other proceedings. For example, on federal contracts, a contractor may be due $1 million on a U.S. Army Corps of Engineers contract, but the contracting officer may have received a notice from the Department of Energy contracting officer stating that the contractor owes the Energy Department $500,000 for breach of warranty or for some other reason. In that case, the contractor will receive only the difference.

## 22.  Pay when paid and pay if paid

These clauses are interpreted literally. The general contractor has no obligation to pay subcontractors unless and until payment is received from the owner—even if the reason for nonpayment is a dispute between the owner and general contractor unrelated to the subcontractor's work. Subcontractors should be extremely careful of contracts with either clause. Subcontracts often contain one or both clauses and eliminate the prime contractor's obligation to act as the owner's banker. The clauses are onerous to subcontractors, who have little knowledge of the owner's financial position, do not have a contract with the owner, have little or no communication with the owner, and have little leverage to force payment.

Pay when/if paid clauses are different from the Prompt Payment Act clauses used on many public contracts. Prompt Payment Act clauses require contractors to pay subcontractors within a specified number of days *after* receiving payment from the owner. Most also require public owners to promptly pay the general contractor, assuming that there are no bona fide owner claims against the contractor.

Some states have refused to enforce these clauses, noting that they are against pubic policy. For example, in 1997 a California court ruled that pay if paid clauses are invalid because they violate a subcontractor's lien rights.

## 23.  Variation in quantities

On unit price contracts, bidders usually spread overhead and profit plus the fixed costs for each bid item over the estimated quantities of work. For example, 100,000 cubic yards (cy) of common excavation is estimated at $2.00 per cubic yard direct costs plus $75,000 for mobilization of special equipment ($0.75/cy), and the contractor adds $0.25/cy for overhead and 10% for profit. The fixed costs are therefore $1.00/cy of the total $3.30/cy price.

If the quantity varies from the estimate, so will the money received for overhead and profit. If the actual quantity is 60,000 cy, the contractor will lose $40,000 (10,000 cy × $1.00/cy) but if the actual quantity is 140,000 cy, the contractor will make an extra $40,000 plus 10% profit.

Variation in quantities clauses are common on unit price contracts. They typically state that the contract unit prices will be adjusted only if the quantities vary more than a specified percentage from the estimate and then only for the amount in excess of that variance, with a typical percentage of 15 percent. In the above example, if the actual quantity of excavation is 85,000 cy (a 15 percent overrun), the work will be performed at the unit price. If, however, the actual quantity was only 65,000 cy (a 35 percent underrun), the contractor would be due the unabsorbed overhead and mobilization costs (plus profit) for the difference between 15 percent underrun and 35 percent underrun. This is $1.00/cy for the 20,000 cy difference between the actual 65,000 cy and the 85,000 cy variance allowed in the contract, plus profit.

Some variations in quantities clauses allow adjustment only if the value of the bid item is greater than a specified percentage of the total contract amount. Contractors also should be aware of the risk of unbalanced bidding (spreading more markup on some bid items than others), either to make more money from expected overruns or to increase early payments to fund mobilization. Deletion of such bid items can cause severe losses.

## 24.  Liquidated damages

Liquidated damages are a preagreed per diem allowance to cover damages the owner will suffer for late completion. They are usually specified in dollars per calendar day of delay. Contractors should prepare a prebid schedule for all bids and include a contingent amount for liquidated damages if the contract time appears insufficient. More likely, however, the additional savings of extended overhead costs will justify acceleration to avoid imposition of liquidated damages.

## 25.  Subsurface conditions and environmental clauses

Subsurface conditions represent one of the greatest risks to a contractor's profitability and often lead to disputes. Extreme caution is advised when encountering disclaimers regarding subsurface conditions, especially if there is no changed condition clause in the contract. Environmental clauses can also pose serious risks to contractors and should be carefully evaluated.

## 26.  Certification

The federal government and some other project owners require contractors to certify the validity of their contract claims and their subcontractors' claims.

Falsifying a claim or intentionally exaggerating the amount can lead to penalties. Contractors must confirm that all claims are valid and accurate, especially when required to certify claims.

## F. Types of Extra Costs and Damages—Based on Entitlement

This section identifies the originating causes of delay and extra cost, and categorizes each cause by (1) how most contracts allocate responsibility and (2) whether the costs are compensable and the delays excusable. Some contracts, and some jurisdictions, may allocate responsibility differently. Verify whether the following types of delays and causes are applicable to your specific project.

For other cost classifications, review:

- Section D.6 of Chap. 11 for the elements of cost (direct, delay and impact costs, markup, etc.).
- Section E of Chap. 11 for categories of direct cost (labor, materials, equipment, etc.).
- Sections F to I of Chap. 11 for the details of nondirect costs (delay, impact, overhead, etc.).
- Section A.3 of Chap. 12 for the reasons for loss of efficiency.

### 1. Nonexcusable and noncompensable delay and extra cost

Nonexcusable, noncompensable delays and extra costs can be classified in three general categories, depending upon contractor responsibility. If skilled in negotiation and on good terms with the owner's representative, try to reclassify some issues from nonexcusable to excusable but noncompensable. For example, the distinction between adverse and abnormal weather and whether the project warranted a time extension may be open to interpretation. Likewise, late fabrication by suppliers due to circumstances beyond their control or failure of a sole source supplier to deliver should be characterized as an excusable delay for the contractor, instead of nonexcusable.

**Contractor error.**  Some problems are clearly the fault of the contractor and are best acknowledged as such to maintain credibility. They include:

- Bid errors, including underestimating production rates.
- Poor scheduling or subcontractor coordination.
- Poor layout or access—if not the owner's fault.
- Insufficiently skilled personnel or supervision.
- Failure to adequately staff the project.
- Inadequate equipment capacity or numbers for the conditions.
- Late shop drawing preparation or submittal.
- Failure to promptly order materials.

- Actions or inactions by the general contractor and subcontractors which are not claimable against the owner, but which the subcontractor and the general contractor can cross-claim against each other.
- Rework to correct defective installation or materials.
- Bankruptcy.

**Misfortune (beyond the contractor's control).** Other problems are the responsibility of the contractor, in accordance with the contract or with industry practice, but are due to bad luck rather than error. They include:

- Adverse (but not unusually severe) weather.
- Work conditions and productivity worse than reasonably expected, but not a changed condition.
- Accidents, unless due to unsafe practices or lack of a comprehensive safety program.
- Acts by third parties not under control of the parties to the contract.

**Uncertain responsibility (beyond the contractor's control).** Some problems are generally considered the contractor's responsibility but not fault and may be claimed as an excusable but noncompensable reason for time extension. They include:

- Late fabrication and materials delivery.
- Improper fabrication or manufacture.
- Restrictive union work practices.
- Labor shortages.

## 2.  Excusable but noncompensable delay

Some causes of delay and impact are excusable but noncompensable. Such delays prevent the imposition of liquidated damages and allow damages for constructive acceleration, if the owner's representative fails to grant timely extensions.

The causes that are normally excusable but noncompensable are:

**Always excusable.**  The following causes of delay are excusable (under most contracts):

- Abnormal or unusually severe weather—rain, snow, wind, heat, cold.
- Epidemics, quarantine restrictions, or other public health measures.
- Acts of God (earthquake, major flood).
- Acts of government *force majeure* (except when the government is also the project owner).
- Commercial impracticality.

- Impossibility of performance.

- Acts of war or public insurrection.

- Concurrent delay, when both the owner and contractor have delayed progress.

Abnormal weather warrants a time extension if the contractor proves it was unusually severe and significantly impacted operations. See Chap. 10 for details. Use caution in claiming weather delays (which are noncompensable) if compensable owner-caused delays also exist. Be aware that owners may use alleged noncompensable concurrent weather delays to protect themselves against compensable owner-caused delay damages and impact.

**Usually excusable.** Most contracts provide for a noncompensable time extension if the project is delayed for any of the following reasons:

- Strikes or labor disputes (slowdown, etc.).

- Interference by other contractors onsite or adjacent to the site, unless controlled by the owner.

- Unavailability of specified materials, if reasonable effort was made to procure in a timely manner.

- Failure of regulatory agencies to issue permits to the contractor, if failure was not due to contractor fault.

- Fire or flood.

## 3. Compensable delay and/or extra costs

Typical owner-caused or owner-responsible delays and extra costs which are normally compensable and excusable include the following:

### Extra work
- Extra work, directed or constructive, that increases the contract scope, including change order work and changes in the contract documents or shop drawings that increase the cost of work, or changes due to interpretations by the owner's representative that increase the cost of the work.

- Owner changes that impact bid work or cause extra work.

### Acceleration
- Constructive acceleration due to the owner's failure to grant timely extensions for excusable delays.

- Owner-directed acceleration by directive to staff the project at a certain level or to finish all or some work by a certain date.

- Contractor-initiated acceleration to avoid inclement weather or other more costly working conditions, originating from owner-caused delay or disruption.

### Changed conditions or differing site conditions

- Differing site conditions are unforeseen subsurface or other hidden conditions, different from those reasonably expected or indicated in the contract documents.

- Changed conditions are a change occurring during the contract, such as a new government standard that increases costs.

- Failure by the owner to disclose necessary or vital information or the provision of erroneous information.

### Design error, change, or uncertainty

- Excessive quantity overruns or underruns on unit price contracts.

- Design errors, omissions, and ambiguities (defective plans and specifications).

- Incomplete design.

- Multiplicity of changes and/or RFIs (requests for information).

- Specifying discontinued or unobtainable materials—provided the contractor made reasonable efforts to obtain or determine availability during bidding.

- Failure of the prescribed design to be constructible.

### Interference or disruption

- Directing the contractor in performing the work.

- Suspension of work, design "holds," and stop work orders.

- Out-of-sequence work.

- Acceleration, directed or constructive.

- Overinspection.

- Partial occupancy prior to completion.

- Project shutdown or termination.

- Rejection of work or materials which conform to the contract requirements.

- Refusal to permit use of more economical methods not expressly prohibited by the contract.

- Rejection of "or equal" material substitutions that truly are equal to that specified.

### Late response or late owner furnished materials

- Slow submittal/RFI review, decision making, or approvals.

- Lack of direction or decision making.

- Failure to provide timely inspection, testing, or construction staking.

- Late owner-furnished materials.

- Untimely payment or nonpayment.

### Access limitations

- Lack of access, late or inadequate access (if less than contract-specified).

- Crowding due to owner changes or the presence of other contractors under the owner's control.

### Coordination

- Failure to coordinate other parties under the owner's control.

- Failure to provide necessary preceding work (utility relocation, etc.).

- Lack of needed permits, if the owner or designer is responsible.

- Failure of adjacent necessary facilities to be completed.

Interference by other contractors, onsite or adjacent to the site and controlled by the owner, is cause for a claim against the owner, unless otherwise identified in the contract. Interference by other contractors not controlled by the owner, when the owner should have anticipated their presence and possible interference, may be cause for a claim against the owner. In some cases, it may be possible to assert changed conditions due to the effect of work by other contractors. If negligence is involved, it is possible to recover damages by pursuing a tort action against the other contractor.

### Other

- Change in start and finish dates, including late award.

- Restricted progress due to limited funding.

- Changes in building codes or other regulatory requirements after bidding.

- Defective owner-furnished materials.

- Improper or ineffective contract administration, unjustifiably low amounts for progress payments, etc.

## 4. General contractor-caused delay and extra cost—compensable for subcontractors

Subcontractors should separate claims against the general contractor from claims against the owner. General contractors need to accept responsibility for their errors instead of claiming damages from the owner for contractor errors, since that destroys their credibility.

Depending upon the subcontract, claims by subcontractors against general contractors may include the following:

- Poor scheduling (insufficient detail, errors, failure to revise or update the schedule when needed, favoring the general contractor or other subcontractors, etc.).

- Insufficient hoisting or other general conditions equipment, if different than specified by the subcontract.

- Inadequate access.

- Insufficient or unreasonably inconvenient storage and/or laydown space.

- Slow shop drawing submittal or RFI processing.

- Accidents or failure of installed materials.

- Damage to installed work by other trades under the general contractor's control.

- Interference by other trades, due to poor general contractor scheduling or coordination.

### 5. Subcontractor-caused delays and extra costs—backchargeable to subcontractor

General contractors may withhold payment from subcontractors or suppliers for defective material or work, delays, or interfering with other subcontractors. If the backcharge amount exceeds the subcontract balances, the general contractor must document a claim against the subcontractor or supplier. In some cases, backcharges are unfairly used as leverage to force a subcontractor to drop claims against the general contractor.

Backcharges against subcontractors (or against sub-subcontractors by subcontractors) may include:

- Late material delivery, which results in project delays, because the subcontractor failed to make submittals or order materials in a timely manner.

- Insufficient workers when not due to regionwide shortages.

- Lower productivity than expected, except if due to accident or unexpected equipment breakdown.

- Any delay to the project for which the general contractor cannot obtain a time extension from the owner.

### 6. Potential counterclaims by the owner

When a contractor files a claim against the owner, owners often make counterclaims against the contractor. These can include the following:

- Nonconforming or defective work.

- Repair of damage to existing or adjacent property.

- Liquidated damages or actual (consequential) damages for late completion, including lost profits.

- Ineffective control of project, resulting in excessive review of submittals or substitution requests, extra construction staking, and additional testing and inspection.

- Interference with the owner's operations.

## G.  How to Proceed if Unable to Negotiate a Settlement

If unable to reach an acceptable resolution through negotiation, there are a limited number of alternatives as briefly discussed below.

### 1.  Appeal to the project (standing) neutral

If unable to negotiate a settlement, the next choice would be to appeal to the project neutral for a recommendation on resolution—if a dispute review board or a neutral expert has been appointed. Section F of Chap. 4 explains how to retain a project neutral and Sec. D of Chap. 7 explains how to make a presentation to a dispute review board or neutral expert.

### 2.  Mediation

If filing a lien doesn't revitalize negotiations, demand arbitration or commence litigation. Mediate first. It is far less expensive and maintains ongoing business relationships. It is senseless to win $20,000 in arbitration if, as a result, you lose $1,000,000 of work per year with that customer. On a large food processing plant dispute, when the owner counterclaimed for $1.6 million, mediation led to the owner's dropping the counterclaim, paying most of our client's $250,000 bill, and awarding a new contract to our client.

Mediation should be requested even if not specified in the contract. It is voluntary for both parties, with the costs of the mediator being shared. Mediation requires thorough preparation and well-planned negotiation to be successful; otherwise the case either will not settle or will result in a much lower settlement than anticipated.

### 3.  Arbitration

If mediation is unsuccessful, propose arbitration. Unless already specified in the contract, negotiate an arbitration clause into contracts as a more reasonable means of settling disputes and forcing payments. However, arbitration is not always beneficial. For example, spending $20,000 to arbitrate a $30,000 dispute if attorney fees are not recoverable is senseless. To avoid this, include an expedited process, provide for the award of the prevailing party's legal fees (under specified guidelines), and include other cost-reducing and settlement-inducing clauses.

### 4.  Litigation for small amounts

Small disputes can be litigated in small claim courts, which usually limit awards to around $2500 to $5000, depending upon the state. In some states, the state contractor's board may provide an economical dispute resolution mechanism for minor disputes.

### 5.  Other options

During construction, project owners depend on the contractor to finish the project as quickly as possible so they can occupy the project. When the owner already enjoys beneficial use, the urgency for meeting the contractor's needs for prompt, fair payment lessens. However, most contracts require the contractor to proceed with disputed extra work pending resolution of disputes. Failing to do so is a breach of contract. However, there is a wide range of acceptable levels of effort and cooperation. Contractors should be aware of how to use reasonable persuasion and partnering strategies to elicit prompt, fair payment.

Before being pushed into insolvency by unpaid claims, contractors should consult with their surety, who will end up completing the project if the contractor cannot. The surety may finance project completion and at the very least will recommend a good construction lawyer.

# 4

# How to Develop and Implement a Change Order Management Program

Carefully planned and executed endeavors are more successful than ad hoc efforts performed in the normal rush of business. Dispute resolution will be improved if contractors implement a program that:

- Identifies and limits contractual risks from onerous contract clauses.
- Ensures that estimating and cost accounting procedures track extra work and impact costs.
- Provides procedures and systems for good communication, recordkeeping, and dispute resolution.
- Trains project personnel to identify extra work, give timely notice, and track extra costs.
- Develops in-house expertise in the preparation of change order requests and claims.
- Is compatible with partnering, yet ensures fair payment for all work.

This chapter includes the following sections:

A change order management program must be implemented companywide. Sections A and B are therefore addressed to the company executive responsible for overall operations. Section C (which describes recordkeeping and reporting) and Secs. D through G (which describe the procedures for each phase of a project) are addressed to both the executive who develops the procedures and the project team members who implement them. If you read the chapter from start to finish, break your review into related sections (A–C, D–G).

## A.  Purpose of a Change Order Management Program

A Change Order Management Program is a set of companywide policies and procedures combined with staff training on managing construction changes, disputes, and claims. It need not be complex and can vary depending on the size of project and the degree of risk, but it should be part of every contractor's standard operating procedures.

A Change Order Management Program is compatible with partnering unless it leads to a focus on change orders instead of doing the work. A well-conceived program actually enhances partnering, as it helps detect and resolve problems early. This requires the following steps.

## B.  How to Establish a Change Order Management Program

Contractors need to *develop a companywide program* of policies, procedures, and systems before implementing a change order management program on individual projects.

---

**How to Establish a Change Order Management Program**

1. Document Goals, Objectives, and Policy.
2. Improve Estimating and Accounting Systems and Procedures.
3. Establish Good Recordkeeping and Reporting Procedures.
4. Train and Motivate Company Staff.
5. Select and Train In-House Claims Expert.
6. Implement Prebid Risk Management and a Loss Prevention Plan.
7. Conduct Preaward Constructibility Reviews.
8. Use Partnering and Retain a Project Neutral.
9. Improve Change Order Management Procedures.

---

### 1.  Document your goals, objectives, and policy

Company executives need to first recognize the need for improved dispute resolution procedures and results. They must then evaluate their goals regarding change orders. Should change orders be used to bid low, expecting to make up the difference on extras? Should they be a profit center? Do you expect to lose money on changes but hope not to go broke from one major unpaid dispute? Or do you just want to be fairly compensated for the work that you do?

After deciding on goals, conduct a needs assessment to determine current practices, results, and needed improvements. Then develop specific objectives that support the goals plus a work plan, budget, and schedule to establish the program.

## 2. Establish effective estimating and job cost accounting systems and procedures

Good estimating and cost accounting systems and procedures are essential to a successful Change Order Management Program. Depending upon the status of your existing systems, this may require major upgrades and extensive input by your estimating and accounting staff and may even require outside assistance. The five steps to achieve a successful program are as follows.

1. *Examine your estimating procedures* for accuracy and whether they use historical data from the job cost accounting system.

2. *Establish an accounting system and procedures* that capture all needed information. They must record and track both budgeted and actual work quantities, costs, labor hours, and productivity. The cost accounting system must *compare planned versus actual costs on an earned value basis* so that prompt action can be taken to correct problems. The system also needs to identify and *record extra work separately from bid work,* which requires a separate cost code for extra work items.

3. *Adjust and convert the project bid estimate into the project cost accounting budget*—for costs, labor hours, and work quantities. Correct bid errors and adjust the work units to facilitate measurement and reporting. During construction, adjust the budget for change orders so that the comparison of budgeted cost to actual is correct.

4. *Record weekly production quantities* in order to *compare actual labor productivity with planned.* This is crucial for using a measured mile analysis to prove inefficiency.

5. *Track the cost of all changes,* even for signed change order work, in order to control costs, to identify further change/delay/impact, and to document additional claims.

A good example of a cost accounting report that generates labor productivity data is given in Fig. 4.1.

## 3. Establish good recordkeeping and reporting procedures

Recordkeeping is the most important element of a Change Order Management Program, as good documentation is essential for establishing entitlement and pricing equitable adjustments.

Bidtek Construction                                                                           SEP 23, 1997  10:45:57
RP00105 - JC LABOR REPORT                                                                                 PAGE    1
                                              JOB COST LABOR REPORT

| | | | ------------ ESTIMATED ------------ | | | ----------- JOB TO DATE ----------- | | | ----- PROJECTED ------- | |
| PHASE | DESCRIPTION | UNITS | HOURS | COST | U/C | UNITS | HOURS | COST | U/C | COST | VARIANCE |
|---|---|---|---|---|---|---|---|---|---|---|---|
| | | | 1000- | FIRST AVENUE HIGH SCHOOL | | | | | | | |
| 01040-200- | PROJECT SUPERINT | 15 | 2,553 | 72,375 | 4825.00 | 2 | 400 | 12,180 | 6090.00 | 91,350 | -18,975 |
| 01510-200- | TEMPORARY WATER | 15 | 30 | 442 | 29.50 | 6 | 13 | 175 | 29.25 | 438 | 3 |
| 01590-100- | JOB OFFICES | 1 | 4 | 73 | 73.28 | 0 | 1 | 14 | 70.00 | 70 | 3 |
| 02510-401- | WALKS FINE GRADE | 4182 | 21 | 272 | 0.06 | 0 | 0 | 0 | | 272 | 0 |
| 02510-402- | WALKS SET SCREEDS | 4182 | 63 | 980 | 0.23 | 0 | 0 | 0 | | 980 | 0 |
| 02510-410- | WALKS PLACE CONCRE | 52 | 36 | 511 | 9.83 | 0 | 0 | 0 | | 511 | 0 |
| 03110-320- | FORM & STRIP CONTI | 1429 | 107 | 1,725 | 1.20 | 1529 | 121 | 1,835 | 1.20 | 1,835 | - 109 |
| 03110-321- | FINE GRADE CONTINU | 1629 | 16 | 214 | 0.13 | 1629 | 22 | 321 | 0.19 | 321 | - 106 |
| 03150-100- | FORM & STRIP WALLS | 16578 | 1,357 | 21,498 | 1.29 | 2486 | 247 | 4,064 | 1.63 | 25,000 | -3,501 |
| 03160-102- | S.O.G. SCREEDS & M | 61500 | 800 | 11,587 | 0.18 | 3075 | 43 | 682 | 0.22 | 11,587 | 0 |
| 03160-105- | S.O.G. EDGE FORMS | 1476 | 44 | 642 | 0.43 | 295 | 8 | 108 | 0.36 | 540 | 102 |
| 03220-600- | WIRE MESH SLABS | 67650 | 203 | 2,669 | 0.03 | 6765 | 25 | 343 | 0.05 | 3,437 | - 767 |
| 03300-101- | POUR FOOTINGS | 46 | 23 | 333 | 7.25 | 0 | 14 | 266 | | 333 | 0 |
| 03300-501- | POUR WALLS | 205 | 105 | 1,508 | 7.35 | 0 | 48 | 768 | | 1,508 | 0 |
| 03300-601- | POUR S.O.G. | 752 | 300 | 4,360 | 5.79 | 22 | 8 | 108 | 4.78 | 4,360 | 0 |
| 03300-704- | POUR PAN SLABS | 656 | 262 | 3,799 | 5.79 | 0 | 0 | 0 | | 3,799 | 0 |
| 06103-160- | 2 X 6 WALL FRAMIN | 9576 | 335 | 5,312 | 0.55 | 0 | 12 | 198 | | 5,312 | 0 |
| 06115-100- | WALL SHEATHIN | 10849 | 162 | 2,579 | 0.23 | 0 | 6 | 114 | | 2,579 | 0 |
| 06115-360- | ROOF SHEATHIN | 65467 | 851 | 1,326 | 0.02 | 0 | 0 | 0 | | 1,326 | 0 |
| 08110-100- | SET METAL DOOR FR | 31 | 47 | 792 | 25.57 | 0 | 0 | 0 | | 792 | 0 |
| 08110-200- | HANG METAL DOORS | 31 | 47 | 792 | 25.57 | 0 | 0 | 0 | | 792 | 0 |
| TOTAL FOR JOB | 1000- : | | 7,366 | 133,799 | | | 969 | 21,179 | | 157,150 | -23,351 |
| | | | | | | | | | | | |
| * * GRAND TOTALS * * | | | 7,366 | 133,799 | | | 969 | 21,179 | | 157,150 | -23,351 |

**Figure 4.1** Typical cost accounting report with productivity reporting. (*Reprinted with permission of BidTek.*)

If problems are encountered or anticipated in specific work areas, record more detailed accounting data, including daily or weekly production rates. Collect productivity figures for both impacted and unimpacted work, so that you can use a measured mile analysis. Use either the existing cost accounting system or special job sampling techniques to collect the data. Verify compliance with procedures, to avoid change order costs becoming buried in standard cost code categories. If costs are recorded in standard cost codes, reallocate labor hours from the cost codes where initially charged to the extra work cost codes. This takes a lot of time for the project manager and supervisor, often cannot be done with reasonable accuracy, and always engenders resistance by the claim reviewer to the resulting data.

If disputed work is ongoing, use timelapse video photography or standard video tape to record the detailed steps of impacted operations. Then use time-and-motion study techniques to deconstruct the operation and determine what the production rates would have been without the impact as described in Chap. 12.

Increase other recordkeeping efforts. Use photographs, additional and more detailed daily diaries, and any other data collection techniques that preserve valuable information.

## 4. Train and motivate company staff

Most claims and change order requests are prepared by project team members with widely varying levels of experience and skill. Training them and providing better tools will improve their performance. This can be done in the following six steps:

1. Train field personnel in the basic principles of contract law and in the company's policy and procedures on change orders. Conduct a series of short workshops at monthly superintendents' meetings, and have personnel read Chap. 3 of this book. Knowledge and training will help them identify contract changes, respond when changes occur, give timely notice, and keep good records.

2. Alert field personnel to critical contract clauses (e.g., the notice provisions) and potential problem areas on their projects. Be sure that they closely review the plans and specifications before starting work. Issue a list of critical concerns and special procedures for their project.

3. Ensure that field personnel use a partnering approach while insisting on fair compensation for changes. Reinforce this by partnering every project. Subcontractors should push the general contractor to partner with them and insist on being included in partnering with the owner.

4. Follow up on conformance with recordkeeping, reporting, and other change order management procedures. Review the daily field reports regularly and respond when appropriate.

5. Support the project team in the preparation of change order requests and claims, including review by a company in-house claims expert.

6. When evaluating personnel performance for pay and promotion, consider conformance to change order management procedures.

## 5. Train an in-house expert in change order request and claim preparation

For the same reason that they prepare virtually all change order requests (economy and availability), project team members prepare most claims. However, the results are often less successful than desired, and settlements are sometimes made at cents on the dollar, or the claim is dropped.

Although claim consultants generally develop more successful claims than project team members, the cost of using consultants is often considered prohibitive. A practical alternative is to train an in-house expert to prepare claims or to supervise the preparation of claims by project team members. The in-house expert can review change order proposals that are likely to be contested. On very large claims requiring more time or experience, the in-house expert can assist consultant experts, thereby reducing costs and providing valuable on-the-job training for the in-house expert.

### 6. Prebid risk management and loss prevention plan

Review every contract bid for unfavorable contract clauses, design errors, ambiguities, and other problems. If problems are found, take additional steps to mitigate the risks—as described in Section D.2.

### 7. Preaward constructibility review

If time allows, conduct a constructibility review of the plans and specifications as described in Section E, either independently or in conjunction with the designer and owner.

### 8. Implement partnering. Use a project neutral and collaborative problem solving

Implement partnering and retain a project neutral—either a dispute review board on large projects or a neutral expert on smaller projects. Collaborative problem solving goes beyond the usual recommendations. It requires full commitment from top management and an extensive implementation effort over an extended period of time. It works only with some owners. Nevertheless, it can pay major dividends. Collaborative problem solving requires strategic thinking to foresee future conflicts, learning new skills in communication and interpersonal relationships, and empowering lower-level personnel to make decisions and resolve problems on their own. For more information, read *Partnering Design and Construction* by Kneeland A. Godfrey, Jr., McGraw-Hill, 1996.[8]

### 9. Construction phase change order management

Continue the program from notice to proceed to receipt of final payment on each project. Emphasize partnering, with continuing contract reviews, improved recordkeeping procedures, timely notice, collaborative problem solving, and the preparation of change order requests. If necessary, prepare claims and resolve them through negotiation, mediation, and arbitration or litigation.

### 10. Vary your procedures for specific owners and projects

Contracting with the U.S. government is considerably different from constructing for a private developer. Contractors working for a variety of owners need to consider each owner's special requirements and culture. Your Change Order Management Program should reflect this diversity and be responsive to government agency requirements.

## C.  Recordkeeping and Reporting Procedures and Forms

Establish simple but complete recordkeeping and reporting procedures with forms and examples for use by the project team. This ensures both consistency and adequate collection of information. Top management must lead the way; any supervisors failing to comply should be censured.

To minimize resistance and the time required, provide fill-in-the-blank forms with simple guidelines for the most common situations.

## 1. General

The crucial elements of a good recordkeeping and reporting system are relatively simple.

**1.  Follow good basic recordkeeping practices.** *Start with clear, concise writing.* Describe issues, items to be clarified, and extra work so that someone not familiar with the topic will understand. Be consistent; different descriptions of the same issue are confusing.

When referring to earlier discussions, note the date, individuals involved, and general or specific discussions and agreements. However, references to written documents need only identify the document.

Identify and report all important information. Record completion of major milestones, the start and finish of scheduled activities, and any significant variations from plan.

**2.  Use rapid written communication.** *Communicate, and confirm verbal communications, in writing.* Fax or E-mail all correspondence to the owner's representative, designer, subcontractors, and vendors. Use speed memos and RFI forms to communicate with the owner's representatives in the field. Verify that field supervisors are using them to request information or to confirm directives and that they are submitting copies to the office.

**3.  Identify and date all documents.** Require project personnel to initial, title, and date all documents—including handwritten notes of telephone conversations, calculations, sketches, etc. Initials identify who created the document and avoid resorting to handwriting analysis to determine the author. The title or description of the document help explain what the document refers to and places the information in context. The date is essential to establishing a chronology. Without a time context, a fact or a statement of understanding or position may be meaningless. Date all the signatures on contracts and change orders. Otherwise, you won't know when the parties agreed to the terms and conditions, which may become crucial to a claim. Also, stamp and date all documents when you receive them.

**4.  Establish standardized descriptions of work areas.** If the design documents do not adequately label the work areas, do so and use the same coding consistently in all records. Mark up a plan view of the work with the coding to create a key plan of the building or site, and post it at the jobsite office so that supervisors and subcontractors will use the same work area descriptions.

**5.   Establish filing procedures.**   Set up a coded filing system for project files and publish it for everyone with access to the files. Cross-file documents by referencing topics or claim issues. Use a routing slip or stamp to facilitate review of incoming and outgoing documents.

**6.   Do not punish supervisors for reporting unfavorable data.**   There is a tendency for field personnel to charge time against under-budget work rather than against items that are over budget. Explain why it is important that cost accounting data be accurate, and do not reprimand supervisors for cost overruns unless they are clearly at fault.

**7.   Use of recommended forms.**   The sample recommended forms displayed in this section are provided as prototypes for modification and use. The computerized versions of the cost forms (e.g., the extra work order form) are designed as spreadsheets to compute values, subtotal columns, and total. Change the company name; add an address, telephone number, fax and E-mail address; and customize the form to meet your company's specific needs. Enter the project name before reproducing, to save the effort of entering repetitive information.

The forms can be retyped or copied from the book and enlarged as a printed form, computerized for use as an on-screen form, or obtained as noted in Sec. F of Chap. 1. They can also be reproduced as two-part carbon-backed forms in order to create copies when completing the form by hand.

**8.   Implement new technology.**   Project-specific websites on the Internet allow all project participants to communicate instantly, access schedules and CAD drawings, maintain RFI and submittal logs, and post or view digital photographs. Sites can be custom-created or built using a commercial package, such as e-Builder, which have built-in menus, forms, and routines.

## 2. Timecards

Timecards are the basic labor records. They are the primary source of data for the accounting system, which generates both payroll and the labor element of job cost accounting. Therefore, timecards must be accurate and must assign the labor hours to the proper cost codes, including extra work cost codes.

Timecards are usually either daily crew cards or weekly individual cards (see Fig. 4.2) although a weekly crew card is possible if they work only a few cost codes. The forms must include the following items.

- The employee name(s) and/or employee number, project number/title, day of the week and date (often the week ending), and a signature line for the supervisor.

- Columns for multiple cost codes, descriptions of the work (especially for non-standard work items), the hours for each cost code, and work area. Provide a total line for each hour column.

- Optional provision for overtime rate, multiple pay rates per individual (e.g., if they operate different equipment), and multiple projects—depending upon the company's needs.

- Space for optional comments on problems (delay, impact, or quality) or extra work performed, discussions or directives by the owner's representatives, equipment and material used, expendables or small tools, etc.

**Procedures.**   Issue condensed lists of cost codes to the foremen with only the cost codes they will use, along with instructions on what to record. Labor hours should be coded to specific cost codes, with a note on the specific work areas. Known or suspected extra work must be listed separately. Encourage foremen to freely annotate the timecards with comments on problems, extra work, etc. The superintendent should review and approve the foremen's timecards for accuracy and completeness, and a payroll clerk needs to check their accuracy before entering them into the accounting system.

**The importance of accuracy and completeness.**   Accuracy in charging time to the correct cost code is essential, and compliance must be verified by the superintendents and payroll clerk. Some foremen are averse to writing any more than essential, and do not break the work down into the proper cost codes. They

## WEEKLY TIME CARD

| JOB NAME OR NO. | KIND OF WORK DONE | S | M | T | W | T | F | S | HRS. | RATE | AMOUNT |
|---|---|---|---|---|---|---|---|---|---|---|---|
| 9258-16 | CLEAN UP | | 6 | 8 | 4 | 3 | 3 | | 24.5 | | |
| 9258-16 | COVER COUNTERTOPS | | | | 3 | | | | 3 | | |
| 9258-16-14 | RETURN PRESSURE WASHER | | | | 1 | | | | 1 | | |
| 9258-CE 267 | SUMP PUMP | | | | | 3 | | | 3 | | |
| 9258-16 | BUSH CONCRETE FOR FLUSHING | | | | | 1 | | | 1 | | |
| 9258-CE 341 | PICK UP JACKHAMMER PW | | | | | 1 | | | 1 | | |
| 9258-03 | MOVE DOORS | | | | | | 5 | | 5 | | |
| | | | | | | | | | | | |
| | | | | | | | | | | | |
| | TOTAL REGULAR TIME | | | | | | | | | | |
| | TOTAL OVERTIME | | | | | | | | | | |

Employee's Name: SMITH, JOHN    NO.    WEEK ENDING 1-29    1994

APPROVED    WITHHOLD | S.D.I. | DEDUCTIONS F.I.C.A. | MEDICARE | STATE WH.    TOTAL EARNINGS
TOTAL DEDUCTIONS
DATE PAID    CHECK NO.    NET PAY

REDIFORM. 4K409

**Figure 4.2**  Typical timecard.

may not know what cost codes to use or what tasks are extra work. Sometimes, they don't list extra work separately from base contract work. Payroll departments sometimes change cost codes without verifying what work was done.

## 3. Field supervisor's daily reports

The superintendent's report is one of the most important documents for substantiating construction claims. The report should contain the essential information needed to determine what actually happened each day and why.

**Sample form.**   Figure 4.3 is a prototype of a one-page daily report. It can be modified for each contractor's specific needs. Expansion to a second page may also be needed, or the addition of more lines for subcontractor activities. If more space is needed for comments, encourage personnel to strike out inapplicable line descriptions and use those lines, write in the margins, or add a second page. Never write on the back, as it may be missed when copying.

**Basic project information.**   Enter the project number and/or an abbreviated project title. Enter the date and circle the day of the week, as a check for the occasional misstated date. The optional report number helps identify when a report is missing.

**Environmental and work conditions.**   Weather information should include precipitation, temperature, wind and visibility (if significant), etc. If important, enter the high and low temperatures for the day. Describe pertinent site conditions (muddy, flooded, frozen ground, water level, etc.) and any impacts from weather or site conditions.

**Schedule activity progress.**   Note the activity numbers of activities being performed (from both the master schedule and short-interval schedule), with start and finish dates of activities that began or ended that date. Also note the expected start and completion dates of other activities. Take particular care to mention activities with intermittent progress, along with the reason why. If the same activities or equipment use are repeated day after day, relist them at least weekly to ensure that there are no oversights. Identify when important milestones are accomplished (e.g., the start and finish of excavation, concrete work, framing, roofing, rough-in, sheetrock, and finish work).

Note tasks scheduled to start shortly and any action needed by the home office—equipment, materials, etc. Also note when work should start but doesn't, with the reason why.

**Work performed, crew size, and work accomplished.**   Enter the size of each crew and the work performed. If two separate crews of the same trade are working

# XYZ Construction Company, Inc. -- Superintendent's Daily Report (SDR)

Job No./Title: _____ Date: _____ (M T W T F S S)

Weather: _____ Temp: high _____ low _____ Site Conditions: _____ Report No. _____

Activity Start/Finish Dates & Numbers: _____

_____

Planned Tasks & Action Needed: _____

| **XYZ Trades Crew  Work Performed:** work area, equip used, if extra work. (Use multiple lines if needed) | **Work Qty** |
|---|---|

Carpenters ___ _____ _____

_____ ___ _____ _____

Laborers ___ _____ _____

Cement Fin ___ _____ _____

Oper Engr ___ _____ _____

_____ ___ _____ _____

_____ ___ _____ _____

Subtotal ___ Equipment Oper/Standby & Mat'l Delivery: _____

_____

| **Subcontractors**    **Work Performed**: work area, equip used, if extra work. (Use multiple lines if needed) | **Work Qty** |
|---|---|

_____ ___ _____ _____

_____ ___ _____ _____

_____ ___ _____ _____

_____ ___ _____ _____

_____ ___ _____ _____

_____ ___ _____ _____

_____ ___ _____ _____

_____ ___ _____ _____

_____ ___ _____ _____

_____ ___ _____ _____

_____ ___ _____ _____

_____ ___ _____ _____

_____ ___ _____ _____

Total _____ Mat'l Delivery: _____

**Problems, Delays & Extra Work:** _____

_____

_____

Owner's Rep, Visitors, Tests, Directives, Discussion: _____

_____

Supervisor's Signature: _____ Reviewer: _____ Date: _____

**Figure 4.3**  Recommended superintendent's daily report form, one-page version.

in different areas or on different tasks, use a separate line for each and enter the foreman's name in the "Trades" column. Enter composite crews separately and periodically describe the composition of the crew. Also enter the work area, equipment used, whether the work was extra work (with the change order/change order request number) and if the crew was impacted and why. Once a week, note the approximate work quantities completed, if these are not recorded daily. Use multiple lines for each crew, and a separate line for extra work.

Subtotal the number of employees working. Identify equipment operating or on standby, if not noted above, and list material deliveries.

List each subcontractor, their crew size, and work performed with the same level of detail as the contractor's own crews. This information can be summarized from the subcontractors' daily reports, which should first be verified for accuracy and completeness.

Total the workforce on the project and list significant material deliveries to subcontractors.

Periodically list the equipment on the project, including the type, manufacturer, capacity, and attachments (or equipment number) and note when equipment is mobilized or demobilized. On equipment-intensive projects, list all equipment onsite and the hours of operation, idle, standby, and downtime for repairs or other problems.

This section provides the most important information needed for the creation of a detailed as-built schedule, should one be required. The section should fully describe the work performed each day, the working conditions, the resources used, and the work accomplished.

**Problems, delays, and extra work.** Describe problems, delays, and extra work being performed, if not fully described under Work Performed. Include the start and finish date, extent, impact, and reason for any delay. Describe labor inefficiencies, the reasons why, and the approximate loss from inefficiency if possible. Describe acceleration or expediting efforts and the response of subcontractors to directives. Descriptions should be sufficient for someone unfamiliar with the project to understand what happened and why. Record when and how the owner was given notice of compensable change or excusable delay.

Describe shortages of labor, equipment, or materials and subcontractor problems, if any, but be careful to avoid exaggerations that may lead to owner allegations of concurrent delay or contractor error. It is generally better to mention action needed by others. Also note safety meetings, labor jurisdictional disputes, and other significant events.

Note when problems are resolved and how; otherwise the claim preparer won't know when an impact ended. Also note accidents, the condition of the injured worker, the cause, if known, and the observed impact to production.

**Others onsite.** List owner or designer personnel on the site, visitors, tests taken, discussions, and directives. Summarize the results of meetings and who attended (if significant, write a memorandum to file with the details). Note directives by the owner's representative or the designer and submit a more detailed record if appropriate.

**Signatures.**  The superintendent must sign the report, and the superintendent's supervisor should review and initial it.

**Procedures.**  *Daily reports must be completed daily* and be either in a bound book or signed by the author. To be most effective, daily reports must be kept in the normal course of business, rather than documenting a specific claim. They must *contain only statements of fact,* not opinions or conclusions. Otherwise they may be found inadmissible in court.

If not on the project for a day, supervisors should summarize the activities for that day as reported by others and note that they weren't there. If no work was done, they should note that fact and the reason why.

Daily reports should be on two-part forms, with one copy submitted to the home office and promptly reviewed by the project manager and operations manager. Action on reported problems should be taken immediately. Otherwise, problems will not be corrected, and supervisors will realize that their efforts are wasted. The results will be inadequate documentation and lower morale. Consider providing superintendents with a pocket recorder to make notes in the field, as playing it back helps jog their memory when writing up their field notes.

**Subcontractor reports.**  Require subcontractors to submit daily reports (per their subcontract), and attach their reports to the superintendent's report. Provide the subcontractors a copy of the company's standard form and require them to use either it or their own form, but to include all of the company's standard information.

**The importance of accuracy and completeness.**  Field supervisors' daily reports are commonly accepted as being more reliable than documents prepared by office personnel. The supervisors actually observed and supervised the work and are generally inclined to be accurate and complete. However, some supervisors are too busy or disinclined to paperwork and don't maintain complete records. A few tend to exaggerate problems and others tend to understate them. Either tendency should be discouraged.

The failure of field supervisors to comment on impacts and problems in their diaries and daily reports can result in claims being rejected. Supervisory personnel should record all impacts in their diary or daily reports, without overdoing it. Managers should verify that all recorded information is without inappropriate comments or posturing. When noting problems, supervisors must state whether the problems are actually impacting their work or just threatening to do so in the future. Otherwise the comments may be used by the owner as a defense for concurrent delay or contractor error if, for example, the problem was due to a subcontractor.

## 4. Daily diaries by project manager, project engineer, and others

All project participants should maintain a daily diary, noting the facts pertinent to their role and knowledge of the project. The guidelines should be similar to those for field supervisors' daily reports, but without the detailed form, and should not duplicate the supervisors' daily reports. The daily diaries are usually not reviewed by management but should be spot-checked to ensure that everyone is conforming to company policy.

As with the superintendent's daily report, all diaries and other records are subject to discovery, and their authors must use caution in recording information that may cause a problem if obtained by the owner.

## 5. Original schedule, schedule updates, short-interval schedules, and progress reports

*Prepare schedules for all projects* using CPM methods or bar charts on smaller projects. Every week, mark up the master schedule (which should be timescaled) with a status line as explained in Chap. 10. In addition, generate monthly computer updates, and compare the current schedule with the original or target schedule.

In addition to monthly updates of the master schedule, the superintendent should prepare a weekly short-interval schedule for use by the foremen and subcontractors, and distribute it at the weekly subcontractor's meeting. The short-interval schedule (Fig. 4.4.) should show the previous 2 weeks' completed activities and the next 2 to 6 weeks' scheduled progress. It *must* be based on the master schedule, with each task tied to a master schedule activity. Otherwise the short-interval schedule may miss a critical task or fail to follow the critical path, and it will be difficult to create an as-built schedule to support a scheduling claim.

For details on recommended scheduling techniques and procedures, see Chap. 10.

## 6. Photographs and videotapes

Take photographs before starting work (of the entire site), regularly (weekly and monthly), and whenever problems occur. Regular photographs often substantiate a problem that wasn't recognized until later, and are excellent for establishing progress. Always take photographs of disputed work, and especially of defective work before repair or replacement, if there is any possibility of a dispute. Also photograph difficult working conditions such as mud, rain, standing water, and crowded conditions.

A sequence of photographs is more effective than scattered photos. Scan the project to create panoramic views, take "clips" of an operation every few minutes or hours to give a sense of progress and explain how the work was done, and regularly photograph the same work areas to provide a visual record of progress.

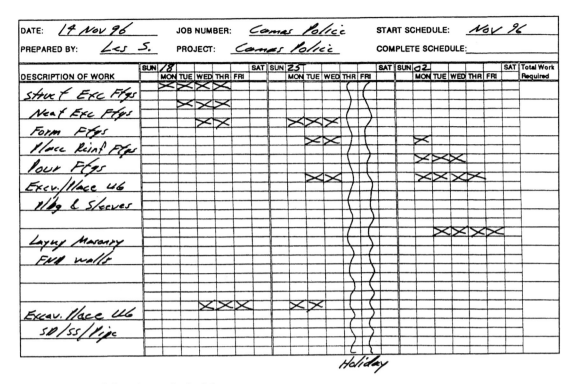

**Figure 4.4** Typical short-interval schedule.

If practical, photograph the entire site from a higher elevation at least once a month. If possible, take periodic aerial photographs of large civil works sites. The cost is minimal, yet the photographs can be invaluable. On one case, I was able to compute the quantity of rock used on a change by measuring the base of the rock pile (by scaling off parked equipment) and the angle of repose (by field measuring a freshly disturbed slope of similar material).

*Use a camera with date stamping,* and make sure that the date is set properly. Keep a log of photographs, the work area, work item, who took the photograph, and why. Issue disposable cameras to field supervisors for recording unanticipated problems that would not otherwise be recorded. Log in, date, and identify their photographs. Use standard descriptions of locations where the photographs were taken.

Instruct field personnel how to take usable shots by introducing a recognizable object in a close-up to provide perspective, adjusting for backlight, using composition and angles to emphasize, etc. Develop film as it is taken, rather than wait until the end of the project, to verify that your procedures are correct with enough detail, etc.

**Videotape.** Videos can be helpful to show operations and inefficiencies, especially with an oral narrative. For usable results, have the camera person practice and check their results. Videos are especially useful for drive-by photography of roadways and easements before construction. They can also be studied with time-and-motion study techniques to analyze inefficiency.

## 7. Agendas, rosters, and minutes of meetings

Schedule weekly jobsite meetings with field supervisors and all active subcontractors and weekly or monthly meetings with the owner's representative and key subcontractors. Keep the meetings brief by focusing on the agenda. Several short meetings are more effective than one long meeting.

Prepare and distribute an agenda for all formal meetings. Remind anyone who needs to prepare for the meeting. At the meeting, if someone isn't known, ask them to introduce themselves, their organization, and their role in the meeting. Get the correct spelling of their name, their company name, mailing address, E-mail address, and fax and telephone numbers. Ask everyone to enter their name, organization, and telephone number on a roster. To ensure timely arrival, set up a coffee and doughnut fund for contributions by late arrivals, and take immediate action to locate anyone whose presence is crucial but who doesn't show.

The general contractor should chair the meeting. Start on time and stick with the agenda. First confirm the accuracy of the prior meeting's minutes. Then address old business before proceeding to new. When consensus or an impasse is reached, summarize the discussion and decision and move on to the next topic. Allow discussion, but don't let it wander too far off the agenda. At the end of the meeting, summarize the main points discussed, agreements, action required, by whom and when.

Always have a large white board or flip chart with markers available for meetings. People can follow the issues better, evaluate multiple alternatives, and reach more effective decisions if the speaker (or someone else) writes the major points for all to see. Information that is spoken *and* written is retained far longer than strictly oral information.

Flip charts have the advantage of being permanent. When discussing problems and their solutions, list the problems in one column and the agreed action (by whom and when) in the second column. Post the list at the job trailer, mark off completed items, and display the remaining list at the next meeting.

Have someone, not the chair, keep minutes. They should be brief and to the point but clearly describe the issues. Number each set of meeting notes sequentially, to help when referencing them, and distribute minutes to all attendees.

If maintaining order is a problem, set some basic ground rules for discussion: one person speaks at a time; everyone takes their turn; quiet members are encouraged to speak out; long-winded speakers are gently advised to get to the point; everyone must be treated with respect; and heated arguments are a reason for a short break.

*Keep extensive notes of meetings chaired by others,* particularly owner/designer conducted meetings. Initial, date, expand them immediately after the meeting while your memory is fresh, and file. Check against the official minutes when issued, and notify the author of any errors or items missing.

---

**Topics for Meetings Minutes**

- Project title, the date, and subject of the meeting.
- Register of attendance.
- Summary of the meeting—a paragraph or two on the purpose and key subjects covered.
- Highlights of each major discussion with conclusions, action required, by whom, and when. List old business (with the date initiated) and new business. Describe how issues were resolved and note when. List a scheduled resolution date, interim action, and responsibility for unresolved issues.
- The next scheduled meeting date and the location, if it varies.
- Distribution list—all those attending plus others as needed.
- Name of preparer and date.
- Notice that any errors must be reported prior to the next meeting.

---

This can be crucial in pursuing a claim later if the minutes don't report critical information regarding an issue or report it incorrectly.

### 8 Notes of conversations

Keep notes of all informal meetings and telephone conversations. Record the project title, date, who met, the subject, initials of the preparer, and (at a minimum) a summary of discussions and agreements. A cryptic note may be sufficient to jog your memory later, although a clearly written summary is far better. Use a colored preprinted form to differentiate telephone conversation notes from other documents.

**Header information.** Enter the date and time of the conversation, and the project number or abbreviated title. Or create a form with that information already printed (Fig. 4.5). Record the phone number, if not a commonly called number, in case you need to call the individual again.

Enter your initials and the name of the other party. Optionally, enter the organization of the other party and check off who called if that is important.

Briefly state the subject of the call.

**Record and summarize the discussion.** Take notes of the conversation, trying to record the key points. When complete, add clarifying comments, write out abbreviated words, etc., so that you (or someone else) will know what transpired if reading it months later.

### 9. Correspondence and transmittals

For successful communication, correspondence must be clear and concise. Limit each letter to one topic and clearly describe the topic on the reference line. Do not exceed one page unless absolutely necessary. If your letter is

| XYZ Construction Company, Inc. -- Telephone Conversation Record | | | |
|---|---|---|---|

☐ a.m.

Date: _____    Time: _____ ☐ p.m.   Phone No. _____   Job No. _____

☐ To                              ☐ To

☐ From _____        ☐ From _____

Subject _____

**DISCUSSION**

**Figure 4.5** Recommended telephone conversation note form (without main portion).

longer than three pages, reduce it to one page and attach an exhibit with the details. Serial number all letters to the owner, ask the other party to serial number theirs, and use a different letter prefix to indicate the source of the letter. This greatly simplifies referencing letters.

Maintain a log of all correspondence. Retain copies of all correspondence in to/from files plus a copy in the appropriate subject file (i.e., concrete, soils, windows, claim, etc.). When a potential change is identified, set up an issue file and post copies of all relevant correspondence.

Transmittals must be dated and must clearly describe attachments. Unless unusually voluminous, staple all attachments to the file copy of the transmittal.

## 10. Facsimiles

Faxes save time over letters in preparation and transmittal and usually get a quicker response. Follow up with a printed copy only if needed or if attaching something physically transmitted. Fax all correspondence to the owner's representative, designer, subcontractors, and vendors. You can broadcast fax to selected or to all subcontractors, with automatic copies to the main office, etc. This saves a great deal of administrative and clerical labor, postage, and expense. If using a fax modem in your computer, you can compose and transmit while printing a hard copy for your files.

If each party on a project programs their fax with their company name and telephone number as their terminal identifier, faxes provide proof of transmittal and receipt. The transmitting terminal identifier (yours) prints on the top of outgoing faxes and the remote terminal identifier (theirs) prints on your fax log. Set up your terminal identifier and ask all parties involved on the project to also set up their terminal identifiers. Periodically print and file your fax logs, in case they are needed to substantiate a transmission.

## 11. E-mail

E-mail can be even more convenient than faxes. They are transmitted immediately and are usually read on a priority basis. Make and file printed copies of all incoming and outgoing E-mail.

## 12. Speed memos

*Issue speed memo forms to all field supervisors* and require their use to document all significant conversations and to confirm verbal instructions by the owner's representative. Instruct the supervisors to list (1) the date and subject, (2) who to and from, (3) a summary of the discussion, and (4) what was agreed or directed.

Speed memos are especially useful when a written response is needed and are sometimes used in lieu of an RFI form.

## 13. Requests For Information (RFIs)

Request For Information (also called Design Clarification/Verification Request or DCVR) forms (Fig. 4.6) are used for obtaining clarification on the design or other issues. *Issue RFI forms to all field supervisors.* Ensure that they are used to request clarifications or to confirm oral questions, and are recorded in a log. RFIs should ask clear questions and reference the appropriate specification sections or drawing details. If a timely response is critical, note the consequences if late. Forward relevant responses to all affected subcontractors and vendors.

**Partnering approach.** RFIs should not be used as a weapon to establish a claim. They may do that, but that should not be their purpose. They are for *requesting information.* Excessive RFIs, intended solely to establish a claim or asking questions that could be resolved by reviewing the plans and specifications, detract from the value of the RFI files and log.

To minimize paperwork, ask minor questions orally in the field or at the weekly meetings instead of issuing an RFI. If the response indicates a potential change, include it in the minutes of the meeting or follow up with written documentation. If an RFI will impact work, note that fact in the RFI log. Also note which RFIs had no effect on the work. Then the RFI log can be a powerful tool for claims analysis.

**Basic information.** Enter the project title, date, and RFI number. The RFI number can be a letter, such as R, followed by a sequential number. Resubmittal of an RFI on the same or closely related question can have a different number, but preferably the same number plus a letter to indicate a revised or second request (e.g., R0023A for the second submittal of RFI number R0023).

For the title enter a brief unique statement that describes important elements of the RFI, the work area, and trades impacted. If possible, the title should indicate the scope of work.

**References.** Note the drawing number and/or specification section describing the work. If the RFI is related to other RFIs, notices of change, change orders, etc., list them in the next field. Also note the work area and any subcontractors affected. Optionally note the subcontractor's RFI number.

**Description of the request.** Enter a clearly written, detailed question or statement to be confirmed. Attach sketches, letters, and other documents, if warranted, as noted at the end. Draw a rough sketch on the RFI if helpful.

**Date needed, priority, and impact.** Note when the response is needed. Avoid "asap" unless a response is truly needed immediately; note "today" if truly needed the same day. Note the relative priority and comment on the current or probable impact if a response is late. The impact can be cost or time and may require more space to be fully described. List the activity numbers affected, if known. Continue on the next line down if more space is needed.

**Author and recipient.** The writer's name, the name of the recipient, and the recipient's fax number or E-mail address should be entered next. Enter who received copies on the next line, and note the total number of pages. The RFI is designed to be faxed or E-mailed and does not need a cover sheet.

**Owner/designer's response section.** The owner's representative or designer completes the next section before returning. They should be asked to use the columns noted or to suggest changes to better meet their needs and yours.

The respondent can note that the following information is their response to the question. Or, if the RFI was a request for approval, they can approve the request, approve it with conditions as noted below, reject it, or request more information as noted below.

The respondent should note the date of the response, and (hopefully) answer the question. The attachment box should be checked if additional pages are attached.

The respondent should enter his or her name, and either leave the "To" field blank if returning to the originator, or enter another individual's name and their fax number or E-mail address. The respondent should note if anyone received a copy and enter the total number of pages (if more than one) on the line below.

**Contractor's resolution section.** The originator of the RFI optionally completes the next section before filing or forwarding the returned RFI to others.

Note whether the RFI is closed (if the answer was complete and acceptable) or if it is still open (e.g., if a change order request must be prepared). If open, note the action needed, by whom, and when. Also enter any additional comments, list schedule activity numbers affected, reference another document (e.g., another RFI or COR), etc. The RFI log is described in more detail in Sec. C.16.b.

## XYZ Construction Company, Inc. -- Request For Information (RFI)

Job No./Title: _____ Date: _____ RFI No: _____

Title of RFI: _____ Ref Dwg/Spec: _____

Related Issues (RFI, NOC, CO): _____ Work Area/Sub: _____

Request/Description: _____

_____

_____

_____

_____

_____

_____

_____

_____

_____

_____ Attachment ❑

When Response Needed: _____ Priority: _____ Impact If Late: _____

From: _____ To: _____ Fax/E-Mail: _____

Copy To: _____ Page 1 of ____

## RESPONSE

❑ See The Following   ❑ Request Approved   ❑ Approved As Noted   ❑ More Information Needed   ❑ Disapproved   Date: _____

Response/Comment: _____

_____

_____

_____

_____

_____

_____

_____

_____

_____

_____

_____ Attachment ❑

From: _____ To: _____ Fax/E-Mail: _____

Copy To: _____ Page 1 of ____

## For Use By Contractor After Return

RFI is: ❑ Closed   ❑ Open   Action Needed: _____

Action By: _____ When: _____

Comment/Reference: _____

**Figure 4.6**   Recommended Request For Information form.

## 14. Extra work order (daily force account/time and materials) records

Force account work is directed extra work without a price agreement, due to either undefined scope or inability to negotiate price, and is paid on a time and materials basis. The work is documented daily on preprinted forms, which are signed by both the contractor's supervisor and the owner's representative (Fig. 4.7). When all work is complete, the labor and equipment hours and materials used are priced, subcontractor charges are added, and costs are subtotaled, marked up, and paid as a change order.

Force account forms can also be used when the work is not clearly acknowledged as extra work. If so, it may be better to title the form a work order form. The use of a preprinted form, the title Work Order or Extra Work Order instead of change order request, and the flat assumption that the work is reimbursable may avoid disputed entitlement.

Work order forms can also be used to document disputed work, in which case the owner's representative will usually decline to sign the form. If so, modify the form so that it acknowledges the work was done without acknowledging entitlement. Then ask the owner's representative to sign it to eliminate one potential issue from dispute if a claim is made. Even if the owner's representative refuses to sign, *submit force account records every day.* Note that the named representative observed the work and refused to sign or was asked to observe the work and declined. If you do so, the owner will find it hard to contest the amount of work performed if the dispute proceeds to arbitration or litigation.

The work order form can also be used for change order requests. Simply complete the form using either actual or estimated labor and equipment hours, materials, subcontract costs, and markup. Otherwise see Sec. B of Chap. 5 for the recommended Change Order Request form. The work order contains the following information.

**Basic project and work order information.** Enter the project number and/or abbreviated title, the date, and an assigned work order number. The work order "number" should consist of a clearly identifiable letter prefix (e.g., WO or EWO) plus a sequential number. If modified, use an ending letter or dash and suffix number to indicate an additional phase of the work order.

Enter a unique title for the work order, preferably describing the type of work, work area, etc. This will be computerized for the work order (extra work) log, with a 50-character field recommended. Enter a sequential record number each day the work order is under way if the work continues longer than one day, to facilitate control and to ensure a report is not missing. When a work order is complete, enter the total number of records (e.g., days worked).

Generally the field supervisor enters the quantities and descriptions of the resources used, and the field office or others enter the unit rates, compute values, subtotal, and optionally add markup.

**Description of the work.** Clearly describe the work performed, and the reason, if known and relevant. If referring to another document, provide at least a

## XYZ Construction Company, Inc. – Extra Work Order (EWO)

Job No./Title: _____  Date: _____  Work Order No: _____

Work Order Title: _____  Report No: _____ of _____

Description of Work: _____

_____

_____

### MATERIAL

| QUANTITY | Unit Size | DESCRIPTION | UNIT PRICE | TOTAL |
|---|---|---|---|---|
| | | | | |
| | | | | |
| | | | | |
| | | | | |
| | | | | |
| | | | | |
| | | | | |
| | | | | |
| | | | | |
| | | | | |
| | | | | |
| | | | | |
| | | | | |
| | | TOTAL MATERIAL | | |

### LABOR

| EMPLOYEE | ST | | OT | | TOTAL |
|---|---|---|---|---|---|
| | HRS | RATE | HRS | RATE | |
| | | | | | |
| | | | | | |
| | | | | | |
| | | | | | |
| | | | | | |
| | | | | | |
| | | | | | |
| | | | | | |
| | | | | | |
| | | | | | |
| | | | | | |
| | | | | | |
| | | | | | |
| TOTAL LABOR | | | | | |

### EQUIPMENT

| HOURS | DESCRIPTION | UNIT PRICE | TOTAL |
|---|---|---|---|
| | | | |
| | | | |
| | | | |
| | | | |
| | | | |
| | | | |
| | | | |
| | TOTAL EQUIPMENT | | |

### SUBCONTRACT

| SUBCONTRACTOR OR TRADE | TOTAL |
|---|---|
| | |
| | |
| | |
| | |
| TOTAL SUBCONTRACT | |

### SUMMARY

| | |
|---|---|
| DIRECT COST (L, M, E & S) | |
| MARKUP @ ____ % | |
| GRAND TOTAL | |

Certified: _____  Approved: _____  Title: _____  Date: _____

Superintendent                Owners' Representative

**Figure 4.7**  Recommended extra work order (force account) form.

summary description of the work and clearly define the referenced document. Do not refer to "...our previous conversation" or other undocumented descriptions.

**Material cost.**   Enter the materials used: the quantity, unit of measure, size, and description of each material line item. If known, enter the unit price; otherwise the office must add that value. Or identify a vendor with a brief description of the materials purchased and attach an invoice. If necessary, refer to an attached sheet for a detailed quantity takeoff. Optionally enter values for consumables and for safety supplies as a percentage of labor costs, or enter detailed costs. Total the material costs. If using a computerized spreadsheet, it will multiply the quantity times the unit price to compute the subtotal and then total for all materials.

**Labor cost.**   Enter the employee names (or employee numbers), the straight time (ST) and/or overtime (OT) hours, and hourly rates. Compute the line item totals. Subtotal the labor costs, optionally add a percentage for supervision, plus travel or subsistence charges. Then total all labor costs.

**Equipment cost.**   Enter the hours of use for each piece of equipment, the description, and the unit rate. Optionally enter a percentage of labor costs for small tools. Compute the line item total and total all equipment costs.

**Subcontract cost.**   Enter subcontractor names (or trade description, if sub-contractors haven't yet been selected) and attach work orders for each subcontractor. Enter the value of the subcontractor's work and total all sub-contract costs.

**Cost summary and markup.**   Add the totals for each cost category (materials, labor, equipment, and subcontract) and post to the "Direct Cost (L, M, E & S)" field. Optionally, enter the percentage markup for overhead and profit, compute the markup, add to the direct cost, and enter the result under grand total. Otherwise the markup can be added on a summary sheet for the change order request.

   If a different markup is used for some cost categories (most frequently sub-contracts), modify the form to show a subtotal and mark up those costs separately from the others.

**Signatures.**   The form must be signed daily by the supervisor, and two copies must be provided to the owner's representative (regardless of whether or not they acknowledge entitlement). If the owner's representative has agreed the work is extra work, the owner's representative should sign the owner's representative line, enter their title, note the date, and return one copy.

If the work is not accepted as extra work, change "Approved" to "Acknowledged" and add, on a line below, "Entitlement is not acknowledged" or some similar phase. Then ask the owner's representative to sign as "acknowledging" that the work was done, while not accepting responsibility.

### 15. Extra work files

Set up a separate file for each issue, as soon as it is identified as a potential extra work item. Start with the document first identifying the change. This may be a notice of change letter, RFI, request for proposal from the owner's representative, architect's supplemental instruction, or other document. Include sketches by the designer or the project team, requests for proposals from the owner, completed extra work order forms, the change order request with all supporting documentation, correspondence and other negotiation records, force account records, and copies of cost reports with the cost code(s) assigned to the change, etc.

Extra work files simplify change order request and claims preparation, especially if a brief narrative is prepared for each that notes the background, summarizes conversations and other unwritten communications, references supporting documents, and includes computation of delay and costs.

### 16. Information logs

Information logs are essential to effective change management, as they fulfill the following functions.

- Tracking of submittals, requests for information, change order proposals, etc., so that you know when action is needed and can act to correct problems and prevent delays.

- Communicating the status and importance of administrative actions to the owner's representative, designer, subcontractors, and others.

- Providing basic information needed to prepare claims for late submittal review, impact from multiple design errors and omissions, late response to requests for information, etc.

Create information logs using a spreadsheet, word processing software, database management software, or a commercial package such as ProLog Manager™, Primavera's Expedition™, etc. Record all event dates: when received from the subcontractor or vendor, submitted to the owner, returned by the owner, and forwarded to the subcontractor or vendor. Enter a clear descriptive title for the item, the specification section and drawing number or detail affected, the location on the project, who originated it, and what the resolution was. Note the date a response was requested and the priority given.

Consider using 11 × 17-inch paper, in order to display more information per sheet. Suppress unneeded columns when printing on 8 ½ × 11-inch paper for faxing or for submitting to the owner's representative.

Although the contractor's logs are the most important, the owner's representative and designer must also maintain submittal, RFI, COR, and other logs. Otherwise, they can't adequately control turnaround time. Emphasize the importance of prompt turnaround of contractor requests in the partnering meeting, and continually stress its importance. Some designers dislike maintaining logs and may not have internal procedures to ensure timely processing. Contractors, on the other hand, are sometimes apt to submit an RFI before trying to figure it out themselves, or they sometimes fail to clearly describe the problem. Be certain to review the available information before submitting an RFI; otherwise it will damage your credibility and result in slower response to important questions.

You will find it most effective to maintain the following logs.

**a. Submittal schedule log.** Submittal schedules are crucial to materials control and *must* be established on all projects. They identify all required submittals, who is to prepare them, when they are due, and where in the specifications they are described. *Require subcontractor submittal schedules within 30 days after notice to proceed,* as preparing them often reveals design errors or material availability problems (Fig. 4.8).

**Form contents.** The suggested contents of the submittal schedule/log are as follows.

- Item number: A sequential record number, assigned when preparing the submittal schedule.

- Specification section: where in the specifications and plans the item is described. This information and the description of the item are often provided by the designer in a required submittal schedule.

- Description: a brief description of the item.

- Type of submittal: optional designation as a shop drawing, catalog cut (manufacturer's data), physical sample, certification, test report, or other type of document.

```
PO8O   TYPICAL DESIGN/BUILD PROJECT                              PMS80                        Update: 02  Page:  1
PROJECT MANAGER:  S. Pinnell   CONTRACTOR:             REPORT 65 - MATERIALS CONTROL LOG        Update date: 1MAR90
PROJECT FILE:PO8O  DBM:PO8OMATL SORT:NOT SORTED    PINNELL/BUSCH INC. - PORTLAND, OREGON        Print: 16:37 22SEP97
--------------------------------------------------------------------------------------------------------------
RECRD SPEC      DESCRIPTION                SUB VENDOR      TRANS DATE SUB REVW DATE REC AP REVI DELV DATE DUE DATE IS  FLO
NUMBR SECTION                              CON             NO  TO A-E  TIME  FR A-E  PR TIME TIME ON JOB   NEEDED   AT
--------------------------------------------------------------------------------------------------------------
   1 1A-3.0     QUALITY CONTROL PLAN                       1  A15FEB86
   2 2A-2.2     PROJECT SIGN LAYOUT                        3  A15FEB86       A25FEB86 A
   3 1B-10.1.1  SAFETY PLAN & HAZARD ANALYSIS              2  A15FEB86
   4 2B-4.22    BACKFILL ROCK SAMPLE       WESTPORT PIT    4  A17FEB86       A21FEB86 A              A24FEB86
  15 15A-3.2    WATER & SEWER PIPE & FITTINGS  ME
   5 3A-8.1     REINFORCING STEEL          MERCER STEEL    5A A17FEB86       A21FEB86 A              A28FEB86
   7 3A-22.4b   CONCRETE MIX DESIGN 2500 PSI  SMITH REDIMIX 6  A22FEB86       22MAR86                 25MAR86
   6 3A-12.31b  CONCRETE MIX DESIGN 3000 PSI  SMITH REDIMIX 6  A17FEB86       A22FEB86 A              A28FEB86
   8 5A-3.1     WELDER CERTIFICATIONS                      7  A20FEB86
   9 5A-3.2     STEEL DETAILS & LAYOUT                     8                                    30
  10 5A-8.1     FIELD CONNECTIONS (BOLT)   MERCER STEEL    8  A25FEB86       A28FEB86 R  31     30
  12 8A-2.11    DOORS AND FRAMES           SEATTLE DOOR                                         70
  14 7B-5.3     BUILT-UP ROOFING           TN                                    20
  16 15B-3.5    FIRE PROTECTION SYSTEM     ME                                               20
  17 15-D4.3    HVAC SYSTEM PACKAGE        ME                                    20           35
  18 16-A1.3    TRANSFORMER & SWITCHES     EL                                            10   45
```

**Figure 4.8**  Typical submittal log.

- Subcontractor or vendor: who is responsible for preparing the submittal.

- Transmittal number: a sequential number, assigned when transmitting the item. Add a letter to indicate a revision when resubmitting the same item. (The transmittal number is blank when initially submitting the submittal schedule to the owner's representative.)

- Document number: optional identification of the document (e.g., a drawing number), provided when the item is submitted.

- Scheduled receipt date: when the submittal is due from the subcontractor or vendor. Suppress this column when providing the log to the owner's representative.

- Scheduled submittal date: the date when the document is scheduled to be sent to the designer. This is superseded by the actual submittal date, when the item is actually submitted.

- Priority: the relative importance of the document, with high-priority submittals to be reviewed before lesser-priority documents. This is needed if a lot of submittals are transmitted at the same time and you want some reviewed first, or an early return is needed.

- Review time: the number of days for review and return as specified in the contract. This can be less than the contract specified time if the item is critical, the requested review time is reasonable, and the designer is notified of its priority. Upon return, this duration is changed to the actual review time.

- Scheduled return date: when the item is due back from the designer. Upon receipt, this is changed to the actual receipt date.

- Approval status: this may be approved, approved as noted, revise and resubmit, or rejected.

- Revision dates and duration: if rejected, the item is usually reentered on another line of the log and goes through the same cycle. Some logs have additional columns for submittal revision time, revision receipt date, resubmittal date, rereview time, and rereturn date. Or revised submittal dates are listed on the next line of the log.

- Fabrication and delivery time: the scheduled/actual number of days for fabrication and delivery of the item after approval.

- Delivery date: the currently scheduled or actual date of delivery to the project.

- Date needed: the date the item is needed, as determined by the early start of the first activity needing the item.

- Float: the number of days from when the item is needed to when it is due. A negative number indicates late delivery will delay the start of the activity (but not necessarily the entire project).

- Comments: a column for comments or additional information.

**Procedures.** Submittal logs are the same as submittal schedules, except that they contain actual dates and durations and are usually the same document.

Most people use the terms interchangeably. Update the submittal log at least monthly. Record actual dates and compare with scheduled dates. Schedule for resubmittal, if required. Update fabrication and shipping times, recompute the delivery data, compare with the date needed, compute the float, and take any required action. Periodically sort the report by review time, selectively print those submittals with excessive review time, and send to the owner's representative with a request for action. Closely track activities with negative or limited float and expedite, if necessary.

The submittal log will help in preparing claims if late material delivery was a problem. Sort and print a report by review time and check those with excessive time that were required for critical path activities.

**b.  Request for information (RFI) log.**   RFI logs track requests for clarification (Fig. 4.9). They should record the RFI number, a title/description, date initiated, who initiated, response date, summary of response, impact, further action, if required (e.g., a change order), the change order request number, if one is submitted, and a comment, if appropriate.

RFI logs can be a powerful indicator of design and administrative problems—by sheer number, the time lapse from initial question to resolution, the relative importance of the issue, and especially by their impact on the schedule. Exclude trivial issues, however, as they cause the owner's representative and designer to disregard future requests and degrade the value of the information for preparing a claim.

When creating an RFI log, consider including the following columns.

- RFI number: a letter, such as R, followed by a sequential number in the order issued. Reissuance of an RFI on the same or a related question can have either a different number or the same number plus a letter to indicate re-request (e.g., R0023A for the second submittal of RFI number R0023).

- Title or description of the RFI: this should be a unique brief statement of the issue.

- Reference specification section or drawing number: where the work is described in the contract documents.

- Work area: an optional field to identify the part of the project affected. This information may be suppressed as nonessential information when submitted to the owner's representative. If a claim develops, it may become very important.

- Subcontractor or crew affected: for internal use by the contractor, as it is nonessential information for the owner's representative unless a claim is being filed.

- Subcontractor's RFI number: optional field for internal use. It aids in responding to the subcontractor.

- Related issues: listing of related RFIs, notices of change, owner requests for proposals, change order requests, accounting cost codes, etc. This may either be included in the report to the owner's representative or suppressed and is very helpful when preparing claims.

- Impact: an optional field with the additional cost and/or time, either currently or if not responded to when needed. Upon receipt of the information, update this field with the activity numbers that were delayed (if there was any impact).

- Date received from subcontractor: an optional field for internal tracking.

- Date submitted to owner's representative.

- Date required: when the RFI must be answered to minimize impact to the project.

- Priority: an optional field noting the relative importance of the submittal, so that it can be reviewed and returned before other submittals.

- Date returned: for tracking response time and impact, if after date required.

- Response: information provided, request approved, request approved as noted, more information needed, or disapproved.

- Status: whether the issue is closed or open, and the action needed if still open.

- Action needed: whether to proceed, obtain more information, prepare a claim, etc.

- Comment: a list of the activity numbers affected, references to another document, etc.

When preparing a claim based on a multitude of impacts, tie each RFI to one or more activities and plot the RFIs together with the affected activities using the ELIPSE schedule described in Chap. 10.

| RFI LOG | | | | | | |
|---|---|---|---|---|---|---|
| **DESIGN / BUILDER:** | | | **PROJECT:** | | | |
| **RFI #** | **REQUEST** | **REPONSE** | **ORIGINATED BY:** | **ORIGINATED DATE:** | **RESPONSE BY:** | **REPONSE DATE:** |
| | | | | | | |
| | | | | | | |
| | | | | | | |
| | | | | | | |
| | | | | | | |
| | | | | | | |
| | | | | | | |
| | | | | | | |
| | | | | | | |
| | | | | | | |

**Figure 4.9**  Abbreviated request for information (RFI) log.

**c.  Extra work logs.**  Create a log of potential and actual extra work items. Enter a record number, a title/description of the work, reference to the document initiating the extra work, the date and form of the notice of change, related document codes (RFI, change order request, construction change directive, etc.), when the work was done, what supervisor or subcontractor did the work, etc.

**d.  Other logs as required.**  Other logs that may be required (Fig. 4.10) include:

- Correspondence to/from the owner's representative and key subcontractors
- Notices of change
- Change order requests and claims
- Construction change directives by the owner's representative
- Change orders
- Architects' supplemental instructions
- Drawings and sketches
- Subcontractor change orders and payments

```
P080   TYPICAL DESIGN/BUILD PROJECT                        PMS80                          Update: 02  Page:  1
PROJECT MANAGER: S. Pinnell  CONTRACTOR:          REPORT 53 - TOOL & EQUIPMENT INVENTORY   Update date:  1MAR90
PROJECT FILE:P080  DBM:P080TOOL SORT:NOT SORTED   PINNELL/BUSCH INC. - PORTLAND, OREGON     Print: 16:38 22SEP97
--------------------------------------------------------------------------------------------------------------
RECRD DESCRIPTION               MODEL AND    EQUIPMENT   RENTAL TOTL UN TO  QTY DATE    TO QTY TO QTY TO QTY TO QTY QTY
NUMBR                           BRAND        NUMBER      RATE   QTY IT WHO  OUT ISSUED  WHO OUT WHO OUT WHO OUT WHO OUT  IN
--------------------------------------------------------------------------------------------------------------
     5 AIR HOSE - 1/2"          (50' length)             2.00 1500 LF SSP  150         LLB 200 JSB  50              1100
     8 BLADE POINT                                       2.50    5 EA SSP    1         LLB   1                         3
     9 CHISEL                                            2.00   10 EA SSP    2                                         8
    10 BUSHING TOOL                                      2.00    5 EA SSP    1                                         4
     1 AIR NAILER               16D PASLODE  N-014      12.00    1      LLB    1 28FEB83                               0
     2 AIR NAILER               6D PASOLDE   N-022       8.00    1      SSP    1 29MAR83                               0
     3 AIR NAILER               6-16 HILTI   N-044      12.00    1                                                    1
     4 AIR STAPLER              SENCO II     N-077      12.00    1                                                    1
     7 CHIPPING HAMMER          GARDNER DEN. N-085      15.00    1                                                    1
     6 CHIPPING HAMMER          INGERSOLL R. N-123      15.00    1      SSP   .1 29MAR83                               0
```

**Figure 4.10**  Typical small tool and equipment inventory log.

## 17.  Cost accounting reports

Cost accounting reports should include weekly labor reports comparing actual productivity (units of work in place per labor hour) and unit cost with budgeted productivity and unit cost. They should also include monthly cost comparison reports by cost category that compare actual unit costs with budgeted unit costs, general ledger reports with detailed transactions, and other reports as required. Equipment intensive projects should have equipment cost reports.

Verify that cost codes are established for tracking the cost of extra work and disputed work and that supervisors are using the codes. Create separate cost codes for subcontractors' change order work.

## 18. Progress payment requests and backup files

File progress payment requests with supporting documentation, which should include surveys or other measurements, quantity takeoffs, subcontractor and vendor submittals, and notes of meetings and negotiations with the owner's representative over what items are approved for the month's pay request.

## 19. Contract documents

Maintain a master copy of the plans and specifications with all addenda and changes posted. Save copies of superseded drawings, specification sections, designer's field sketches, drawing issuance logs from the designer, etc.

Maintain a separate set of as-built drawings with all changes from the design drawings marked in red. These are normally turned over to the owner at the end of the project. Also maintain copies of all subcontracts, purchase orders, bonds, insurance certificates, etc.

## 20. Shop drawings and other submittals

Maintain copies of all submittals including stamped shop drawings, along with the log described in Sec. 16 above. Review and approve subcontractors' shop drawings before submitting them to the owner's representative.

## 21. Other records

Other records that should be maintained include prebid documents, estimate and bidding files, other preconstruction documents (e.g., regarding bonding), field records such as delivery tickets (note on the ticket and in your diary if they are late), safety reports, accident reports, weather records, newspaper clippings, documents received from subcontractors or the owner, etc. See Sec. A.1 of Chap. 8 for a complete list of records normally maintained at the jobsite or home office.

On equipment-intensive jobs, maintain daily records of equipment use. If not used, note if idle, standby, or down for repairs and the reason for any standby. Record the date and effort for equipment mobilization, demobilization, and other special activities.

On weather-sensitive work (e.g., earthwork, paving, painting), consider keeping your own weather records. Rainfall varies widely over a relatively short distance and a gully washer at your jobsite may not reach the nearest official weather station. Temperature and wind velocity also vary widely depending upon the microclimate. High/low recording thermometers, wind gauges, and rain gauges are relatively inexpensive and easy to maintain. Maintain records of any condition that may affect your work (e.g., wind speeds for crane operations, temperature and rainfall for concrete work).

**Organization.** Organize correspondence (letters, facsimiles, speed memos, etc.) with separate folders for each recipient/sender. If extensive correspondence is expected (i.e., with the owner's representative, designer and key subcontractors), set up separate folders for To and From correspondence. If only limited correspondence is expected (i.e., with minor vendors), include their quote, subcontract or purchase order, submittals, test reports, delivery tickets, etc. in the folder. Otherwise, set up separate subfiles for each type of document. Insert the latest correspondence on top and maintain that order to facilitate retrieval.

## 22. Subcontractor records

Improved subcontractor recordkeeping will help document your change orders and improve the subcontractors' recovery for extra work. It will help you avoid problems with unpaid subcontractors who don't perform or default on their contract.

Require subcontractors to submit daily subcontractor reports similar to the daily field supervisors' reports and obtain their input to your schedule. Require that they submit a submittal schedule with material delivery dates prior to their first progress payment and that they update it monthly. Late delivery of subcontractor materials is a frequent cause of project delay but can usually be avoided by close tracking of submittals and material procurement.

## 23. Reports to company management

Improved communication between field personnel and company management is *essential*. Managers need to receive copies of many of the above records, in addition to formal management reports. Management reports should be submitted monthly at the same time as the monthly cost accounting reports and progress payment request. Brief weekly reporting is recommended if company managers don't visit the job weekly. One or two pages is usually sufficient for most projects.

The recommended format for management reports is as follows.

---

**Monthly Status and Progress Report**

Project title _____ Report date _____ Author _____

1. *Summary of Status and Direction.* One or two paragraphs summarizing the status of the project and major pending developments.
2. *Work Accomplished Last Period.* A description of the work completed during the reporting period (or to date) with a discussion of schedule, cost, quality, safety, and changes. It should include the effort expended (costs, equipment use, etc.), major quantities of work accomplished, milestones met, crew size, problems encountered and resolved, etc. Schedule and cost variances should be discussed in detail.
3. *Work Planned Next Period.* The work planned next period, with resources planned or needed, upcoming problems or opportunities, scheduled milestones, expected crew size, etc.
4. *Problems and Opportunities.* A brief discussion of problems, proposed solutions and questions, opportunities for savings, and other general comments.
5. *Action Needed, by Whom, and When.* List of actions needed by others besides the project team (i.e., company management, estimating, subcontractors, utility companies, etc.) with the names of the individuals or companies responsible and when the action needs to be taken.

---

The report should include comments on notice of changes, a description of the change, status of the change work, probable cost magnitude, response of the owner's representative (especially regarding entitlement), and whether support is needed to estimate and negotiate a settlement.

## D.  Prebid Phase Risk Management

To control contract risks, implement the Change Order Management Program on every contract bid. This requires only a few simple procedures and is economical.

### 1. Conduct a prebid review

Every company needs standard bidding procedures for their estimators which identify issues to discuss with management before bidding. They should include:

1. Nonstandard and potentially risky contract clauses.

2. Indications of design deficiencies that might increase construction costs.

3. Scheduling problems that may push weather-sensitive work into bad weather.

4. Environmental windows for working in sensitive areas that could interfere with scheduling

5. Unique features of the site or work that substantially increase risks.

A thorough prebid site visit with a written report of findings (on a checklist) helps identify potential problems and shows due diligence when pursuing a claim for changed conditions.

Prebid activities should include preparation of a preliminary schedule, even if only a bar chart. It will confirm the project duration for computation of job-site overhead costs and determine if scheduling or weather-related problems are likely. Resource-load the schedule with critical labor and equipment to verify the practicality of the schedule.

Subcontractors should pay as much attention to prebid scheduling as general contractors. They should be especially concerned with the amount of time available for their work, whether multiple mobilizations will be required, and whether work will be performed in inclement weather. Subcontractors need to integrate the schedules for all of their projects into a master schedule and resource-load the activities, to verify whether they have enough supervisors and tradesmen to perform.

If the bid is low, estimators should thoroughly review and document their bid for the project team's use when building the project, and any concerns for prompt management action. Their notes and copies of pertinent portions of the bid become part of the change order management files for the project and are turned over to the project team.

## 2. Implement a loss prevention plan for contractual risks

If particularly onerous contract clauses (described in Chap. 3) are found, a loss prevention plan will reduce the probability of the problems occurring and mitigate impacts if problems do occur. Subcontractors faced with onerous subcontract clauses can take the same action against the general contractor as contractors take against the owner.

A typical loss prevention plan includes the following 10 steps.

1. Before bidding:
   - *Attempt to change the clause*—Contractors can negotiate with private owners to change the clause. Contractors can also talk with public works officials or lobby the governing body, or they can ask their contractors' trade association to intervene or to lobby for state legislative changes. Contractors can cooperate with the local bar association to educate and inform attorneys about the unfairness of the clause, possible disadvantages to their clients (in obstructing partnering), and why it is not in the public interest for such clauses to be used.

     Subcontractors have a particularly difficult time with onerous contract clauses, as they often bid without knowing the general contractor's terms and conditions, and then must negotiate more reasonable terms or lose the subcontract. A subcontractor refusing to sign a subcontract containing such terms may be sued if the general contractor relied on the subcontractor's bid. To avoid damages, the subcontractor must prove that the proffered terms are unreasonable.
   - *Alert other bidders to the risk*—Directly contacting competing bidders could be construed as collusion and could expose contractors to severe penalties. However, general contractors can notify potential subcontractors and suppliers of the risks. Undoubtedly, they will inform the other bidders.
   - *Add a contingency*—Adding a contingency will cover a portion of the probable cost of an onerous contract clause, but you may emerge the high bidder. If your bid is low, and the event occurs, the contingency will not cover the actual costs.

     To calculate contingency, compute the "expected value," i.e., the product of the probability of an event occurring times the cost. An event costing $250,000 with a 10 percent probability of occurring has an expected value of $25,000. An expected value computation is a reasonable way to account for the risk but is not suitable if the possible cost is more than the contractor can afford to lose.
   - *Ignore it—not recommended*—Hope for the best. You can also invest your life savings in the lottery or drop your insurance policies to increase profit margins.
   - *Refuse to bid*—If enough contractors refuse to bid contracts with such onerous clauses, changes will occur. The contractors may miss an opportunity but will also avoid a potential disaster.

- *Develop and implement a loss prevention plan*—The best response to onerous contract clauses is a loss prevention plan. The plan includes all the choices above, except ignoring the problem. It minimizes the probability of the problem occurring and the amount of loss if the problem occurs.

2. If you decide to bid, consult with your attorney and increase the markup to cover the additional costs of implementing the loss prevention plan and to provide higher margins to compensate for the additional risk. In extreme cases, some contractors form a separate company, just for a risky project. The separate company must be adequately capitalized to avoid the appearance of fraud but can protect other assets.

3. Analyze the plans and specifications for circumstances in which the problem may occur, and document the risks and appropriate action to be taken.

4. Inform your field personnel so that they promptly identify the problem if it occurs and know how to respond.

5. Try to prevent the problem from occurring.

6. Comply with all contractual notice requirements. Promptly notify the owner's representative in writing if the problem occurs, and *insist* the owner take immediate action to mitigate your loss. Notify the owner of the legal obligation to mitigate your costs, regardless of the contract language, and your intent to pursue reimbursement should the owner fail to respond. If necessary, involve your attorney.

7. As soon as possible, write the owner's representative describing the cause of the problem, specific action by the owner to resolve the problem, how soon action needs to be taken to avoid delay, and the probable cost and time impact if action isn't promptly taken.

8. Continue pressing the owner's representative to resolve the problem. Document your efforts and the owner's response or lack of response.

9. Implement more intensive recordkeeping of costs and events, take corrective action to minimize costs, and promptly prepare a change order request.

10. Aggressively pursue the change order request based on established principles of entitlement, equity, any contract loopholes, your notice, and the legal mandate that each party in a contract mitigate costs and not hinder the other party. Do not wait for the work to complete. If the change order request is rejected, file a claim in accordance with the dispute provisions of the contract and press for resolution. If the claim is rejected, demand arbitration or file for litigation as soon as the contract allows.

## 3. Identify noncontractual risks

The prebid review may identify problems with design errors, poor overall quality of the design, environmental hazards, unreasonably tight schedules, or work that must be scheduled for unfavorable weather, inadequate access, etc. If so, in addition to including the cost for additional risk in the estimate, establish cost codes to track the resulting costs, so that prompt action can be taken if the problems occur.

### 4. Clarify ambiguities before bidding

Attend the prebid conference and clarify ambiguities such as "...the owner may pay for materials stored off site." A clarification at that time is a commitment that can be enforced. Absent a prebid conference, write or fax requests for clarification to the owner's representative or designated individual. Inform them of your interpretation, to establish your position in case a dispute does occur.

### 5. Identify unenforceable clauses

Some clauses are not enforceable under state law (e.g., no damages for delay in some jurisdictions), but a few owners include them either from lack of knowledge or in an attempt to intimidate contractors. Have your attorney identify those clauses and ask the owner to delete them from the contract, as both a partnering commitment and a legal obligation.

## E.    Preaward Phase Constructibility Review and Risk Management

In an ideal world, the owner and designer would perform a thorough constructibility review during design. Unfortunately, this is seldom done. A preaward constructibility review by the contractor is the next best thing. It would be better to perform a thorough prebid constructibility review, but there isn't time and the cost would be prohibitive since you won't necessarily win the bid. Therefore, wait until after bidding, but before starting work, to conduct the review.

The following list of suggestions, although provided for a contractor's preaward review, can be used for a midproject review when a pattern of problems has developed. It can also be used by the owner and designer as a checklist for value engineering analyses and constructibility reviews prior to bidding. For additional information, see Andrew Civitello's book, *Contractor's Guide to Change Orders.*[9]

### 1. Who should participate and what they need to do

The constructibility review team should consist of the estimator(s) who bid the job, one or more senior construction company executives, the assigned project manager and superintendent, any other personnel who may be available and able to contribute, and key subcontractors.

Before meeting, all team members need to review the bid notes and bid estimate (subcontractors review their own bids), the plans and specifications, an instruction sheet identifying areas of concern and assigning tasks for individual investigation, and any other pertinent documents. They should take notes and prepare questions and comments for the meeting.

The joint review session should be managed like any other meeting, with an agenda (that is followed), recorded minutes, and an action plan distributed to each attendee afterward. Look not only for problems but also for opportunities such as value engineering change proposals, alternative methods, etc. The joint session may be followed by a more detailed review of areas with identified high risk or opportunity.

An in-depth constructibility review conflicts with a contractor's goal to reduce overhead but pays major dividends. Finding a problem when you're ready to do the work, or in the middle of it, can be very expensive, and it is difficult to recover the costs. It is easier to negotiate a fair settlement when there is time and much of the cost can be avoided. You also have the psychological advantage of identifying the problem. If identified early enough, address it at the preconstruction meeting or partnering workshop, or shortly afterward, when the owner's representative is more likely to be cooperative.

**Minimum acceptable level of effort.**  Even if a contractor doesn't perform a full preaward constructibility review, project team members should carefully review the plans and specifications and report action required or potential problems to the project manager. This should be done before the preconstruction conference and partnering workshop. Not only does this minimize problems, but it also provides a distinct advantage by initiating action before the owner's representative is aware of the problem.

## 2. The source of design errors

Design is a difficult, complex process, especially the coordination between design disciplines, verification of material availability, and ensuring compatibility. To avoid misunderstanding and design errors, the different design disciplines need to communicate extensively, and be coordinated by the lead designer. However, design fees are often insufficient for adequate research of existing conditions or coordination between design disciplines. Project changes are often budgeted at 3 to 5 percent of construction costs and often cost far more, while the fee for basic design services may be only 5 to 6 percent.

There are numerous causes of design errors:

- The large number of items specified, the complex relationship between elements of the project, and the potential for unidentified conflicts all cause problems. Even home building can be complex, especially in specifying materials which may not be available, may not fit together, etc.

- Money is always short for good design, let alone for detailed specifications of materials and workmanship.

- Different elements of the contract documents are prepared by different people and different companies, as each design discipline is usually a separate design subconsultant. The result is frequent miscommunication.

- The lead designer sometimes fails to coordinate the work of the various design disciplines, creating conflicts and gaps.

- The specifications are often assembled by inexperienced junior designers lacking the experience to recognize inconsistencies or common errors.

- The specifications may be copied from other projects from years earlier and for different conditions, or from standard boilerplate without adjustment for specific project needs.

- When "cutting and pasting" specifications from other projects, an important element may be left out or an inapplicable item included.

- Specifications are written after design, often in haste to meet the bid date.

- Some designers write a material or equipment specification with desired features that may not be available.

- Some specifications are copied from a particular manufacturer's literature, written around a manufacturer's product, or actually written by a manufacturer's representative. The result may be impossible for another manufacturer to meet, and no equal may exist.

**3. Guidelines on where to look during a constructibility review**

The majority of project problems tend to occur in just a few work areas, as noted below. They are a result of the documents:

- Being unclear
- Being conflicting
- Being insufficiently detailed
- Failing to anticipate a condition that either existed and wasn't known or evolved after the design
- Requiring work that can't be accomplished or materials that aren't available

**1.  General conditions.** Start a constructibility review by checking the following areas of the general conditions:

- *Award*—may be late with winter approaching and result in serious weather impacts, even if within the contract-specified time for award. An owner's request to extend the award date may force a contractor to risk losing the contract or suffer weather impacts. In either case, press for early award, expedite your bond and insurance submittals, and closely review your CPM schedule.

- *Project duration and progress requirements*—may not allow enough time to complete the work. Check for scheduling constraints, long-lead materials, interim completion dates, etc., and closely review your CPM schedule. This should have been done prior to your bid.

- *Schedule of values on unit price contracts*—may have incorrect quantities that will result in quantity underruns. This may result in insufficient recovery of overhead allocated to those items, or in extra profit if other items overrun the scheduled quantity. Prepare an independent quantity takeoff of critical items before spreading overhead and mobilization costs to unit price bid items.

- *Exculpatory language*—usually indicates a less than fair approach to risk allocation and the strong probability of problems.

**2. Drawing errors.** Next, review the plans to gain an overview of the project, and detect common drawing errors:

- *Discrepancies*—may exist between large-scale plans and small-scale details, or between the same dimension on different drawings.

- *Poorly detailed drawings*—are almost certain to lead to problems.

- *Differences between the plans and actual conditions*—often cause problems on renovation projects.

- *Numerous drawing changes*—evidenced by more design revisions than normal, or different designers on the same drawing or on related drawings normally done by one individual that indicate the likelihood of problems during construction.

- *Differences between the specified and actual dimensions of specified equipment*—may cause conflicts. This may be due to supplying one of several "or equal" pieces of equipment specified, which has different dimensions and characteristics than the one drawn on the plans.

- *Conflicting dimensions*—may result in a gap or overlap between portions of the project.

- *Assembled systems*—may violate clearance requirements of building, fire safety, plumbing, or other codes.

- *Undersized mechanical rooms*—require relocation of equipment or rerouting lines to avoid conflicts and to meet code clearance requirements.

**3. Technical specifications.** The specifications may have more errors than the plans. You should check the following items in the technical specifications and drawing notes.

- *Nonspecificity of references*—with notes that say "See Structural" instead of "See Section 06321" indicate sloppy design and the probability of more errors than normal.

- *Numerous extensive and last-minute addenda*—are a precursor of problems to come.

- *Missing or duplicated drawing notes and specification sections*—indicate an incomplete cut-and-patch job.

- *Vague specifications*—are almost certain to lead to problems.

- *Performance specifications*—of work that is normally prescriptive, or where means and methods are also partly specified, may create problems.

- *"Or equal" clauses*—when no true equal exists result when a designer wants a specific manufacturer's product and writes the specification so closely around that product that no other product will comply.

**4. Discrepancies between the drawings and the specifications.** Coordination between the drawings and specifications can be a problem but isn't apparent without careful review. Either may reference an item that doesn't exist in the other, and there may be gaps not covered by either. The contract documents usually specify rules of precedence. Items to look for include:

- *Missing or orphan work items*—when an item is deleted or added to either the plans or specifications without a corresponding change to the other.

- *Mislocated drawing notes*—when a work item for one trade is noted in the drawings for another trade. Subcontractors only review the drawings for their trade and will probably miss the item.

**5. Permits and regulatory requirements.** Permits and regulatory agency rules are a major source of delay, impact, and disputes over extra work. They should be checked for conflicts with the design, failure to secure permits (by either the owner or contractor), and unanticipated requirements.

- *Building permits*—can be late owing to slow building bureau review, design errors, owner changes, etc.

- *Building code violations*—are due to faulty design or variations in interpretation of the code by building officials. Take immediate action to resolve the problem and have the designer revise the documents to meet code.

- *System development charges*—may be payable by the contractor under the terms of the contract.

- *Environmental permits and restrictions*—are sometimes not included in the design. Verify the need for a grading or wetlands permit, river crossing, discharge permit, hazardous materials permit, etc.

**6. Project location and site conditions.** Your prebid site visit may uncover the following site problems:

- *Street access*—may be inadequate or not clearly stated, and the availability and distance to disposal sites, aggregate sources, laydown, and office/shop/yard area may be unsatisfactory.

- *Site access, project limits, easement and right-of-way restrictions or inadequacy*—may cause problems.

- *Neighborhood and adjacent property use*—may result in restrictions on noise and truck traffic or hours of operation. Problems may also result from concurrent construction on adjacent sites, or flooding from waterways or from adjacent property.

- *Topography and onsite facilities or landscaping*—may require unplanned demolition or extra costs in working around them.

- *Subsurface conditions*—soils, groundwater, and buried facilities may result in problems. Check for existing utilities not on the plans and mislocation of utilities on the plans that might interfere with the work. Examine the soils report and check for susceptibility of the soils to degradation under rain and the need to gravel access roads or to make other adjustments for winter work. Also check the location of the borings versus the building location as the designers may have moved the building after the soils report was prepared.

**7.  Utilities.**  Utilities, especially underground utilities, are often a cause of problems:

- *Availability of temporary utilities*—may be a problem.

- *Offsite utilities*—may present a problem owing to their location and difficult access from the site.

- *Relocation of utilities by others*—is a frequent source of delay and should be carefully monitored with the utility companies.

- *The location of existing utilities*—may interfere with new construction.

- *Unmarked existing utility lines*—not shown on the documents may interfere with new construction.

- *Deteriorated condition of existing utilities*—may require special protection, relocation of new lines, or replacement of the existing lines.

**8.  Inadequate space, physical conflicts of different trades, and poor design coordination.**  Trade conflicts are a frequent source of problems, which some designers try to cover by requiring the contractor to "...coordinate the trades." You may find any or all of these typical problems.

- *Insufficient space*—to install equipment, to move equipment into its specified location without extraordinary efforts, or to assemble the work.

- *Inadequate ceiling space*—for structural members, mechanical lines, electrical conduit, insulation, light fixtures, sprinkler piping, etc., will increase costs.

- *Mechanical rooms*—are usually too small, with trade conflicts, inadequate code-required clearances, etc.

- *Inadequate structural support*—for mechanical equipment (e.g., rooftop mechanical units or VAV boxes hung from the roof trusses) will require relocation of the equipment or reinforcement of the structure.

- *Conflict between gravity (sewer) lines and other items*—will cause relocation of work items.
- *Excessive piping or electrical conduit to fit in the pipe racks and conduit chases*—will increase installation costs or require relocation of some lines.
- *Structural elements and anything else*—architectural building details, plumbing, HVAC duct, lighting fixtures, construction access, etc., will often conflict.
- *HVAC duct and any other element*—may conflict.
- *Lighting fixtures or heating elements and other features*—may conflict.

**9.  Other.**   Other design items to check are as follows.

- *Changed conditions*—can occur between the time of design/site visit and notice to proceed, as when a building being renovated isn't heated over the winter, or winter flooding causes damage. Document conditions before starting work, especially on renovation projects, with extensive photographs, measurements, tests, etc.
- *Equipment coordination, installation, and interfaces*—can be problems.
- *Specified materials*—may not be available or can create delays through long lead times.

## 4. Take action if problems are encountered

If serious problems are identified, notify the designers and schedule a joint session to review them and initiate corrective action. Use a partnering and collaborative problem-solving approach.

If problems are found in one area of a design discipline, additional problems should be expected in other areas of that discipline. Prepare instructions for the project team on how to respond if the anticipated problems occur, and submit RFIs for any unresolved issues.

## 5. Brief the project team on potential problems

Before starting work, brief the project team (if they didn't participate in the constructibility review). Have the estimators share their notes, opinions, and strategies for accomplishing the work.

## 6. Value engineering change proposal clauses and other opportunities to increase profits

Don't limit your review to problems. Look for opportunities to make changes that would save money. If changes can be made without modifying the contract, make them. However, if they require a modification in the contract, negotiate an agreement with the owner. This may be covered by a value engineering change proposal (VECP) clause, which allows the contractor to propose money-saving changes that perform basically the same function, with the contractor sharing the savings with the owner.

If the contract doesn't have a VECP clause, propose one be added. Bring this up at the partnering or preconstruction meeting. Refer to a federal contract for an example clause and modify it to fit your needs.

### 7. Subcontractors: negotiating subcontract terms

Some general contractors include unbalanced or onerous terms in their subcontracts. These clauses include no damages for delay, pay if (or when) paid, and other provisions that put subcontractors at risk because another party has control over the outcome. Subcontractors should carefully review the subcontract before signing such contracts and consult with an attorney, if necessary.

General contractors can force the low subcontractors to stand by their quote or risk being sued for breach of contract. If the proposed subcontract has unacceptable, non-industry-standard terms, however, the subcontractor may be able to refuse to sign the subcontract and negotiate more acceptable terms. Before doing so, they must check with an attorney. Negotiating better contract conditions isn't possible if the general contractor has previously informed the subcontractor of the terms. To be safe, subcontractors should condition their quote on "...mutually acceptable terms" or on an acceptable subcontract form such as the new AGC subcontract 610. If you ask for a copy of the general contractor's standard terms and conditions prior to bid, read it. Then either accept the conditions, decline to bid the project, or condition your bid. Consult with an attorney if you have any questions.

### F.  Implement Partnering and Retain a Project Neutral

Partnering and a project neutral can help avoid or quickly resolve issues that would otherwise deteriorate into a claim. They will decrease the need for arbitration or litigation.

The use of a project neutral is one of the newest techniques to avoid and resolve disputes. The most common form of project neutral is a dispute review board (DRB), which is utilized on very large projects. The concept of a neutral expert is more recent and not widely used in the United States, but has great potential for most projects.

### 1. Partnering

Partnering is crucial to the change order management program and results in a better response by the claims reviewer. Contractors should therefore press for project partnering, even if it is not included in the contract. Subcontractors should insist on being included in the partnering workshop and press the general to implement partnering with subcontractors just as the general contractor does with the owner. Be prepared, however, to enforce your contract rights if the partnering efforts aren't successful. You must comply with your contract obligations, as partnering does not alter contract requirements (e.g., the notice requirements).

The partnering concept and techniques are integrated throughout this book. The process of initiating partnering with a formal workshop is briefly described in Sec. C of Chap. 2, with more detailed information available in the referenced trade literature in the bibliography.

## 2. Dispute review boards

Dispute review boards (DRBs) are usually a panel of three individuals experienced in the type of construction being accomplished. The contractor and owner each select one board member and the two together pick a third. All three must be acceptable to both parties. The board meets regularly to keep abreast of progress. Whenever there is an unresolved dispute, the board hears presentations and renders a nonbinding written recommendation for settling the disputes.

Although nonbinding, few dissatisfied parties take an unfavorable DRB decision to arbitration or litigation. The DRB decision is admissible evidence. Since the decision was reached by neutral experts in the industry, who were familiar with the project and dispute, few arbitration panels or courts are willing to disagree with their findings.

On 100 underground construction projects with a value of $6.4 billion using DRBs, 98 disputes were referred to the boards and none were arbitrated or litigated. DRBs have also been very successful in other types of construction, although there have been a few cases of litigation.[10]

DRBs are suitable only on large contacts, as board meetings are relatively expensive. The total cost of DRB programs typically ranged from 0.04 to 0.5 percent of final project costs.[11]

For a detailed description of DRBs, their costs, and how to maximize their usefulness, see *Construction Dispute Review Board Manual* by R. M. Matyas, A. A. Mathews, R. J. Smith, and P.E. Sperry, published by McGraw-Hill, 1996.[11]

## 3. Neutral experts

The use of neutral experts is new in the United States but has been used elsewhere for years. The neutral expert is retained jointly by both the contractor and owner to determine the facts, develop a recommended solution, and present the solution to the parties. The parties need not accept the findings and recommendations of the neutral expert.

The use of neutral experts is similar to the use of DRBs but can be applied to smaller projects owing to the lower cost. In addition, neutral experts may be more proactive than DRBs and can help mediate disputes. Neutral experts also may be hired by a DRB to independently investigate and report on disputed technical issues outside the expertise of the DRB members.

See Sec. D of Chap. 7 for suggestions on how to work with neutral experts and dispute review boards.

## G.  Construction Phase Change Order Management

The change order management program continues from notice to proceed through construction to final payment.

### 1.  Use the preconstruction conference to initiate partnering and change management

Come to the preconstruction conference prepared to move the conference toward partnering and in the direction that you want with the following steps:[12]

- Thoroughly review the contract documents before meeting. Identify your objectives, the specific questions you want to ask, and the issues to be resolved. Being forearmed with knowledge is a powerful factor in shaping a positive outcome.

- Prepare at least a 60-day look-ahead schedule and a summary schedule to completion, or a completed schedule ready for submission if possible. Plan to cover the highlights of the schedule at the meeting and ask for an early review.

- Ask the owner's representative to include your specific issues on the agenda and to be prepared to discuss them.

- Clarify any contract ambiguities, stating them in a manner that will lead to a favorable interpretation. Argue for your interpretation if the response is unfavorable and ask for reconsideration if necessary.

- Schedule the partnering session, or request that one be scheduled.

- Establish procedures to better manage changes. For example, request agreement that the owner's representative will sign force account sheets of disputed extra work to confirm the work was done, without acknowledging entitlement.

- Request that a project neutral be selected—either a DRB on a large project or a neutral expert on smaller projects.

- Identify who has authority to approve change orders, any limits on their authority, and the process for submitting and obtaining approval of change order requests and claims. Also discuss allowable markups and extended overhead costs (if appropriate).

- Request establishment of a two-part change order process for large modifications. This allows partial payment for extra work for which a price hasn't yet been negotiated, up to the value the owner concedes, with the balance to be negotiated and paid in the future.

### 2.  Communicate with subcontractors, the owner, and the designer

Better communication is important to effectively managing change orders, but it requires attention to detail and a willingness to share information.

1. *Brief subcontractors* on company policy on changes, partnering, the need for a partnering attitude even when pursuing claims, potential problem areas, contractual notice, recordkeeping and reporting requirements, and action to be taken when a problem is identified.

2. *Obtain subcontractor input* as part of your scheduling process.

3. *Notify subcontractors and suppliers when changes may affect their work.* Require a timely response and insist that they track and submit costs in a timely manner.

4. *Maintain frequent open communication with the owner's representative* on the status of all change order requests and claims.

### 3. Ensure that extra work is identified and notice given as required by contract

Check the notice requirements in each new contract and alert the project team to any special requirements. For example, issue a plastic laminated sheet with the most important requirements to supervisory personnel, to protect it from damage and ensure it isn't mislaid. Make sure that field personnel give timely notice of change to the owner's representative, in addition to reporting to the home office.

The designer's review notes and requirements on shop drawings and other submittals may constitute changes to the work. These changes are not usually identified as such by the designer. Contractors must be alert to the practice and place the owner on notice when appropriate.

**Change-monitoring procedures.**  Effective change-monitoring procedures will ensure that extra work and impacts are promptly identified so that notice can be given and adequate documentation maintained. Monitor the following.

- *Cost overruns,* as reported by the cost accounting system. This requires an adequate job costing system, a detailed and reliable estimate, conversion of the estimate to a budget and input of the budget into the accounting system, recording of costs against the proper cost code, and regular review of cost reports. Most importantly, it requires that work accomplished be recorded, so that you can perform earned value analyses. Otherwise you won't really know where you are costwise.

- *Schedule delays,* as identified by the schedule updating procedures. This requires an accurate, thorough schedule, regular updating and revisions when necessary, and comparison of plan versus actual to determine delays and their causes.

- *Frequent or major questions or changes,* as evidenced by the number and significance of RFIs, RFPs (requests for proposals), CDs (change directives), architect sketches, etc.

- *Changes in construction sequences,* with work being done differently than planned, not because of contractor decision to reduce costs.

**Notice of change.**  There are four basic methods of giving notice, all of which are satisfactory if fully contract-compliant. The notice must clearly communicate that the issue is a change to the contract and, if known, schedule or cost impacts.

1. *Request for information* (RFI), which should provide information on the cost and schedule impact if known, or at least indicate there may be impacts.

2. *Discussion in meetings,* if recorded in the minutes and clearly constituting a notice.

3. *Correspondence,* with clearly stated notification that the issue is a change in the contract.

4. *Conversation* with the owner's representative, *if* confirmed in writing by a speed memo or other written document.

See Sec. E.12 of Chap. 3 for a discussion of the notice provision.

**Project audits.**  If a project appears to be in difficulty because of clearly identified cost overruns, delays, or other factors, a project audit is recommended. This consists of a physical inventory of the work accomplished, comparison with the budgeted work to be accomplished, adjustment for errors in the budget or changes to the contract, and projection of the cost to complete based on the productivity to date. This may reveal significant problems either for the overall project or for specific portions.

A project audit should also include close review of the schedule status and revised projection to complete. Delays should be carefully investigated to identify pending impacts due to work being pushed into inclement weather, crowding due to the need to complete on time, etc.

## 4. Avoid paper wars

If possible, avoid paper wars with the owner. They waste time and engender an adversarial approach and reprisals. Document the facts and provide notice of changes as required by the contract, but avoid exchanging accusations and allegations.

Personalize notice of claims and other correspondence establishing your side of a dispute, instead of using a form letter. Set a positive, friendly tone that matter-of-factly complies with the contract requirements and establishes your position without arguing the point. Save the arguments for later. If warranted, call before sending the notice to let them know it is coming and that you don't want to escalate the dispute.

## 5. Resolve disputes when they occur—fairly, timely, and firmly

When disputes do occur, verify that the change order management procedures are followed. This limits delays and costs, initiates resolution, and leads to timely preparation and submission of change order requests and claims. To summarize:

- *Monitor all projects for problems and possible disputes,* to identify and resolve problems before costs become excessive.

- *When problems are encountered, respond promptly,* give timely notice, implement expanded recordkeeping efforts, mitigate damages, initiate discussions with the owner's representative, and start preparing a change order request.

- *Prepare change order requests in a professional manner,* submit them promptly, and negotiate a settlement as established by the change order management program.

- *If negotiations are unsuccessful and the change order request is denied, start preparing a claim,* as provided for in the disputes resolution clause of the contract.

Even if a project is ongoing and disputes already exist, implement a Change Order Management Program. A weekend session with the project team and key personnel from the home office will usually identify most problems. Present the results to the owner's representative and designer in a proactive working session, with specific recommendations and initiate a positive joint review.

## 6. Speed up project closeout

Project closeout, the time from substantial completion to final completion, is an important step toward a successful project and the resolution of outstanding issues and disputes. Unfortunately, the project team is usually transferred to other projects, and must try to "clean up" the old project in their spare time. The priority is low and the work drags out—costing more money and endangering customer relations. Meanwhile, pending change order requests and disputes remain unresolved.

**Standardize closeout procedures.** Organized closeout procedures will result in quicker, less-expensive completion, and better resolution of disputes and change order requests. Checklists help supervisors and their managers ensure all issues are resolved as quickly as possible.

**Consider changing personnel for project closeout.** For success in project closeout, the superintendent must work closely with the project manager or in-house claims/change order expert to resolve open change order requests and claims. Greater emphasis on paperwork and client relations is needed than normally required of construction supervisors. Closeout requires a different attitude and skills, with an emphasis on cajoling and pressuring subcontractors to complete or repair their work. It is a tedious task, and many superintendents become frustrated with the aggravation, preferring to build a new project. They also have problems balancing the demands of their new project with cleanup on the old.

Some contractors and subcontractors assign a closeout supervisor to wrap up nearly completed projects. They take over all projects at or near substantial completion to complete punch list work, preparation of O & M manuals, and warranty work.

**Improve customer relations.**  Don't allow slow punch list resolution to mar a good experience for the customer. Use better closeout procedures to speed the process, and resolve lingering disputes and change order requests.

# 5

# Preparing and Settling
# Change Order Requests

Changes are common in construction. Most are resolved by the contractor's project team working with the owner's representative, within the provisions of the changes clause of the contract. This chapter addresses the process of preparing and settling change order requests (CORs)—also called change order proposals (COPs) or requests for equitable adjustment (RFEAs). It shows how training and improved procedures lead to better results, with less effort.

This chapter covers the following topics:

## A. The Changes Clause and Related Contract Provisions

Most contracts have a changes clause and notice provisions that govern the process of initiating and resolving change orders. Your first step when preparing a change order request should be to reread the changes clause and related provisions.

### 1. The changes clause

The changes clause specifies the procedures for authorizing and paying for extra work. It usually specifies what cost elements are reimbursable—direct labor, materials, equipment, and subcontract costs. Some clauses pay for supervisors but not for superintendents and specify a flat percentage markup for jobsite overhead, home office overhead, profit, bond, insurance, etc. The common markup is 15 percent, with a lesser percentage being allowed on subcontract work. For details, see Sec. E.13 of Chap. 3.

Payment for extra work is characterized as a contract adjustment, if prescribed under the contract, or as an equitable adjustment if not. The request for payment is called a change order proposal, change order request, or request for equitable adjustment.

**Notice requirements for changes to the contract.**   Contracts usually specify how soon and in what form the contractor must notify the owner of changes in the work. Failure to comply with these provisions may result in a lengthy dispute or rejection of the change order request. See Sec. E.12 of Chap. 3 for details of the clause and how to prevent rejection of your change order request for failure to give timely notice.

**Authority to direct a change and the requirement that authorization be in writing.** Two contract provisions related to the notice clause require that directives to change the work (1) be in writing and (2) be issued by authorized individuals. These provisions prevent misunderstanding the intent of the owner representative's conversations or the scope of directed extra work. They also ensure that owner's junior personnel don't exceed their authority. See Sec. E.13 of Chap. 3 for details.

## 2. The disputes clause

If the change order request is rejected, the contractor can proceed with a claim, under the disputes clause if one is included in the contract. For details, see Sec. E.14 of Chap. 3.

Both the AIA and the EJCDC (engineering societies') standard contracts provide for review of claims by the architect/engineer. This is usually a condition precedent to filing for arbitration or litigation.

## B.   How to Prepare a Change Order Request (COR)

This section addresses how to prepare a change order request (COR) for which entitlement is acknowledged (or not yet contested), and only price must be negotiated. The process is summarized in Fig. 5.1. If entitlement is disputed, refer to Chap. 7 for a basic explanation of how to prepare a claim, or to Chaps. 8 through 13 for a more detailed treatment.

## 1. General

Preparation of a change order request (COR) depends upon (1) whether the owner acknowledges entitlement, (2) the size and complexity of the change, (3) the expectations of the reviewer, and (4) the contractor's available records and internal procedures. Change order requests can be quite simple, consisting of merely a lump sum price quote, or very complex with pages of computations.

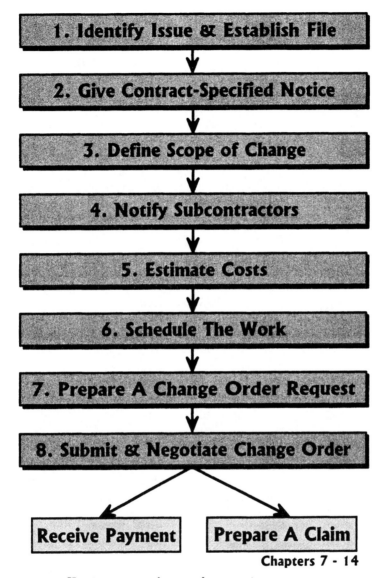

**Figure 5.1**  How to prepare a change order request.

They can be rigorous, following a carefully described process with detailed analysis of all pertinent facts, or they can be simplistic, focusing on one or two major issues.

**Develop and follow standard procedures.**  Most contractors have established procedures for preparing and negotiating change order requests. If successful, these should be continued. However, modifying your typical procedures, and standardizing them on a companywide basis, will probably result in larger settlements.

**Make partnering integral with the change order request process.**  Agree with the owner's representative and designer to use a partnering approach for resolving changes and avoiding disputes. Agree to work together to define the scope, examine entitlement, prepare cost estimates, schedule, and accomplish the work. Obviously, the contractor has the lead in most of these tasks, but the owner's representative and the designer must be involved and committed to a joint effort.

Periodically review the results of completed changes, to determine if you followed the agreed procedures and whether the result met all parties' needs.

**Agree to job-specific procedures and rates.**  Agree on COR procedures at the preconstruction conference or partnering workshop. Establish labor and equipment rates and attempt to negotiate markups based on your actual costs instead of contract-specified markups. This will minimize later disputes or delay in issuing change orders. The joint sessions will reduce adversarial attitudes and promote a fairer settlement.

Jointly developing change orders requires trust and a major shift in attitudes. It won't be possible with some owners and designers but should be attempted whenever possible.

**Entitlement.**  Entitlement is normally assumed for CORs, and only minimal justification is provided, usually in a statement of the "Reason for Change" (see Fig. 5.2, p. 137). In some cases, however, you may need to provide additional justification similar to that provided for a claim.

When stating the reason for the extra work, don't try to educate the owner's representative on the nuances of contract law. Provide a simple, reasonable rationale that is reasonably accurate and most likely to be accepted. Start with the premise that entitlement is obvious, and focus on computing costs. If a claim must be filed later, modifying your theory of recovery will be much easier if your change order request isn't too specific on entitlement issues. The most acceptable rationale for entitlement is differing site conditions—no one is at fault. Use it whenever reasonably applicable, in lieu of design error, interference, etc. Other nonconfrontational rationales are as-built discrepancies and scope changes.

## 2. If possible, prepare change order requests and negotiate based on the estimated cost

It is generally better to negotiate a fixed price before doing the work. That provides an opportunity to be more efficient and to increase your profit margin. It also avoids detailed recordkeeping for proving claim damages, although cost records should be maintained.

When pricing extra work based on estimated costs, use the following checklist.

### Prepare and negotiate a change order

1. *Identify the extra work, set up a file, and log in the extra work item.* Clearly describe the extra work and assign a sequential change order request (COR) number for tracking. Set up a file folder and collect copies of all related documents. Log the change order request into the COR log for tracking and control. Consistently use the same number and description on all future documents.

2. *Comply with contract-specified notice requirements.* Check the contract notice requirements and notify the owner in writing unless the work is already acknowledged as extra work.

3. *Define the scope of the change.* Owners often fail to define the scope of extra work and contractors sometimes forget to ask enough questions before estimating the cost of changes. The result is inadequate compensation or disputes over what work items are included.

Fax or hand deliver RFIs to the owner's representative to clarify the scope of work. If appropriate and if time is short, concurrently fax a copy to the designer. Log the request and response into the RFI log for tracking and control. If the change is large or may have severe impact on other work, meet with the owner's representative, the designer, and key subcontractors. Discuss the owner's needs, define the scope of the change, and jointly develop optimum solutions.

Determine the work schedule, probable impacts, and subcontractors that will participate or may be affected.

4. *Notify subcontractors and request quotes.* Fax notices to all potentially affected subcontractors and vendors. Request response in a specified format by a specific date, and note the consequences if late. Log and track responses, and take further action as necessary. If a subcontractor fails to respond in spite of numerous requests (and if allowed by your subcontract), estimate their costs and delay for inclusion in the COR. Notify the subcontractor that a unilateral change order will be issued, or a backcharge assessed if appropriate.

5. *Estimate the cost of the change.* Prepare an estimate using a standard COR estimating form. Use a crew/productivity/unit rate type of estimate whenever practical, as they are generally preferred by owner's representatives over unit price estimates. Include impact and delay costs unless excluded from the agreement.

Include the costs of preparing the change (scope definition, detailing, estimating, procurement, materials handling, etc.), if excessive. If conventional overhead costs have been reallocated to "direct support costs," as described in Sec. C.2 below, include those costs in the direct cost of the extra work.

Provide vendor quotes, detailed calculations, applicable pages from estimating manuals such as a *Means* estimating manual, and other backup for your estimates. Require the same of subcontractors. Review subcontractors' CORs and mark them up to correct errors. Forward their CORs to the owner with the corrections (after confirming with the subcontractor), to indicate that they have been reviewed and are therefore correct. Submit a professionally prepared and professional-looking estimate. If you have reallocated conventional overhead costs to "direct support costs," include those costs in the direct costs of the extra work.

Consider documenting and proposing the use of actual jobsite overhead and home office overhead costs, plus profit, insurance, and bond—instead of using the contract-specified markup. Evaluate risks and develop a descriptive and quantified contingency for possible delay, weather, and other unexpected conditions. The markup and contingency proposals will almost certainly result in a dispute, which can be the first step in ultimately negotiating a compromise.

Complete any backup documents and obtain internal approval. Clear, solid details with an understandable estimate and a summary of costs make settlement easier.

6. *Schedule the extra work.* Scheduling the extra work can be more important than the estimate. In many cases, original bid work is dependent upon the change order work, and some activities or the entire project may be delayed. Even if sufficient time is available to perform the extra work without delaying the project, waiting too long for approval may result in delay and cost impacts. Determine the required dates for authorization to proceed, return of submittals, etc., and include that information in the COR. Then carefully track all actual dates. If delayed by the owner, notify the owner's representative that another change order may be required.

To schedule the extra work use the work quantities from the change order estimate and the crew size to determine how long the work will take, after adding time for setup, interruptions, and cleanup. If the crew is already committed to critical path work, note the effect on the project schedule, and consider mobilizing a separate crew and equipment for the change order work.

Examine the project schedule, create a "fragnet" of the effect of the added work on the overall project and individual activities, and prepare a Time Impact Analysis as described in Sec. E.8 of Chap. 10. Attach it to the COR.

7. *Prepare the change order request.* Using a standard printed form or a template with word-processing software, enter the pertinent cost information and write up the description of the work in the format described in Sec. B.5 below. Attach relevant documentation.

Include your detailed estimate with the change order request. This allows the owner's representative to prepare their estimate using the same format, thus facilitating resolution of differences between the two estimates. It ensures that you agree on what work items are included. If delays are identified, either include an estimate of delay and impact costs or specifically exclude such costs from your proposal. Attach a narrative of your estimate assumptions, sources, and methodology.

8. *Submit and negotiate a change order.* Present the change order request in person at a brief meeting, unless the COR is very small and noncontroversial. Refer to Chap. 14 for recommendations on presenting and negotiating claims. The same techniques are applicable to change order requests, including clearly stating how soon a response is needed to avoid delay and increased costs. Remember to "sell" your proposal to the owner.

9. *If forced to proceed without a change order.* If work must start on an unapproved change, to avoid delay and impact, notify the owner's representative in writing. State that you are constructively required to proceed, in order to mitigate damages, and ask for confirmation or other directions in writing.

**Perform the work.**  The recommended steps to performing the extra work and getting paid are listed below.

1. *Upon approval, adjust the budget and schedule.* Add new activities to the schedule and modify the project budget to account for the extra work. If the work can be differentiated from original bid work, set up a separate cost code to track the cost.

2. *Prepare shop drawings, submit, and obtain approvals.* Prepare sketches or other details of the work, obtain shop drawings and other submittals from subcontractors or vendors, and submit to the owner. Log the submittal into the submittal log for tracking and control. Clearly state on the submittal how soon a response is needed and the priority of the request. Track the submittal process through to approval.

3. *Procure materials.* Upon approval, purchase the materials. If time is critical, track fabrication, shipping, and delivery. Implement procedures for effective materials handling, storage, and temporary protection.

4. *Layout and mobilize for the work.* Ensure field supervisors have the latest information so that they can prepare for the work. Survey or lay out the work area, notify subcontractors, move materials to the work area, and schedule the crews.

5. *Perform the work.* Install the work as designed. Keep detailed records of the time and materials used, noting any variances and problems. Perform testing, startup if applicable, and coordinate inspection and acceptance.

6. *Get paid.* Include the change order amount in the next payment request, and in interim pay requests, if applicable.

### 3. If unable to negotiate a price before proceeding, keep force account records

Many contracts provide for force account work and require detailed daily records of the work performed and effort expended. Even if entitlement is questioned, work can be documented with daily extra work order forms as described in Sec. C.14 of Chap. 4.

Require extra work order forms from subcontractors performing force account work. Require daily certification by the subcontractor's supervisor, confirmation, and authorizing signatures from your supervisor and the owner's representative. Require subcontractors to accompany monthly payment requests with a summary sheet of force account work for convenience in checking the totals, hourly rates, labor classifications, etc.

If the extra work lasts only one day, use the extra work order (EWO) form as a COR form, but assign a COR number for tracking and control. Normally, several work order forms are attached to a COR, which totals the amounts from the individual work orders and adds markups.

## 4. Document and submit actual costs, if pricing after-the-fact change orders

If extra work is not priced before the work is done, or if detailed force account records aren't maintained, you will have to price the work based on other records.

Hopefully, separate cost codes were added to the cost accounting system for the added work and the extra work was charged against those codes. If not, or if the costs include impact (which won't show up on the cost accounting reports), a detailed cost analysis will be necessary. This may require determining crew size and equipment used from daily reports, computing inefficiency using the measured mile or other techniques, and other computations described in Chaps. 11 and 12.

## 5. Use a standard form for change order requests

Using well-designed standard forms for change order requests will speed their preparation, minimize overlooked items, and improve acceptance by the owner's reviewer. Either handwrite or type the information on printed forms or create a template on word-processing software. Deliver or fax the completed form to the owner's representative with attached documentation.

The example below (Fig. 5.2) is recommended, as it includes the most important data. However, it may need modification to match company procedures, to add or delete information, to comply with the contract requirements, or to satisfy the preferences of the owner's representative.

**Extra work order form.**   The work order form in Chap. 4 (Fig. 4.7) can be used as a simple change order request form. Either use the form as it is or change the title from "extra work order" to "change order request" and adapt to suit your requirements.

**Change order request (COR) form.**   The recommended COR form (Fig. 5.2) consists of (1) a scope of work section, (2) a section summarizing the proposed cost and time, and (3) a cost summary and markup section. Optionally, it may include a change order agreement section that can be used as the agreement between the contractor and owner in lieu of the cost summary and markup. The form is designed to be supported by one or more pages of cost and scheduling data. It can be created with spreadsheet software to automatically post the costs from the cost estimate sheets.

- *General information.* Enter the job number and/or abbreviated title, the date the COR is prepared or submitted, and the sequential COR number assigned.

Enter a brief unique COR title to simplify references to the COR and avoid confusing it with another. The title should be descriptive of the work, and everyone must use it. A letter prefix is suggested for the COR number (e.g., COR-001), which helps identify the type of document. Reference the drawing number and/or specification section describing the work.

The COR author's name and title should be entered and the form signed. Enter who it is addressed to, provide their fax number or E-mail address, note everyone who is to receive a copy, and enter the number of pages included with

---

**XYZ Construction Company, Inc. – Change Order Request (COR)**

Job No/Title: _____  Date: _____  COR No: _____

COR Title: _____  Ref Dwg/Spec: _____

From: _____  Signature: _____  Title: _____

To: _____  Fax/E-mail: _____  Copy: _____  Page 1 of __

**SCOPE OF WORK**

Description of Change: _____
_____
_____
_____
_____  Attachment ☐

Reason for Change: _____
_____

Related Issues & Impacts (RFI, activity #, EWO # ) _____

Assumptions / Conditions: _____

Exclusions: _____

**PROPOSED COST AND TIME**

Change Contract Amount? ☐ Yes ☐ No  Amount: $ _____    Impact Costs? ☐ Included ☐ None ☐ Deferred

Change Contract Time? ☐ Yes ☐ No ☐ Deferred  Calendar Days Change _____    Time Impact Analysis: ☐ Attached ☐ Not Done

Work Is: ☐ Completed ☐ On-going ☐ Pending.  When Agreement Needed _____  Impact if Late _____

**CHANGE ORDER COST SUMMARY & MARKUP**

| | | | |
|---|---|---|---|
| 1. Labor (from page __) | | 10. Home Office Overhead @ __% of Jobsite Cost | |
| 2. Materials (from page __) | | 11. Extended Home Office Overhead (from page __) | |
| 3. Equipment (from page __) | | 12. **Subtotal – Project Costs** | |
| 4. Subcontract (from page __) | | 13. Profit @ __% of Project Cost | |
| 5. **Subtotal – Direct Costs** | | 14. Subtotal – Project Cost + Profit | |
| 6. Jobsite Overhead @ __% of Direct Cost | | 15. Bond and Insurance Premiums @ __% of line 14 | |
| 7. Extended Jobsite Overhead (from page __) | | 16. Subtotal without tax | |
| 8. Direct Support (from page __) | | 17. Sales and/or Use Tax @ __% of line above | |
| 9. **Subtotal –Jobsite Costs** | | 18. **TOTAL AMOUNT OF CONTRACT CHANGE** | |

**Figure 5.2**  Recommended change order request form.

the COR. The supporting data pages can include a direct support cost form, one or more cost estimate forms, and other data as required.

- *Scope of work.* Clearly describe the work. Note where it is on the project, the general work quantities, and how it will be or was done, if pertinent. Make (or attach) a sketch, if needed, and note any other attachments.

State the reason for the change, if known. If there are several alternate theories of recovery, use the one that best supports entitlement and is the most likely to be accepted by the owner's representative.

Note related issues and impacts. Include any RFIs referring to the work, the notice of change letter, any schedule activity numbers affected, letters from or to the owner discussing the work, etc.

Note assumptions made and exclusions. These can be very important later if problems are encountered. Be certain to document both.

- *Proposed cost and time.* Note whether there is a cost increase or decrease, and the amount. Also note if impact costs are included in the amount, if none were incurred, or if determination of impact costs is being deferred until later.

Note whether a time extension is needed or if it is deferred until a schedule analysis can be performed. If a time extension is requested, note the number of calendar days. Unless the change in contract time is deferred, prepare a time impact analysis and note that it is attached. See Sec. E of Chap. 10, especially Subsection 8, for guidance on preparing a Time Impact Analysis.

Note if the work is completed, ongoing, or pending; when agreement is needed to avoid further impact; and the impact if an agreement is delayed.

**Alternative final sections.**    The final section can be either a cost summary and markup section or a change order agreement as shown on Fig. 5.3 below.

**Cost summary and markup section.**    The separate cost categories (labor, materials, equipment, and subcontracts) can be subtotaled and markup can be added on the estimate sheets, in which case this section can be deleted from the form. Some contractors will find the recommended form too complicated and may wish to simplify it. For example, a single markup percentage can be used that includes jobsite overhead, home office overhead profit, and bond. Or it can be modified to comply with a specific contract's allowed markup percentages and procedures.

The percentage for markups will vary depending on how they are computed. Industry practice is to add markups to markups as indicated in Fig. 5.2.

- *Estimated cost by category—labor, materials, equipment, and subcontract— lines 1 through 4.* These costs are posted from separate cost estimating sheets or from extra work order forms. The cost summary and markup section assumes that all direct costs are marked up an equal amount. If not, modify the form to allow for a different percentage markup for different cost categories, as is required by some contracts for subcontract costs.

- *Impact, acceleration, escalation, and other delay costs—add another line, if listed separately.* Impact costs (excluding extended overhead costs from delays) can be either (1) broken down by cost category and combined with the direct costs in the labor, materials, equipment, and subcontract costs above, or (2) computed and listed separately from the direct costs by cost category. If the latter, create one or more additional lines and enter the values.

- *Direct cost subtotal and markup for jobsite overhead—lines 5 and 6.* Subtotal the direct costs on line 6. If the average jobsite overhead has been computed as a percentage of direct costs, enter that percentage. Compute the jobsite overhead allocable to this change and list on line 6.

  Many contracts specify a flat markup, which is difficult to avoid using unless the change was clearly beyond the contemplation of the parties when they signed the contract. However, it is worthwhile to propose actual overhead costs if they are unusually high or the change order processing costs are high. Start by computing actual overheads as described in Chap. 11 and documenting your rationale for using them.

- *Extended jobsite overhead—line 7.* If the project has been delayed and if the COR includes delay costs, compute the extended jobsite overhead costs on a separate page and enter the amount on line 7.

- *Direct support costs—line 8.* These are costs that directly support the change order but are normally classified as indirect (i.e., overhead) costs. If accounting procedures have recategorized some of those costs that are conventionally overhead to be direct support costs, identify these costs as a direct cost of the change order. For details, see Sec. C.2.

  Claim these costs even if the accounting system hasn't been changed if extraordinary overhead costs have been incurred for change order work. However, you may have considerable difficulty in negotiating their acceptance.

- *Jobsite cost subtotal and markup for home office overhead—lines 9 and 10.* Subtotal the jobsite costs on line 9. If you have computed a percentage for home office overhead costs, enter that value, compute the amount, and post it to line 10.

- *Extended home office overhead—line 11.* If there has been delay and if the COR includes delay costs, compute the extended home office overhead costs on a separate page (using the modified Eichleay formula as explained in Chap. 11) and post to line 11.

- *Project cost subtotal and markup for profit—lines 12 and 13.* Subtotal the project costs on line 12. Enter the percentage markup for profit (traditionally 10 percent), compute, and post the markup for profit to line 13. If you have computed the markup for profit and wish to attach documentation supporting your profit markup, reference a page number after the percentage.

- *Subtotal and markup for bond and insurance premium—lines 14 and 15.* Subtotal and enter the value on line 14. Enter the total percentage markup for bond and insurance premiums, compute the additional markup, and post to line 15.

- *Subtotal and markup for sales and/or use tax—lines 16 and 17.* Subtotal and enter the value on line 16. Enter the markup for sales and use tax, compute the additional markup, and post to line 17.
- *Total amount of contract change—line 18.* Add lines 16 and 17 and post the total amount of the COR on line 18.

**Alternate change order agreement section.** As an alternate to displaying the cost summarization and markup, use the remainder of the COR form as a change order agreement (Fig. 5.3). This has advantages and disadvantages, with the largest disadvantage being that change orders normally incorporate multiple CORs. Having a separate change order for each COR results in more paperwork.

If using the change order agreement section on the COR form, enter the change order number and a title (brief description) of the change order on the first line. Enter the original contract amount, changes by previous change orders, the amount of this change, and the new contract amount. Enter the original contract duration in calendar days, the number of days for previous changes, the time required for this change, and the new contract duration. Also compute and enter the new contract completion date.

Note whether the change order includes or excludes delay and impact costs. Unless you have fully analyzed the effect of the extra work on the schedule and other activities, exclude delay and impact.

The change order will normally be signed by the contractor, the owner's representative, and the authorized owner's executive. Each signatory must enter the name of the organization and their title, sign, and date the agreement. If there is not a separate owner's representative, that signature block can be deleted.

**CHANGE ORDER AGREEMENT**

Change Order No ________ Title ________________________________

| | | | |
|---|---|---|---|
| Original Contract Amount | ________ | Original Contract Duration (calendar days) | ________ |
| Net Change by Previous Change Orders | ________ | Net Change by Previous Change Orders | ________ |
| Amount of this Change Order: ☐ Add ☐ Deduct | ________ | Time for this Change Order: ☐ Add ☐ Deduct | ________ |
| **New Contract Amount** | ________ | **New Contract Duration** ........................ | ________ |
| | | New Contract Completion Date | ________ |

This change order (☐ does ☐ does not) include impact costs from delay or the effect of this change on other work. XYZ Construction Company, Inc. reserves the right to be compensated for the additional costs and extensions of time for the result of such impacts.

| | | |
|---|---|---|
| **XYZ Construction Company, Inc.** | Owner's Representative | Owner |
| Signature | Signature | Signature |
| Title | Title | Title |
| Date | Date | Date |

**Figure 5.3** Recommended alternate change order agreement section for COR form.

**Supporting data pages.** Standard forms and customized worksheets can be attached to the COR with supporting data. The number depends upon the size and complexity of the change, whether it is an estimate or based on actual costs, whether it includes only direct costs, or whether delay and impact costs are included. These supporting pages may include the following.

- *Analysis of entitlement.* Entitlement analysis, if needed, can include a more detailed description of the reason for the work, a narrative, review of the contract language, and analysis. For details, see Chap. 9.

- *Cost summary and markup.* A cost summary sheet can be either a standardized form or a customized worksheet.

- *Overhead computations.* These would normally be provided only once to establish the overhead rates. They would include home office overhead percentage, jobsite overhead percentage, daily extended jobsite overhead computations, and daily extended home office overhead (using the Eichleay formula).

- *Impact, delay, and acceleration cost analyses.* These costs can be computed on the standard estimating forms or special spreadsheets.

- *Direct support cost.* See Fig. 5.4 for a list of direct support costs.

- *Cost estimating sheets.* Use the standard or detailed cost estimate forms (Figs. 5.5 and 5.6), customized estimating forms, or the extra work order form (Fig. 4.7).

- *Other costing data.* If warranted, include quantity takeoffs, subcontractor and vendor quotes, excerpts from cost estimating manuals or catalogs, and other costing data as needed.

- *Time analysis.* Include a time impact analysis that determines the days of compensable, noncompensable but excusable, and nonexcusable delay. Optionally include a schedule fragnet and delay narrative and computation. For details, see Sec. E of Chap. 10.

At a minimum, there should be an estimate or listing of actual costs, a summarization with markup, and attachment of source documents supporting the estimate and analyses.

**Subcontract cost proposals.** Attach subcontractor change order proposals, or refer to them and provide separately. Exclude subcontractor costs if some affected subcontractors haven't submitted claims or sign-offs. However, this is likely to delay or prevent resolution.

## 6. Direct support costs

Direct support costs are defined here as those costs that are normally or often classified as overhead (jobsite or home office) but which can be reclassified as direct costs of the extra work and reimbursed under certain circumstances.

**Reallocation of overhead costs.** If the accounting procedures had been modified to reallocate some home office overhead costs directly to projects and some jobsite overhead costs to field direct costs, many of the following direct support costs will already be allocated to the extra work. For details, see Sec. C.2.

**Extraordinary costs.** Some change orders result in extraordinary support costs that greatly exceed the markup. A relatively small dollar value change can require weeks of effort for investigation, development of a solution, soliciting quotes from vendors and subcontractors, and preparing a change order proposal. Then a difficult owner's representative can extend negotiations beyond any reasonable expectation. Purchasing and expediting materials delivery, scheduling and tracking progress, engineering, and shop drawings can consume even more time. Finally, the supervision time required can also be excessive owing to the nature of the work or late clarification and authorization.

Although not normally claimed as a separate line item in a change order request, these costs can and should sometimes be claimed.

**Checklist for field activities.** Fig. 5.4 lists four categories of direct support costs. The first two are conventionally categorized as overhead costs and included in the markup. They are sometimes claimed, however, either by reallocation of some overhead costs or as extraordinary costs due to specific claims or a large number of small claims. The cost types listed are for a typical building contractor. Civil works contractors will have a somewhat different list.

The ancillary and wrap-up costs on the attached list should be included as direct costs of a change order request, with Fig. 5.4 serving as a checklist.

## 7. Cost estimating forms

Cost estimating for change order requests can be in any format and practically any style that is readable and understandable by the claim reviewer. However, a standardized, well-organized format will save time and engender more confidence.

**Standard cost estimate form.** Figure 5.5 is an example form for simple, small change order requests. It is designed for a one-page estimate or actual cost record.

Enter the job number and/or title, the date the estimate was prepared, the COR number it supports, and the COR page number. Also enter the title of the COR, the drawing number or specification section governing the work, and the name or initials of the estimator.

The data fields are similar to the extra work order form (Fig. 4.7).

- *Material costs.* Enter the quantity, units, size, description, and unit price of the material items. Then compute and total the materials costs. Note the

## XYZ Construction Company, Inc. – Direct Support Costs

Job No/Title: _____  Date: _____  COR No: _____  Page ____ of ____

COR Title: _____  Ref Dwg/Spec: _____  Estimator: _____

| DESCRIPTION | LABOR HRS | RATE | SUBTOTAL | EXPENSES (MAT'L) | TOTAL |
|---|---|---|---|---|---|
| **Change Order Preparation** | | | | | |
| Research and design of solution | | | | | |
| Estimating and preparation of change order request | | | | | |
| Negotiation of change order | | | | | |
| **Administration** | | | | | |
| Purchasing and expediting | | | | | |
| Scheduling, schedule update, and tracking | | | | | |
| Office engineering, design, and drafting | | | | | |
| Shop drawing preparation, review, revision and processing | | | | | |
| Accounting and recordkeeping | | | | | |
| Clerical, supplies, office fixtures & furnishings & equipment | | | | | |
| Postage and shipping | | | | | |
| Photographs | | | | | |
| Telephone charges, faxes, cellular phone | | | | | |
| Additional office, shop, yard, and toolsheds | | | | | |
| Special travel and subsistence | | | | | |
| Permits and fees | | | | | |
| Financing costs | | | | | |
| Special insurance | | | | | |
| **Ancillary to Prosecution of the Work** | | | | | |
| Supervision and subcontractor coordination | | | | | |
| Mobilization and setup | | | | | |
| Utility hookup and use: power, water, heat, lighting, sanitary | | | | | |
| Field Engineering, layout and surveying | | | | | |
| Coordination of utility relocation / installation | | | | | |
| Freight, delivery, pickup, and cancellation/restock charges | | | | | |
| Materials receipt, warehousing (on/off site), and handling | | | | | |
| Crane, elevator and other yard equipment use | | | | | |
| Equipment fuel, oil, grease, fueling, and maintenance | | | | | |
| Temporary facilities: staging, scaffolding, lighting | | | | | |
| Special construction: access roads, parking | | | | | |
| Weather protection and cold/wet-weather operation | | | | | |
| Inspection and quality control | | | | | |
| Small tools | | | | | |
| Consumables | | | | | |
| Cutting and patching | | | | | |
| Fencing, security and allowance for theft and vandalism | | | | | |
| Safety – supplies, equipment, facilities and procedures | | | | | |
| **Wrap Up** | | | | | |
| Startup | | | | | |
| Testing and final inspection | | | | | |
| Punch list completion | | | | | |
| Demobilization and cleanup | | | | | |
| As-builts, O&M manuals, certification, and guarantees | | | | | |
| Warranty work for specified warranty period | | | | | |
| SUBTOTAL DIRECT SUPPORT | | | | | |

**Figure 5.4**  Direct support costs estimating form.

| XYZ Construction Company, Inc. – Cost Estimate |
|---|

Job No/Title: _____  Date: _____  COR No: _____  Page ___ of ___

COR Title: _____  Ref Dwg/Spec: _____  Estimator: _____

### MATERIAL

| QUANTITY | Unit Size | DESCRIPTION | UNIT PRICE | TOTAL |
|---|---|---|---|---|
| | | | | |
| | | | | |
| | | | | |
| | | | | |
| | | | | |
| | | | | |
| | | | | |
| | | | | |
| | | | | |
| | | | | |
| | | | | |
| | | | | |
| | | | | |
| | | | | |
| | | | | |
| Consumables & Safety Supplies @ __% of labor | | | | |
| | | TOTAL MATERIAL | | |

### LABOR

| EMPLOYEE | ST HRS | ST RATE | OT HRS | OT RATE | TOTAL |
|---|---|---|---|---|---|
| | | | | | |
| | | | | | |
| | | | | | |
| | | | | | |
| | | | | | |
| | | | | | |
| | | | | | |
| | | | | | |
| | | | | | |
| | | | | | |
| | | | | | |
| | | | | | |
| | | | | | |
| SUBTOTAL | | | | | |
| Supervision @ __% of Labor | | | | | |
| TOTAL LABOR | | | | | |

### EQUIPMENT

| HOURS | DESCRIPTION | UNIT PRICE | TOTAL |
|---|---|---|---|
| | | | |
| | | | |
| | | | |
| | | | |
| | | | |
| | | | |
| | | | |
| | | | |
| Small Tools @ ___% of Labor | | | |
| Fuel, oil, grease & maintain rented equipment | | | |
| | TOTAL EQUIPMENT | | |

### SUBCONTRACT

| SUBCONTRACTOR OR TRADE | TOTAL |
|---|---|
| | |
| | |
| | |
| | |
| | |
| Subcontractor Bonds | |
| TOTAL SUBCONTRACT | |

| SUBTOTAL DIRECT COSTS | |
|---|---|
| MARKUP @ ___% | |
| GRAND TOTAL | |

**Figure 5.5** Recommended cost estimate form, standard version.

optional line item for consumables and safety supplies (normally two separate line items) that are estimated as a percentage of labor costs.

- *Labor.* Enter employee names if reporting actual costs or the trade and journeyman status for estimated labor. Enter the number of hours and hourly rate for straight time, overtime, or both. Compute the values and subtotal. If extra work caused bid work to be worked on an overtime basis, charge all the hours at the overtime rate, which is the incremental rate for added work. Enter a value for supervision as a percentage of labor costs, multiply by the labor subtotal, and enter the total. Add the two together for the total labor cost.

- *Equipment.* Enter the hours of equipment use, description, and the unit rates. Compute values. Optionally enter a factor for small tools as a percentage of labor costs, compute, and enter the value. Enter a value for fuel, oil, grease, and maintenance of rented equipment. Equipment rates for owned equipment generally include these costs, but rental equipment is normally paid on the basis of invoiced costs, and adjustment is needed for these costs.

- *Subcontract.* Enter the names of the trades expected to perform extra work or the names of subcontractors, if they have been selected. Also enter the estimated or actual cost for each and attach appropriate backup. Add a value for bonds (a percentage of subcontract costs) and total the subcontract costs.

- *Factors on labor costs.* The form has lines for supervision, consumables and safety supplies, and small tools as a percentage of labor costs, as might be used by a small contractor or specialty subcontractor. Many estimators account for consumables (fasteners, welding rod, etc.) as part of direct costs and separately estimate supervision and safety supplies as a monthly cost of jobsite overhead.

- *Summary and markup.* Sum the total direct cost for all cost categories, enter a percentage for markup, compute the markup, and enter the grand total for the estimate. Attach referenced documents.

**Detailed cost estimate form.** The detailed cost estimate form (Fig. 5.6) is a standard estimating spreadsheet for unit price or crew/productivity rate estimating.

- *Job and COR information.* The form header is identical to the standard cost estimating form.

- *Cost element—contract document reference and description.* Enter a reference to the specification section or drawing number governing each major element of work comprising the COR, and enter and underline the description of the element. Then list the components for that cost element below it.

- *Line item—specification section and/or drawing number.* Optionally refer to the specification section number and/or drawing number governing each line item.

- *Line item—description, size, unit, and quantity.* Enter a description of each line item comprising the cost element. Include the size, quantity, and units.

- *Line item—unit price or crew/production estimated costs or actual costs.* If preparing a unit price estimate, enter the estimated unit costs for each cost category (labor, materials, equipment, and subcontract) of the line item.

Optionally, enter the estimated labor hours and the hourly labor rate, which will enable computing the unit labor costs. From this, compute the costs for each line item of work.

Alternately, list the trades included in a crew, the number of each trade, the hourly rate for each, etc. Add equipment, materials, subcontract costs, and consumables to develop a composite hourly rate for the crew. Then develop a production rate for the crew, determine the units or hours of work, and compute the total costs (by cost category) for the element of work.

If entering actual costs, enter material descriptions and quantities, labor by trade and skill level or name with their hours and unit rates, equipment use hours and rate, and subcontract unit costs. Then compute the costs by category for each line item.

- *Cost element—subtotal by cost category.* Subtotal the cost of each cost category for the cost element and total.

- *Page totals (for manual use).* When a page is full, enter the estimate page number (which will be different from the COR page number). Subtotal the cost for each category and total for the page as a math check on the total by cost element.

- *Additional columns.* Many contractors subtotal consumables and temporary materials in a separate column from permanent materials, which would require an additional column. Heavy equipment contractors may add a separate column for fuel, oil, and grease separately from other equipment costs. Users should customize the form to meet their needs.

Use as many pages of detailed estimate as required. Attach referenced material, quantity takeoff sheets, vendor quotes, etc. When done with the detailed estimate, sum the values on a single sheet and post to page 1 of the COR form.

**Use factors to simplify estimating and ensure including all miscellaneous costs.** The use of factors for estimating miscellaneous costs will simplify estimating, ensure including all miscellaneous costs, and result in higher recovery. The factors that can be included are as follows, with the cost element they are based on.

- Small tools as a percent of labor (can be 5 percent or more of labor cost)
- Supervision
- Consumables
- Materials handling (can be 5 percent or more of labor cost)
- Safety (can be as high as 3 percent of labor cost)
- Freight, drayage, and pickup
- Testing and cleanup
- Nonproductive labor (breaks, tool pickup, etc.; can be 10 to 15 percent)

## XYZ Construction Company, Inc. – Detailed Cost Estimate

Job No/Title: _____  COR No: _____  Page ____ of ____

COR Title: _____  Ref Dwg/Spec: _____  Estimator: _____  Date: _____

| SPEC/ DWG | Size | DESCRIPTION | UNIT | QTY | UNIT COSTS | | | | | LABOR COST | MATERIAL COST | EQUIPMENT COST | SUBCONTRACT COST | TOTAL COST |
| --- | --- | --- | --- | --- | --- | --- | --- | --- | --- | --- | --- | --- | --- | --- |
| | | | | | LABOR | | MAT'L RATE | EQUIP RATE | SUB RATE | | | | | |
| | | | | | HRS | RATE | | | | | | | | |
| | | | | | | | | | | | | | | |
| | | | | | | | | | | | | | | |
| | | | | | | | | | | | | | | |
| | | | | | | | | | | | | | | |
| | | | | | | | | | | | | | | |
| | | | | | | | | | | | | | | |
| | | | | | | | | | | | | | | |
| | | | | | | | | | | | | | | |
| | | | | | | | | | | | | | | |
| | | | | | | | | | | | | | | |
| | | | | | | | | | | | | | | |
| | | | | | | | | | | | | | | |
| | | | | | | | | | | | | | | |
| | | | | | | | | | | | | | | |

**PAGE TOTALS**

Estimate Page ____ of ____

**Figure 5.6** Recommended detailed cost estimate form.

147

- Cleanup (can be 2 to 5 percent of labor costs)
- Foremen (often 20 to 25 percent of journeyman hours/costs)
- General Foreman (assume 12.5 percent of journeyman hours)
- Laborer supporting mechanical, electrical, and other specialty trades (can be 5 percent or more of labor)
- Inefficiency factors for overtime, cold/wet weather, fatigue, trade stacking, morale and attitude, too small or too large crew size, concurrent operations, dilution of supervision, learning curve effects, early owner occupancy, limited site access, etc.

**Record your assumptions, cost data sources, and methodology.**  Footnote your estimates extensively with references to a narrative listing any assumptions, data sources not clearly noted on the document, and reasons for your choice of methodology. The narrative should explain *why* costs are required and why certain procedures are assumed. Attach it to the estimate for review by the owner and your own use.

## 8. Time impact analysis

Most changes will affect the schedule, and some may delay the project. A time impact analysis (as described in Sec. E.8 of Chap. 10) should be made for all changes, while preparing the change order request. This requires, however, that the schedule be updated regularly and revised when necessary. Otherwise the schedule will be out of date and you will be unable to evaluate the effect of the change.

All too often schedules aren't current, due to either contractor failure to schedule or an excessive number of changes that prevent timely rescheduling. No one knows the true impact of a change until the project is completed (late and over budget) and a claims expert analyzes what happened and determines that extra work delayed the project. By then, it is too late for the owner to take alternative action and it is difficult for the contractor to be compensated for the impact costs.

**Pricing based on schedule.**  In some cases, CORs should be priced for acceleration to avoid delay, by adding money for overtime labor premium, extra workers, expedited material delivery, multiple shifts, etc. This will avoid impacting existing work, escalation, and extended overhead costs, and may be cheaper than delay.

**Least cost expediting and cost-time tradeoff analysis.**  As explained in Sec. A.2 of Chap. 10, accelerating to reduce project duration increases direct costs but also reduces indirect (overhead) costs. Also, some critical path activities are less expensive to accelerate than others. Whenever extra work delays the project, analyze the time impact and take appropriate action.

- Estimate the daily extended overhead, equipment standby (if applicable), and escalation costs to determine the savings from one-day acceleration.
- Examine the critical path and identify those activities that can be accelerated for less than the savings from reduced overhead.
- Accelerate the least expensive activities to accelerate until the costs of acceleration are equal to the savings from reduced overhead. In some cases, you may have to accelerate two concurrent critical paths to save a day of time.

The project may have been accelerated beyond the time required to prevent delay from the extra work. If so, this indicates a similar analysis of other projects would probably result in savings from reduced project duration.

**A modest proposal.** Contractors should budget and staff projects for monthly schedule updates and for revisions due to weather delays or contractor errors. The owner should pay for schedule revisions due to extra work, and contractors should promptly mobilize additional scheduling capabilities to prepare time impact analyses to accompany CORs. Contractors also need to sufficiently staff for prompt preparation and negotiation of CORs, and these costs should be reimbursed by the owner.

Owner reimbursement of COR preparation and time impact analysis will result in more detailed, accurate, and timely response by contractors. Late submittals will be eliminated, total costs will be reduced, and disputes will be less frequent.

## 9. Include delay, impact, and acceleration costs

Cost may not be the only issue. The project may be delayed if the added work is on the critical path or if it uses limited resources that are needed on critical path activities.

**Analyze the schedule delay.** First analyze the schedule to determine which activities will be delayed and whether those activities are on the critical path. If so, determine the amount of delay and the impact.

**Price extra work to avoid delay, if possible.** If the extra work would normally be performed by a crew engaged on critical path activities, price the work to avoid delay. This can be done by (1) bringing in a separate crew or (2) working the existing crew overtime and adjusting for inefficiency from fatigue.

**Include delay costs in the change order proposal.** If the extra work is on the critical path because of job logic and it isn't possible to bring in a separate crew or to work overtime, include delay costs in the change order proposal. The delay costs will include extended overhead costs, including home office and jobsite overhead, for the period of delay. The delay costs may also include escalation, interest, and impact. Many owners will resist paying these costs (although they know "the time value of money" for their own operations), and you may have to file a claim to get paid. However, if you put the owner on notice and eventually end up in arbitration, the arbitrators will consider the notice when making their award.

**Exclude delay and impact from the change order.**    An alternative to pricing delay and impact in the change order is to defer the issue until later. This requires that the change order specifically state that the issue will be addressed later.

## 10. Add markup

One problem in negotiating change orders is that the markup specified in some contracts is inadequate. The AIA General Conditions, AIA Document A201, requires that the contractor "...present...an itemized accounting together with appropriate supporting data...," but many other contracts specify a fixed percentage of markup. The markup is often 15 percent or less for jobsite overhead, home office overhead, profit, bond, and insurance. That may be adequate for a large contract modification with little paperwork but is often grossly inadequate for the smaller changes normally encountered.

In many cases, the administrative time to estimate the cost of the change, obtain subcontractor quotes, schedule the work, and administer the contract change order exceeds the markup. If this happens, propose an equitable alternative.

**Propose actual overhead rates with actual administrative costs.**    The first step is to assemble documentation proving actual jobsite and home office overhead costs (described in more detail in Chap. 11) and to submit these costs in lieu of the contract-specified markup. Many contracts allow the parties to negotiate an "equitable adjustment," requiring the specified markup only if the parties cannot agree. Invoke the partnering spirit to help convince the owner's representative to accept the actual overhead rates.

Some contractors try to collect excessive "guesstimated" administrative costs for the change order on top of the contract-specified rates—with little or no success. A better approach is to record actual administrative costs and to propose them, using actual administrative costs from prior change orders to justify costs that haven't yet occurred.

**Focus estimates on direct costs.**    Another approach is to place more attention on estimating direct costs, using composite crew costs and estimated productivity rates. Be conservative in estimating these costs, and include an adequate contingency as noted below in Sec. 11.

**Equipment rates.**    Equipment-intensive projects don't have a serious problem, as equipment rates (especially Blue Book rates) are usually more than adequate and will compensate for insufficient contract-specified markups. That may not be true on U.S. government projects using agency-determined equipment rates.

**Minimize overhead costs for changes.**  Since superintendents' time is usually included in the contract-specified markup, use foremen to plan, lay out, and manage the work. Their time is considered a direct cost and is reimbursable.

**Deductive changes.**  Many owners want a credit on markup for deductive changes. Although superficially logical, this argument fails to account for the cost of estimating and negotiating a credit for the reduced direct cost, let alone the costs for canceling materials procurement, etc. Nor does it consider unabsorbed overhead. A credit for reduced overhead on deductive change should be considered only on very large changes where overhead costs are actually reduced.

**Net change for add and deduct changes.**  When deducting some work items and substituting other work, owners will attempt to compute overhead on the net balance of direct costs. If the value of the work added is close to the value of the work deleted, the net change in the contract amount will be very small or even negative, yet the administrative and estimating costs may be substantial. A reasonable method of determining overhead for this type of change is to estimate the expected, or record the actual, "direct office support" cost for the change.

## 11. Use expected value to compute contingency

Estimators instinctively use the "most likely" cost to do the work. If the spread of probability were a bell curve, this would be satisfactory. Unfortunately, the curve is skewed, and the probability of things going wrong is far greater than the chance that things will go better. Therefore, adjust for this in the estimate, either by using conservative estimates of productivity or by including a contingency for the unexpected.

One means of handling unknowns is to include a contingency for the possibility of an event or condition happening. For example, there may be a 15 percent probability that heavy rainfall will flood a cofferdam during the extended time that a change has made it vulnerable to damage. If the estimated damage from overtopping is $100,000, the expected value of the event occurring is $15,000.

## 12. Don't try too hard when forward pricing change order requests

Do not be overly optimistic or premature when forward pricing change order requests. There may be many unknowns that can't be quantified. In your eagerness to resolve all issues and maintain a partnering relationship, don't settle for too low a price for extra work. If necessary, work on a force account basis—at least until enough cost data are accumulated to price the balance of the work.

## 13. Submit your change order request and negotiate a change order

The final step is to submit the change order request and negotiate a change order. This should follow the procedures outlined in the contract, or as agreed to with the owner's representative. It is important, however, to follow the basic steps outlined in Chap. 14, although you can be more informal and less detailed. Make a brief oral presentation of CORs that may be contested, follow up with additional information, and confirm progress is being made. Insist on another meeting before they reach a final decision, as it is difficult to convince someone to change their mind once they have reached a decision.

## C.    How to Improve Settlement of Change Order Requests

Too many change order requests are rejected. Too many others are accepted only after tedious negotiation and excessive compromise. Those rejected will encounter more resistance when submitted as a claim.

### 1.  Train project personnel in how to prepare change order requests

Train key project personnel in the basics of contract interpretation, how to identify extra work, when to give notice, and how to prepare change order requests. Include instruction in the company's policy and procedures, the use of standard forms to prepare the requests, how to submit and negotiate change orders, and when to request help from the main office. Detail these instructions in the company's Change Order Management Program as described in Chap. 4. Include review by the company's in-house expert of all change order proposals over a specified dollar value.

Because it is often too difficult to train project personnel in good writing techniques, give staff standard forms and letters to aid them in preparing clearly written proposals.

### 2.  Reallocate conventional "overhead" costs to be "direct job support" costs

Traditional accounting systems allocate overhead costs on the basis of a single factor—usually the direct costs for each project. Unfortunately, this may result in an inequitable allocation, with some projects bearing an unfair proportion of the overhead costs and some bearing less. Inevitably, *troubled projects bear less than their fair share of overhead,* as management time (the essential element of cost control and profitability) is focused on those projects to the detriment of others. Although management time is "free" in the sense that management costs are expended regardless of whether expended on a claim or other activities, the reduced management time spent on other projects or nonimpacted activities results in higher costs for those projects and activities. The troubled project or activity should bear its fair share of those losses.

There are two approaches to a more equitable allocation of overhead to extra work. One is through using a relatively new concept called "activity-based accounting," which allocates overhead on multiple factors rather than only on project direct cost. The other is to allocate as many overhead costs as possible to specific projects and to specific issues or problems when possible. For example, home office project managers should charge their time to specific projects instead of a general overhead account. This takes very little effort and can result in a substantial transfer of overhead to impacted projects. For a more detailed discussion, see Sec. C.4 of Chap. 11.

### 3. Support and review by in-house claims expert

If you have a designated in-house claims expert, the expert can assist the project team members in preparing change order requests. This will improve the quality of proposals while increasing the skills of the project team members.

### 4. Provide more detail

If the reviewer has confidence that the proposed prices are fair, the change order request will be accepted with little or no discussion.

One means of getting a price proposal accepted is to provide more detailed cost information, as suggested in Sec. A of Chap. 11. Use a crew composition and production rate analysis, which owner's representatives can analyze, rather than unit prices, which they cannot analyze.

### 5. Be timely

Don't wait too long to submit CORs, especially if the owner's representative is aware of the forthcoming change and has generally accepted entitlement. It is more difficult to negotiate a price after the problem is solved and the work is complete. You have a psychological advantage if you discovered the problem and make a timely effort to resolve it, especially if presented in a partnering mode.

Delaying a change order request can have significant interest costs. Six months' delay in a $90,000 change is $4500 if interest is 10 percent per annum. That is your profit and some overhead, if markup is limited to 15 percent.

The most important steps to ensure timely submission are to adequately staff projects and to add staff if projects are in trouble. This goes against the grain of most contractors but will pay large dividends and prevent a possible disaster.

After inadequate staffing, the next most common reason for delaying submittal is late submission by subcontractors. To avoid delays, notify all affected subcontractors and vendors of the time frame required, track their progress, and insist on timely submission.

### 6. Do it right the first time

Be especially careful to be thorough, accurate, and reasonable with your first change order request. This is where you establish procedures and start building trust.

### 7. Initiate the disputes process

If unable to reach an equitable settlement and forced to file a claim, tell the claim reviewer that you will use a partnering approach and win/win negotiation.

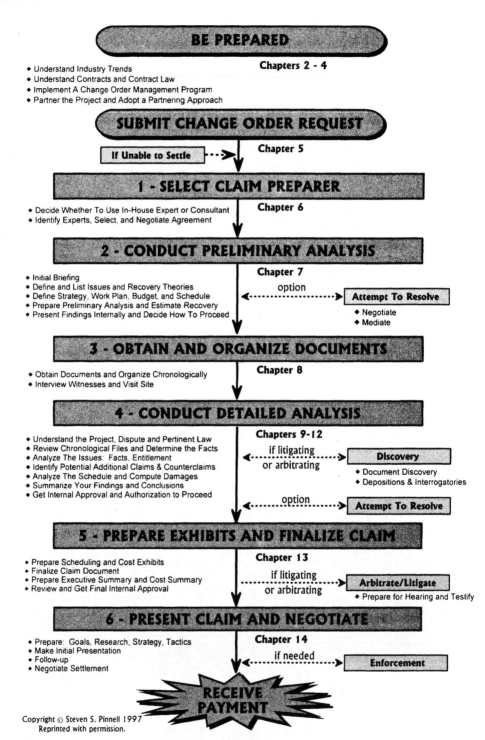

**Figure 5.7**   Flowchart of how to get paid for claims.

Then prepare and submit a claim in accordance with the disputes process specified in the contract. *The balance of HOW TO GET PAID describes how to prepare and negotiate a successful claim.*

Be assertive. Do not wait until the work is done to file a claim. Initiate the disputes process as soon as possible, using estimated costs. Prepare and submit a claim, try to negotiate entitlement, and revise the damages estimate when actual costs are available. Then, if the claim is rejected, request mediation immediately. If mediation is unsuccessful, consider filing for arbitration or litigation without waiting for the project to complete. There are risks in this approach, however, as the inspectors and owner's representative may retaliate.

Figure 5.7 describes the process of preparing, presenting, and negotiating a claim. It starts with being prepared, as recommended in Chaps. 2 through 4. Next is submission of a change order request as described in Chap. 5. Only if unable to settle do you proceed to the claim process.

As noted on the figure, there are six steps to preparing and resolving a claim. The first is to select a claim preparer, as explained in Chap. 6. That can be a project team member, an in-house claims expert, or a consultant. The second step, as described in Chap. 7, is to conduct a preliminary analysis to define the effort required, the probable recovery, and a work plan, budget, and schedule. It may be possible to settle the dispute at that point, with the information developed.

If the decision is made to proceed, the third step is to obtain and organize the documents. Then conduct the detailed analysis as described in Chaps. 9 through 12. Chap. 13 describes how to prepare the claim document and exhibits. Chap. 14 explains how to present the claim and negotiate a settlement.

# 6

# Phase 1—Selecting Claim Preparers

The most important step in developing a successful claim is selecting the right person to prepare it. If the right person is selected, all else will follow. This chapter is addressed to the construction executive who is responsible for a claim and needs to select and manage the individual(s) preparing a claim, whether a consultant or a company employee. It includes:

## A. Qualifications for a Successful Claims Preparer

Successful claim preparation is difficult. The claim reviewer must be convinced that the request for additional time and money is both fair and contract-compliant. To do this requires considerable construction experience, certain basic abilities, and a number of specific skills.

### 1. Individual knowledge and skills required for an effective claims preparer

Look for these characteristics and experiences when selecting someone to prepare claims:

1. **Broad construction experience.** Construction claim preparers need expertise in scheduling, costs, construction management practices, and the specific type of construction in dispute. For example, a claim preparer who has only building construction experience may miss critical issues when preparing

a civil works claim. That claim preparer will require more support, time, and expense to get up to speed, and his or her testimony will not be as effective as that of an expert with extensive civil works experience. However, everyone's experience is limited. It is better to retain an outstanding claim preparer supported by a second expert (e.g., a subcontractor) in a specific trade than to use a less qualified claim preparer who is familiar with that type of work.

**2. Scheduling expertise.** If the claim involves scheduling (most claims do), the preparer must have expertise in critical path scheduling. An "old hand" with years of experience, but comfortable only with bar chart scheduling, will not do well on a major scheduling claim. Scheduling experts must be experienced in the industry. They must know construction methods, construction industry practice, and sometimes engineering/architectural design practices and standards. Extensive field experience is needed to understand and to communicate the impact of winter weather, overcrowding, uncertainty, constant change, acceleration, and other factors affecting morale and productivity.

If a contractor's personnel have expertise in critical path scheduling, an expert is needed only to provide independent verification. In-house scheduling is seldom done at the expert level, but a review of Chap. 10 can help provide the necessary scheduling knowledge and some powerful new techniques.

**3. Estimating expertise.** Although contractors' estimators are expert at determining direct costs, many are not aware of the estimating formats preferred by claim reviewers. Contractors' estimators may lack experience in estimating impact costs or computing other claim costs using the Eichleay formula, the measured mile, learning curve adjustments, etc. To learn those techniques, they should read Chaps. 11 and 12.

**4. Contracts and claims expertise.** Claim preparers must know construction contract law and the procedures for preparing and presenting claims, which are presented this book, starting with Chap. 3. If expected to testify, experts should have litigation and arbitration experience, although a claims preparer inexperienced at testifying can learn courtroom procedures.[13] Experience as a mediator or arbitrator is helpful if the forum is mediation or arbitration. If the dispute will be resolved through negotiation or mediation, it is best if the expert is known and respected by the other party and has a reputation for fairness.

**5. A good memory.** Claim preparers must review and absorb large quantities of information and grasp the essential elements affecting progress and costs. Standardized procedures and techniques for breaking the work into phases and step-by-step tasks help reduce the amount of information that must be simultaneously considered. These procedures are described in Chaps. 7, 8, and 9.

**6. Personality.** The personality of the claim preparer is important. Some are good at research and painstaking analysis but crumble under cross examination. Others are very convincing witnesses and do well on broad issues but are not sufficiently organized or efficient to handle detailed investigations. Experts should have an imaginative approach to strategy and tactics. Personality should be matched to the task and the audience. A convincing speaker and a good presence are crucial for a jury trial. A team approach, using two individuals with complementary skills and personalities, may work best.

**7. Written communication skills.** The ability to write clear, concise, and convincing statements of facts and analyses is essential. Use narrative text to organize the material, as described in Chap. 9. Writing skills can be improved through practice, reading references on the subject, and reviewing Sec. B.2 of Chap. 13. Effective graphics, as explained in Sec. B.3 of Chap. 13, complement good writing.

**8. People skills.** Claim preparers need to understand human behavior and motivation to differentiate between factual and self-serving statements or documents. Negotiating a change order request or claim requires one-on-one people skills. These are discussed in Chaps. 13 and 14. Evaluation of documents is discussed in Sec. D. of Chap. 9.

The claim preparer can be either a construction company employee or an independent consultant. If an employee, the preparer may be a home office staff person or a senior project manager between projects and available to prepare the claim. However, a senior manager is more valuable running projects and is seldom satisfied with claims analysis. In either case, the claim preparer must be impartial and willing to inform management that a case is too weak to pursue.

A knowledgeable construction attorney is often needed on the claim support team and is essential if the dispute goes into mediation, arbitration, or litigation, or if there are significant legal questions.

## 2. Advantages of a team over an individual

A team is more efficient and economical than a single individual on all but the smallest claim preparation effort (under 40 hours). On most claims (under 200 hours), the team consists of an expert and a clerical assistant (to organize the files, make copies, etc.) plus a project team member who helps identify files, answers questions about the project, communicates with the contractor's decision maker, etc. The clerical assistant can be an experienced construction secretary, although it is more efficient to use an experienced claims technician, perhaps the same secretary but with some training and experience in claims preparation support. On larger claims, additional but less senior experts, scheduling and computer specialists, bookkeepers or accountants, and other specialists can provide support in specific areas more economically, and more effectively, than a single expert.

## B.  Deciding Who to Use to Prepare a Claim

The first decision is whom to choose: a project team member, an in-house expert, or a consultant expert.

### 1. Project team member or independent expert (in-house or consultant)

The most common reasons for using an expert to prepare claims are the project team's lack of time and insufficient experience in claims. Project team members may have insufficient familiarity with construction law to identify all the important issues. However, even if they are experienced in claims and available, a contractor should not rely solely upon the project team's view of the facts, as their involvement in the project may cause them to overlook some issues. Many construction claims, as presented by a contractor's project team, miss major issues that significantly affect the dispute. In some cases, the initial thrust of the claim changes to other issues with greater cost impact, which had been obscured while the project was being built.

The biggest advantage of using a project team member is the lower cost, owing to their salary structure and their familiarity with the project and the dispute. If they prepare the claim while running the project, there are no out-of-pocket costs. Compare, however, the savings of using project personnel to prepare claims with the higher profit achievable if they spend more time supervising construction.

If project team members prepare claims, training and use of this book will greatly aid them in preparing successful claims. Support by an in-house expert will also help.

### 2. Selecting and training in-house experts

The preferred solution for most claims is to use an in-house expert. Employees are less expensive and can be available immediately. They are familiar with company procedures and the strengths and weaknesses of the project team members. They are also available to coach and support project team members in preparing change order requests and small claims. One disadvantage, however, is that they may leave the company.

Selecting an employee for training as the in-house expert is a matter of finding someone who expresses an interest, is not committed full time to project work, and does well on change order requests and claim preparation. The individual should be generally available part time to work on claims. Potential candidates could be a project engineer, an estimator, or even a project manager.

If the selected in-house expert lacks some necessary skills or has limited experience in an area, the individual needs to learn those skills. Start with on-the-job training in support of a consultant expert on one or two claims, and continue by attending seminars and reading trade journals and books on claims. One of the best guides to being an expert witness is the booklet *A Guide to Forensic Engineering and Service as an Expert Witness* published by the Association of Soil and Foundation Engineers, 1985.[13] For other suggestions on retaining and managing experts, see Martell and Poreth's paper, "Using Experts to 'Prove Up' Your Construction Case" in the April 1988, issue of the American Bar Association's *The Construction Lawyer.*[14]

### 3. In-house or consultant expert

The important items to consider when deciding whether to use a consultant expert or in-house expert to prepare a claim are:

- Availability of an in-house expert.
- The skill level of the in-house expert relative to the needs of the claim analysis.
- The greater credibility of an independent expert if in arbitration or litigation.
- The possibility of using a consultant expert for advice and peak workloads.

A highly qualified expert is needed if the contractor's attorney is inexperienced in construction law and disputes. The expert can help counsel identify potential recovery theories and strategies and therefore avoid extensive research time by the attorney to learn the applicable construction law.

If a consultant expert is used, they should work with the in-house expert. Not only will this save money, but it will also contribute to the training of the in-house expert.

## C.   Selecting a Consultant Expert

If a consultant expert is used, selecting the right claims consultant is crucial. Although several consultants may be available, they may not be matched to your specific needs.

### 1. Procedure for selecting a consultant expert

See the recommendations in Fig. 6.1.

### 2. Hire the expert as early as possible

Too often in litigation, consultant experts are retained late in the process. Discovery may be complete, and trial or arbitration hearings may be fast approaching. The result is a rushed effort, additional costs, possible failure to pursue some valid claims, and a less than desirable settlement.

**During construction.** If possible, assign the in-house expert or retain a consultant expert while construction is ongoing. The expert can mitigate damages by helping resolve the cause of the problem and ensuring that the project team maintains adequate documentation. The expert will also become

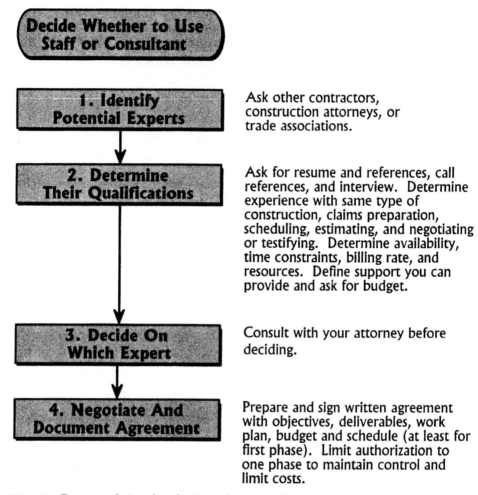

**Figure 6.1**  Recommendations for selecting a claims consultant.

familiar with the project as it is being built and will be able to testify firsthand about the conditions and facts. The expert should review the schedule for possible oversights and errors and can also help negotiate a settlement.

**At the same time or immediately after retaining an attorney.**  If a construction attorney is retained, select the expert at the same time (and with the approval of the selected attorney).

**For preliminary claims analysis.**  If selected early, the expert's preliminary claim analysis, as described in Chap. 7, can be used as a briefing document for the selected attorney and claim support team in deciding how to prepare a claim and resolve the dispute.

**Prior to discovery.**  If in arbitration or litigation, it is essential that an expert be retained prior to discovery. The expert needs to identify what documents to copy, prepare questions for depositions and interrogatories, and attend depositions of opposing experts (as discussed in Sec. C of Chap. 8). This requires familiarity with the dispute and having conducted at least a preliminary analysis.

### 3. Experts for technical issues

In addition to needing an expert in cost/schedule/management issues, technical experts may be needed to analyze means and methods or construction defects for specialty construction trades and materials. A subcontractor usually provides the specialty trade expertise during claims preparation, but an outside expert with experience and proven credibility may be needed for testimony if the claim is headed to arbitration or court. If design issues are involved, a design expert is needed to advise during claims preparation and to testify if necessary.

The procedures in selecting a technical expert are similar to selecting a construction cost/schedule/management expert. Contacts can be found through the American Institute of Architects, the engineering societies (American Society of Civil Engineers, American Consulting Engineers Council, American Public Works Association, etc.), or contractor trade associations (AGC, ABC, NECA, MCA, etc.). The role of the technical expert is more limited, as their testimony addresses the reasons, responsibilities, and cures for specific construction deficiencies. If involved, the attorney normally selects the expert or at least approves the selection. In some cases, the cost/schedule/management expert identifies technical experts, especially if the management expert has a design background or extensive experience in the type of construction in dispute.

### 4. Cost of an expert

Fees of consulting experts vary depending on their reputation, whether they are a free-lance consultant or a senior partner of a prominent claims consulting firm, and the ability and willingness of their clients to pay. Project team members or in-house experts will seem more economical, but their cost should include the lost opportunity cost of not using that experienced construction manager to manage a current project.

## D.  Overseeing the Expert and Managing Claim Preparation

Regardless of whether a consultant or in-house expert is used, the contractor needs to oversee their efforts and ensure they set and meet clearly defined objectives for the cost, time, and quality of their finished product.

## 1. Preparation

First, define what you want to achieve and the role of the expert. Do you want a quick, informal claim for one-on-one friendly negotiation with your counterpart on the owner's side? Or do you want a large, formal document to dump on their desk to show you "mean business"? Or do you prefer something in between? What will the expert do and what will you and your staff do?

Next, prepare a briefing paper and assemble documents for the expert's use. If a draft claim has not been prepared, describe the project, what has happened, and the disputed issues in a memorandum. Cover all important issues to be investigated. If expecting litigation or arbitration, address the memo to your attorney, control its distribution to limit discovery, and present the most sensitive information orally. Regardless if the expert is a consultant or a project manager or staff expert, every effort should be made to facilitate their work.

Inform the expert up front of any problems; don't hold back. The expert will probably discover it anyway and lose confidence in your statements. It will be far worse for both of you if the expert learns about the problem from the other side during negotiations or hearings.

Clearly communicate your expectations to the expert, and confirm they are understood.

## 2. Kickoff meeting and preliminary analysis

Next, hold a kickoff meeting to discuss the project and objectives. Follow the meeting format described in Sec. C.1 of Chap. 7, and require a preliminary analysis as explained in Chap. 7 before proceeding with the majority of the claim preparation work. This ensures efficiency and effectiveness.

## 3. Organization of documents and analysis

Organization of the documents and full-scale analysis begins after completion of the preliminary analysis and confirmation of your objectives and the expert's work plan, budget, and schedule. This can take from a few weeks for a small claim to months for a larger claim, and includes the tasks described in Chaps. 8 through 12.

## 4. Preparation of the claim document and exhibits

It is critical that your expert reports interim findings, to permit directing the effort and deciding whether and how to proceed. The timing, extent, and content of meetings depend upon the claim and continue through review of the final claim document and exhibits.

## 5. Presentation and negotiation

Many construction executives take over after the claim is prepared, to make the presentation and negotiate a settlement, with the expert playing a supporting role. It may be more effective for the expert to make the presentation and conduct preliminary negotiations. This keeps the executive once removed from the direct negotiations, making it easier to escalate negotiations to the next tier of the owner's management, if necessary. A detailed discussion of claim presentation and negotiation appears in Chap. 14.

### 6. Using experts (and narrative text) to counter punitive discovery requests

Some owners' attorneys use overly aggressive legal tactics to delay and raise cost barriers to the prosecution of legitimate claims in arbitration or litigation. A favorite tactic is punitive but "entirely proper" discovery devices that burden your company management. This may include contention interrogatories and deposition of corporate officers and key employees demanding the details of all facts forming a basis for each statement in the complaint, the names of individuals personally knowledgeable of those facts, all documents relied upon, specific dates of all events, etc. This can distract the management team from their primary responsibilities, severely disrupt company operations, and cost a lot more than the use of an expert, either in-house or consultant.

Using a highly qualified expert with sufficient budget to fully investigate the facts and prepare a thorough report (in the form of narrative text), instead of using key employees and company management, reduces demands on company personnel from punitive discovery requests. The expert will be the most knowledgeable individual of the facts and will have the information needed to respond to interrogatories. Although the hourly rate may be higher than the salaries paid to company officers and key employees, the expert's involvement will (1) save their time, (2) focus the effort on one individual who isn't needed for ongoing operations, and (3) require only management oversight and limited involvement. If adequate documentation exists, the expert can give testimony on those documents. Management and key employees will be needed only for confirming testimony and for added credibility. This won't eliminate demands on company management but will substantially reduce their involvement.

# Phase 2—Preparing Preliminary Claims Analyses, Dispute Review Board Presentations, and Small Claims

This chapter describes how to prepare a *preliminary analysis, the recommended first step in claims preparation.* It consists of an initial review and analysis to confirm the validity of a claim, the approximate value, your priorities, and a strategy and work plan for preparing the claim.

The procedures described are similar to the process of preparing a small claim or one that is weakly contested and merely needs reasonable justification. This chapter can therefore be used for preparing a small claim, in lieu of the detailed procedures in Chap. 9. The results can also be used for preparing a presentation to a dispute review board, as discussed in Sec. D below.

This chapter and the remainder of the book address the claims preparer (the expert) who may be a project team member, an in-house claims expert, or a consultant. This chapter includes the following sections:

## A.   Why Make a Preliminary Analysis Before the Detailed Analysis?

Making a preliminary analysis of a claim saves time and money, and results in a better claim than proceeding immediately with a full-scale effort. A preliminary analysis is recommended even if the claim preparer was a project team member.

### 1.   Advantages of making a preliminary analysis before proceeding with the claim

A phased approach to claim preparation has numerous advantages over launching an all-out effort:

- Dead-end approaches that don't pan out are reduced or eliminated, thereby reducing cost and time.

- More control can be exercised over costs and the overall effort, owing to the opportunities to evaluate progress and to redirect efforts if required.

- Contractors using a consultant need not commit a lot of money to a relatively untested individual and an uncertain result. Instead, they commit for a fixed limited cost and a specific work product. Should the contractor decide to proceed, the work product can be used by anyone to complete the claim.

- For a consultant claims preparer, a preliminary analysis avoids nonreimbursable initial reviews, extended negotiations, and possibly inadequate budgets.

- It may be possible to negotiate a settlement based on the preliminary analysis, thus saving the cost and time of preparing a complete analysis and claim.

The greatest advantage of a preliminary analysis is not wasting time and money on a weak claim or walking away from a good one.

### 2.   Goals and objectives

The goal of a preliminary analysis is an overall assessment of the dispute, the claims preparation effort required, and the method and likelihood of resolution. Alternately, the analysis may focus on a single issue, such as the facts relevant to a crucial legal issue on which the entire claim depends.

Your objectives during preliminary analysis are as follows:

- Identify and describe all the issues in dispute, to ensure that the final analysis includes all issues.

- Obtain and briefly review the most important records documenting the facts of the dispute.

- Discuss the issues with project team members and others so that everyone understands them.

- Determine the key facts of the dispute, including the significance of the contract language.
- Develop tentative findings and recommendations.
- Identify what needs to be done to complete the analysis.
- Determine the documents needed, and describe how to obtain them.
- Estimate the probable size of the ultimate recovery and the probability of recovery.
- Describe a strategy, work plan, budget, and schedule to complete the claim.
- Help decide how to proceed.

Your objectives may include requesting information from the owner's representative and opening negotiations to settle the claim.

### 3. An opportunity for strategic review and risk assessment

The preliminary analysis objectives should include review of business issues, in addition to the factual legal and technical issues of the claim. For example, on one large public works project, we identified preserving a 10-year business relationship as a primary goal. The solution to presenting a large controversial change order request without overly adverse reactions was to (1) better communicate the magnitude of the cost overruns to the owner's senior management so that they accepted it on an emotional level, (2) redefine everyone's perception of the problems as a "joint" problem for the contractor, owner, and designer, (3) call for a partnering/value engineering workshop to deal with owner-generated scope creep and designer changes, and (4) characterize prior changes as "changed conditions" while requesting better notice and documentation of future changes. The difference between our approach and normal procedures was slight, but the difference in how it was received was significant.

On another case, while defending an owner from a $600,000 claim (after the contractor had refused a $200,000 offer), we determined that not only was the claim worth only $100,000 but that there were almost $400,000 in construction deficiencies. Had the contractor evaluated the down-side risks of pursuing arbitration, they would have been half a million dollars ahead.

In a third case, while preparing the claim document for a small general contractor against a federal agency, we decided to clearly assign responsibility to the designer instead of skirting the issue. Not only did the majority of the problems originate in poor design, but many others resulted from poor contract administration. The designer was no longer a filter to the owner and we had no reason or need to placate them.

### 4. Level of effort and costs

A preliminary analysis generally takes less than 10 percent of the total claim preparation effort, although it can take more on a very small claim. It may be as little as a one-day briefing and work session with a consultant expert, with the contractor's staff completing the claim.

The cost is considerably less for an in-house expert or project team member than for a consultant expert, as employees are already on the payroll and are familiar with the project. A consultant expert typically charges $2500 to $5000 for a preliminary analysis of a small to medium-sized claim. This allows 20 to 40 hours time plus limited staff support and expenses. Additional costs to complete the claim can vary from $5000 to well over $50,000.

## B.  How to Prepare a Preliminary Analysis

The steps for preparing a preliminary claim analysis vary widely, depending upon the size and complexity of the claim, the available documentation, the expected level of resistance by the reviewer, and the level and quality of support received. Preparing a preliminary analysis requires the following 10 steps.

### 1. Initial briefing

The first step is the initial briefing, the kick-off meeting for the claims preparation effort. If a consultant expert is used, the kick-off meeting may also serve as a final selection interview to determine whether to hire the expert.

**Obtain and briefly review key documents.**  Prior to the meeting, request an initial information packet containing a summary of the dispute(s), any prior change order requests or claim documents with referenced exhibits, the contract, the plans (reduced size if available) and specifications, the original schedule and all updates, other important documents, and all readily available overview information on the project. If not provided earlier, be sure they are available at the meeting so that you can briefly examine them. Request copies rather than originals, so that they can be marked with comments.

Ask for a site plan to be marked up with work areas and disputed items identified (as referenced in the daily logs). Also request photographs showing the progress of the work and the issues in dispute. Ask for all the photographs, organized chronologically, not just the best ones.

**Identify the location and attendees.**  If practical, hold the initial briefing at the jobsite in order to tour the project and view specific problem areas. Request that the claim support team and all key project team members attend, as they may have different views on the problems and how to approach the claim.

**Conduct oral briefing and discussion.**  The quickest way to learn about the project and the issues, while ensuring that everyone agrees on the issues, is an oral briefing and discussion with the claim support team. Cover the following.

- A general description of the project and construction elements, including the disputed work items.

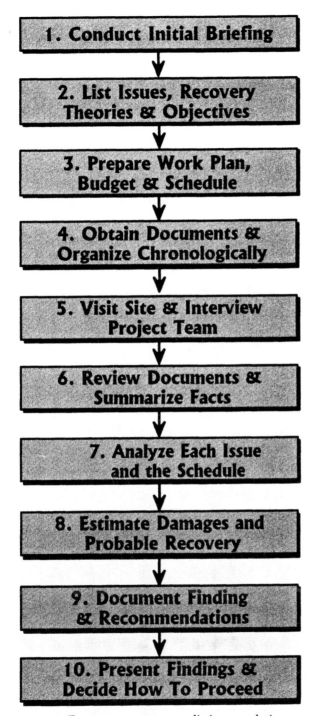

**Figure 7.1**   Ten steps to prepare a preliminary analysis.

- A chronology of key events and activities during construction, comparing actual progress to plan. If scheduling is an issue, briefly review the critical path and impacted critical path work.

- A description of the significant disputed issues, with quantification of cost and time, if known.

- Identification of the key project participants and their role in the project and the disputed issues.

- A discussion of crucial portions of key documents, with a brief opportunity to examine them.

- The current strategy to complete the project, if work is ongoing, the completed or ongoing claim preparation efforts of others, legal theories of recovery, and settlement strategies.

- All expectations regarding the analysis, including the time frame, fees (if a consultant), support required, anticipated results, work product, etc.

Keep extensive notes and highlight and tab documents for future reference. Ask questions as the briefing progresses or take notes and return to the issue later. Clarify all ambiguities. Start a list of all the issues in dispute and an action list identifying what needs to be done, by whom, and when. Note taking on a laptop computer is the recommended method of making notes.

**Identify and prioritize the issues.**   List all of the significant issues that need to be investigated. Give each a descriptive title, categorize them by group (type of claim, chronological, CSI code, subcontractor, entitlement theory, initiating cause, responsibility, etc.), and prioritize by the probable recovery value, relative difficulty of investigation, and order of investigation.

**Check key contractual issues.**   Briefly investigate key contractual issues that could derail the claim. These include:

- Notice requirements, actual notice, and the consequences for noncompliance.

- Whether contract rights to impact and delay have been abrogated through accepting change orders for the direct costs without qualifications.

- Contract requirements for timeliness and documentation of change order requests and claims.

- Potential owner defenses and counterclaims.

**Give initial impressions, action plan, and agreement.**   After the presentation and discussion, consider giving your initial *impression* of the case to the claim support team with caveats, available alternatives, and how you recommend proceeding. Caution the claim support team that this is only your initial impression and that you cannot commit at this stage to the outcome of full investigation. Reach a verbal agreement on what action is to be taken, and confirm it in writing. Develop an action plan with objectives, deliverables, tasks to be accomplished, budget, and schedule. Always do this, even if working as an employee on a company claim.

**Conduct in-depth schedule review—if not completed prior to meeting.** If scheduling is an issue, review the schedule in depth with the project scheduler or project superintendent. Review the critical and near-critical path activities and any impacts to those activities. Ask for copies and mark them up during the review. A joint review is far more efficient than trying to determine the job logic yourself. An additional benefit is learning their assumptions and strategy.

**Suggestions for consultant experts.** If you are a consultant or a construction executive planning to use a consultant, but without a consulting agreement at the start of the briefing, break the meeting into five phases:

1. A brief presentation describing the issues in dispute.
2. Questions and a brief discussion of the issues and possible actions, to establish the consultant's credibility and how reasonable he or she will be to work with.
3. Review of the consultant's qualifications relevant to the specific issues in this dispute.
4. Negotiation of a verbal agreement (to be confirmed in writing).
5. A more in-depth presentation and discussion as described above.

## 2a. Use narrative text to collect the facts, and organize and communicate your findings

Narrative text is the term which describes a simple but powerful method of recording and organizing the facts and your analysis and communicating your findings. It uses word processing software and a few simple procedures summarized below and described in detail in Sec. B of Chap. 9.

**Advantages of narrative text.** Narrative text improves communication with the claim support team. It enables the claim preparer to record and organize factual information, analyses, and conclusions. It stores and provides easy access to key facts, names of the individuals and firms involved, telephone numbers, and any other vital information. When claim preparation is delayed or if there is a long period between preparing the claim and giving testimony, narrative text is crucial for quickly refreshing your memory. You can easily resume where you left off and finish the analysis with little lost effort. When the narrative text is complete after the detailed analysis, it is converted into the claim document as described in Chap. 13.

**Start your narrative text file at the initial briefing.** If available, use a laptop computer to record your notes of the initial briefing. Create a narrative text file. For convenience, use a standard template similar to the example outline below, and then edit it to create the narrative text for your preliminary analysis.

**Create document footer.** To help protect the confidentiality of privileged information, create a page footer that titles the document, alerts the reader that the information is privileged, and helps identify the latest version.

---

Preliminary Dispute Analysis of ABC Project for XYZ Constr.—
Confidential Attorney Work Product—date—page #

---

### TABLE OF CONTENTS FOR NARRATIVE TEXT

*A.* **Administrative Matters**
 1. Summary of Status and Direction of the Analysis
 2. Scope of Work, Deliverables, Budget, and Schedule
 3. Work Accomplished Last Period (or to date)
 4. Work Planned Next Period (or to completion) with a brief work plan, budget, and schedule
 5. Questions and Comments (reference those embedded in the issue analyses)
 6. Recommendations
 7. Action Needed, by Whom, and When
 8. How to Use the Narrative Text, Definitions, Abbreviations, and Miscellaneous Information
*B.* **Summary of Tentative Findings and Conclusions**
 1. Summary
 2. Factual (including project background and history)
 3. Contractual (with legal questions for response or action by counsel)
 4. Preliminary Conclusions
*C.* **Issue Analyses**—each issue has a subsection with five sub-subsections:
 1. Issue 1
  *a.* Summary
  *b.* Description
  *c.* Analysis
  *d.* Conclusions, Unresolved Items, and Recommendations
  *e.* Chronology of Summary Notes Related to the Issue
 2. Issue 2, etc.
*D.* **Schedule Analysis**
*E.* **Cost Analysis**
*F.* **Review Notes**—nondated documents, telephone conversation notes, meetings, and interviews
*G.* **Chronology of All Summary Notes**—optional, in addition to chronology for each issue
*H.* **Appendixes**

**Organization of a narrative text file.** The following outline works well for a preliminary analysis. Modify it to meet the specifics of each claim.

**Contents of each section.** Section A, Administrative Matters, contains notes of work accomplished and action needed, information for communication with and action by the claim preparer or claim support team, and a record of progress. Update it as work progresses, and before distributing to the claim support team.

Section B, Summary of Tentative Findings and Conclusions, condenses the analysis and conclusions for each of the individual issue analyses from section C and combines them into an overall summary for all claims on the project. Although coming before section C in the narrative text, it is prepared afterward.

Section C, Issue Analyses, addresses the individual issues in dispute with a subsection for each issue. Each subsection has a summary that presents the highlights for a quick review. The description describes the issue for the reader. Each subsection also has analysis and conclusion sub-subsections, where you collect and organize your thoughts for presentation or further analysis. The subsection ends with a chronology of summary notes related to the issue, which were generated while the chronological source document files were reviewed.

Sections D and E, Schedule and Cost Analyses, contain the schedule analysis and computation of costs. These are usually prepared just before section B (the summary).

Section F, Review Notes, contains nondated information such as pertinent provisions of the contract, notes of meetings with the claim support team, and telephone interviews. The notes are created while reviewing the documents, attending meetings, or recording telephone conversations, and are one of the first parts of the narrative text to be completed.

Section G, Chronology of All Summary Notes is optional, as it duplicates the summary notes for each issue. The notes summarize actual events and progress as recorded when the chronological source document files are reviewed. They contain the pertinent time-relevant information, paraphrased for brevity except when a direct quote is desirable. The notes consist of the date of the event, a summarization of the information, a code for the issue(s) discussed, and a reference to the source document. The summary notes are also entered in a chronology within the subsection for each issue and are used during analysis of the issue.

Section H, the Appendixes, contains exhibits, detailed analyses and computations, and subcontractor claims.

**How and when to create and use narrative text.** If possible, type your notes on a computer. If not, write longhand notes and have them transcribed, or use voice-recognition software, as the work must eventually be entered into a computer in a word-processing program.

If discovery is a potential problem, read Sec. B.13 of Chap. 9, consult with legal counsel, and use caution in what you record to avoid embarrassing data being obtained by the opposing attorney.

The process of generating narrative text of a claim is generally as follows:

---

### THE GENERAL PROCESS OF CREATING NARRATIVE TEXT

1. Set Up Narrative Text and Enter Initial Administrative Information—Section A.
2. Convert List of Issues from Initial Briefing into Outline of Issue Analyses—Section C.
3. Create Review Notes of Meetings, Telephone Conversations, and Nondated Document Information (Contract Documents, Prior Claim Documents, etc.)—Section F.
4. Expand Issue Analyses with Information from Document Review—Section C.
5. Review Source Documents and Create Chronological Summary Notes. Alternately, Post Chronological Summary Notes Directly to Relevant Issue Analyses—Section G or C.
6. Expand the Issue Analyses Based on Chronological Summary Notes—Section C.
7. Review and Analyze the Schedule—Section D.
8. Review and Analyze the Costs—Section E.
9. Expand the Issue Analyses Based on Schedule and Cost Analyses—Section C.
10. Periodically Update the Administrative Section and Use the Narrative Text to Communicate with the Claim Support Team—Section A.
11. Draft and Update The Tentative Findings—Section B.
12. Finalize the Issue Analyses—Section C.
13. Finalize the Tentative Findings, Conclusions, and Recommendations—Sections A and B.

---

#### 2b. Review your notes, any existing claim documents, and key contract clauses

Review notes from the initial meeting, and all interviews and telephone conversations regarding the claim. Verify you fully understand the dispute and your objectives. If necessary, clarify the objectives.

Review the change order request that was submitted and rejected by the owner, along with the owner's reasons for rejection. If a preliminary claim document has been prepared, review it and note questions, comments, and action needed. Understand the dispute as viewed by both parties.

Briefly review the contract documents. Examine pertinent contract clauses and flag potential problem areas for review by legal counsel. The most important clauses, as described in Chap. 3, are:

- The changes and disputes clauses.

- The notice provisions.

- The contract duration and starting date, if scheduling is an issue.

- Any onerous risk-shifting clauses in the general and special conditions.

- The technical specifications governing any work item in dispute.

When reviewing the contract, create review notes of pertinent portions for later review. These notes should be recorded in section F of the narrative text.

#### 2c. Define and list the issues, tentative recovery theories, and your objectives

Break the disputes into their basic elements and list all the disputed issues. Identify and organize the issues by major category (type of dispute or type of cost, responsibility, work area, construction trade, etc.). Prioritize each issue by dollar value, the relative ease of claims preparation, and the probability of settlement. This list, started at the initial briefing and updated periodically, usually includes:

- Specific claims by the contractor and subcontractors.
- Potential claims not identified by the project team.
- Probable owner defenses or counterclaims.
- Legal questions, including recovery theories, needing review by counsel.
- Questions of fact that are unknown, uncertain, or not readily provable.
- Other items of importance.

After completing the list of issues in dispute, create a separate subsection in section C of the narrative text for each issue. Then collect all the pertinent facts on each issue within that subsection of the narrative text.

Write down the tentative recovery theories for each issue, your comments, additional information required, and whether to seek legal advice. Determine if the contract supports your tentative recovery theory and what claim defenses must be overcome.

### 3. Prepare a work plan, budget, and schedule for the preliminary analysis

Write down specific objectives and expected deliverables. Create a work plan for accomplishing the objectives and producing the deliverables. The work plan can be a bar chart or timescale arrow diagram. It should be detailed for the preliminary analysis and in outline form for completion of the claim. At this point, it should identify all the tasks required to complete the analysis, their sequencing, and the estimated hours for each person assigned to each task. The estimated hours times the person's hourly cost, plus expenses and a contingency, constitutes the budget for this phase and can be used to control claim preparation costs.

The work plan, budget, and schedule are contained in Sec. A.2 of the narrative text.

### 4. Organize and review critical documents chronologically

The most important documents to review usually include:

- The contract, plans and specifications, especially the general conditions.
- Any previously prepared, unresolved claims and change order requests.
- All documents referenced in the previously prepared but unresolved change order requests.
- The original schedule and pertinent updates, especially for delay or acceleration claims.
- The bid summary and any relevant backup.

- Copies of all change orders with supporting documentation and record of negotiations.

- All information logs: submittals, change order requests, RFIs, notice of change, etc.

- Copies of progress payments, at a minimum the first and last ones.

- Correspondence and other documents related to the disputed issues, especially those giving notice.

- Daily diaries and other field reports, at least for periods of disputed work.

- Photographs—request all, preferably sorted by date.

- Any other documents the claim support team believes will be useful, while reasonably limiting the information to be reviewed, considering the limited budget and time available.

To facilitate review, have a staff person *organize the documents into one or more chronological files,* with the oldest on top. Group all correspondence items in one file. Superintendents' daily logs and other field reports, also organized chronologically, may be in either the main file or a separate file.

## 5. Visit the site and talk with key project team members—if practical

Visit the site accompanied by one of the project team members. If work is ongoing, interview job personnel. Failure to visit the site can result in an incomplete grasp of the project.

Interview key project personnel by phone or in person to obtain their perspectives and any additional information. Later, during the detailed analysis, with full understanding of the issues, interview them in depth. Guidelines on interviewing job personnel and conducting site visits appear in Secs. D and E of Chap. 8.

## 6a. Briefly examine the schedule and cost reports

Briefly examine the as-planned schedule and the as-built schedule (the latest update) to determine the time frame of the project and major milestones, and how actual progress compared with planned. If delay or acceleration is an issue, review the schedules in detail.

Examine the latest cost comparison report that lists the budgeted and actual cost for each cost code with a breakdown by cost category (labor, subcontractor, materials, etc.). Identify the most important issues in terms of cost overruns, and use this to prioritize the analysis of the high-value issues.

## 6b. Review the documents and summarize the facts—in the narrative text

The next step is to review the chronological document file, determine the facts, and summarize them as summary notes in the narrative text.

**Take review notes of nondated information.** Summarize *nondated* information (e.g., the contract documents) in section F of the narrative text—as in the following example:

---

3. Specifications—Volume I—Architectural

<u>Division G - F.1.5—Markup</u>
Specifies 15% on labor, 10% on equipment and materials, and 5–10% on subcontract work under/over $2000.
C: PROPOSE ACTUAL OVERHEAD & REASONABLE PROFIT.

---

**Review the chronological files and highlight important text.** Review the documents chronologically, looking for references to the identified issues or to possible new issues. Highlight important text with a nonreproducible (e.g., yellow) marker to facilitate finding the information when reviewing that page later.

**Tab pages with important text.** Place Post-it note tabs extending beyond the right margin of documents with important text. Label the tabs with a code for each issue. Stagger the tabs down the side by issue so that the documents related to each issue are readily accessible. Since only five tabs can be placed without overlapping, use the staggered tabs for the most important issues and use the tab space at the top of the page for multiple less important issues.

**Take chronological summary notes of dated events.** While reviewing the documents, create a summary note of each significant *dated* reference to an issue and enter it into either section G or the chronology subsection for that issue within section C of the narrative text—as in the following example:

---

14Aug98   Formed and poured E footing slab. Had problem with last minute changes by owner's rep and had to shut down for 4 hours to make adjustments. CONC DELAY [DailyLog]

---

Entering the summary notes into a Sec. G is more efficient than entering into separate sub-subsections for each issue, but it requires (1) coding each summary note with a code for each issue and (2) selectively printing by the code for each issue. Since some summary notes refer to multiple issues, this saves copying the summary notes to each of the issue chronologies.

For details, see Sec. B.6 of Chap. 9. If a note references two or more issues, copy it to the chronology sub-subsection of the narrative text for the additional issues.

**Enter comments or questions.** If a question arises or comment seems appropriate when creating the summary notes, note it in capital letters (so it stands out), after the summary note with a search code (Q: or C:) for retrieving it later. Optimally, copy and post the comments and questions to the narrative text in subsection A.5, under a sub-subsection for the individual responsible for the question or comment, to ensure the individual reads them when you distribute the text.

**Ask questions, obtain more documents when needed, etc.**    If the facts or issues in dispute aren't clear, or if you need additional documents to complete your analysis, ask for clarification or for the missing document.

## 6c. Optionally, create a detailed as-built schedule while reviewing the documents

In addition to summary chronology notes, consider creating a detailed as-built schedule, described in Sec. G of Chap. 10. The detailed as-built schedule merges hundreds of daily events onto a single sheet of paper and defines what actually occurred on the project. It is a very powerful, but simple, tool that determines when activities were completed, under what conditions, and why they took so long.

Although not normally prepared during a preliminary analysis, the detailed as-built is relatively easy to create and can take as little as one or two hours per month for a limited review of key facts.

## 6d. Selectively print the summary notes by issue

If the summary chronology notes are entered into a common chronology in Sec. G, selectively print a separate chronology for each issue.

## 7a. Analyze each issue separately—in the issue analysis section of the narrative text

*Analyze each issue separately.* Gathering all the relevant information, organizing it chronologically, and eliminating superfluous details makes patterns obvious and the essential facts discernible. This is especially true for large, complex claims when the facts are so voluminous that a claim preparer cannot retain every detail in working memory.

**Review the chronology and related documents to determine the facts about each issue.**    Read the chronology for each issue and any other relevant documents (e.g., daily diaries, photographs, the detailed as-built schedule, etc.). Take notes, and start the analysis using the following checklist.

- Note obvious conclusions and develop a list of subjects to be analyzed.
- Verify or refute asserted facts regarding the issue.
- Check for facts supporting your tentative recovery theory and whether other theories may apply.
- Check for facts that might support or refute potential owner defenses or counterclaims.

- Check for possible new issues and claims.
- Identify missing documents that are referenced by the summary notes.
- Mark quotes that may be printed in the analysis.

**Analyze each issue.**   Type the facts and your observations into the computer, organize them, and record the analytical thought process by which you reached your conclusions in the Analysis sub-subsection for each issue. Then expand the analysis, noting the facts supporting your conclusions and referencing each fact to a source document. Also note if additional information or legal review of contractual questions is needed.

- Determine the facts. Describe and reference the initiating factual event, action, or condition to the supporting documents.
- Deduce the essential facts from those available, if the needed information isn't found in the documents.
- Identify the applicable contract provisions or law—the theory of recovery.
- Examine liability. Confirm whether the proposed recovery theory appears viable and note the facts supporting your tentative conclusions. Comment on the impact of pertinent contract clauses and contract law, state uncertainties, and note the need for legal review by counsel.
- Examine causation. Trace the linkage from the initiating event or condition to the resulting costs.
- Compute damages, even if only a guesstimate, to determine whether to proceed with the claim.
- Analyze the schedule, if the issue involves impacts to the schedule (i.e., if there was delay or acceleration), using the same process.

**Reference each statement of fact to a source document.**   To document the facts or to support your analysis, it is essential that you reference every statement to a source document, if possible. Place references after each statement in square brackets with a letter code for the type of document, abbreviations for who from/to, and the date (e.g., [L-GC/OWN-14Aug98]). If referencing to an individual's comment, enter his or her initials and the date the comment was made. This enables you or someone else to verify each statement in the narrative text.

**Develop your tentative findings on the issue, identify unresolved items, and make recommendations.**   Record your findings on the key facts, entitlement, and damages for each issue in a separate subsection for the issue. Identify weaknesses in the claim, documents needed, questions and possible resolutions, additional information or other action needed, and your recommendations on how to proceed. Identify exhibits and other work products to prove entitlement and damages and to convince the eventual reviewer or fact finder of your position.

**Examine potential additional claims and issues, owner defenses, and counterclaims.** Project teams often miss or misunderstand at least one major issue of a dispute. The most important issues sometimes turn out to be substantially different from those identified when starting the analysis. Therefore, briefly check for additional issues, including owner defenses or counterclaims.

**Evaluate the reliability of the documents and performance of the project participants.** As in all endeavors, the human element is of paramount importance. If you were a member of the project team, you will know what documents are unreliable and which parties tend to exaggerate. If not a member of the project team, and if your review of the documents is limited, you must rely on the opinions and memory of the project team members to a large extent. Their statements and position must be accepted unless contradicted by the immediately available evidence. However, if the project team is inexperienced or weak in management skills, some of the problems may be a result of contractor errors. Therefore, be cautious in accepting unsupported statements. When recording them, always reference them to the individual making the statement to facilitate correcting them if contradicted by other evidence.

**Look for inconsistencies.**    Most activities in construction have a pattern that with experience is recognizable but hard to define. Therefore, identify and investigate inconsistencies to determine if they affected the issue.

**Finalize your findings, describe and summarize each issue.**    Finalize the analysis of each issue by revising the tentative findings and recommendations sub-subsection. Write up a brief description of the issue and a summary of your investigation, findings, and recommendations regarding the issue.

## 7b.  Analyze the schedule

Schedule analysis is similar to the analysis of any other issue, but there are specific procedures and techniques, described in Chap. 10. Apply these tools to the preliminary analysis.

## 8.  Estimate damages, the probable recovery, and downside risks

Cost analysis is usually completed last, after entitlement has been established, although some cost analysis is recommended concurrently while examining entitlement. As noted above, this helps prioritize your efforts and may establish entitlement. You also need to estimate the amount of recovery (which may be significantly less than damages, owing to the difficulty of establishing entitlement and proving damages) and the probability of recovery.

Identify and quantify the risk of counterclaims and other potential losses. The downside risk is normally small but in some cases may be so large that the claim should be dropped.

## 9a. Develop your tentative findings, note unresolved items, and make recommendations

Although it is important to keep perspective and avoid being locked into an opinion or strategy that proves untenable, a hypothesis is necessary to guide continued research. Write up an overall summary of the claim(s) and a discussion of the important issues in section B of the narrative text. Share this with the claim support team without raising unrealistic expectations. If discovery is possible, communicate critical issues orally, and revise the narrative text later with your final conclusions.

## 9b. Prepare a work plan, budget, and schedule to complete the claim

Prior to presenting your preliminary analysis to the claim support team, establish a tentative work plan, budget, and schedule for expanding the preliminary analysis into a complete claim.

## 10. Present your findings to the claim support team and help decide how to proceed

The ultimate objectives of the preliminary analysis are to present and discuss your findings and recommendations with the claim support team, and to decide whether and how to proceed. If you are both the claim preparer and the decision maker, present the results of the preliminary analysis to several others to help evaluate the claim and decide how to proceed.

When making the presentation:

1. Verbally present your findings and recommendations to the claims support team and/or decision maker. An oral presentation greatly aids understanding and retention, especially if supplemented by graphic exhibits.

2. Identify and discuss legal issues that need review and analysis by counsel.

3. Discuss the issues, recovery theories, strategy, and probable tactics.

4. Discuss any potential problems with entitlement, contract clauses, notice, or proof of damages.

5. Present a tentative work plan, budget, and schedule for completing the claim preparation.

6. Participate in making any necessary decisions, including whether to proceed with a claim.

7. Obtain authorization to proceed and identify actions required by each party.

## C. Optionally, Attempt to Settle the Claim

It may be possible to settle the claim upon completion of the preliminary analysis. Make the attempt, as even if you don't settle, the discussions may lead to eventual resolution as the claim reviewer is now involved in the claim. Another benefit is that you will have learned how the claim reviewer responds and what is needed to approve the claim.

## 1.    Prepare a preliminary claim document

The findings and conclusions section of the preliminary analysis is easily converted into a claim document. Organize the preliminary claim and address the issues as follows:

- Introduction, following a cover letter and table of contents.
- Executive summary of one or two pages that includes the time extension and amount claimed.
- Brief description of the project and disputed work.
- Statement of applicable contract terms (and contract law if provided by counsel).
- Description of the contractor's original strategy, schedule, and expectations— if it is a delay claim.
- Detailed description, chronology of summary notes, and analysis of each issue.
- Schedule analysis, including computations of delay.
- Computation and summary of damages.
- Overall conclusions.
- Appendix with exhibits and key source documents.

## 2.  Initiate settlement discussions with the owner's representatives

Present the preliminary claim document to the claim reviewer, if that person has been identified.

It might be possible to settle all issues at this time—especially if a partnering approach is being used. Or you could reach a partial settlement. Resolve the easy issues first: retainage that could be reduced under the terms of the contract, acknowledged change order work without final approval, extra work for which the disputed amount is small, etc. If successful, continue and attempt to resolve everything. If entitlement is acknowledged but agreement can't be reached on the cost of a large change, ask for a two-part change order. Request payment of the amount the reviewer agrees to, leaving final payment until resolution. This helps if the contractor is in a financial bind. It also protects the owner from additional interest charges, which can be high in a state with a prompt payment act.

Informal discussions can identify the individual(s) who will be reviewing and approving the claim, their initial position, and what theory of recovery they are most likely to accept. Initiate discussions at this time and provide limited preliminary information to involve them psychologically in the claim preparation while attempting resolution.

Settlement discussions also enable you to request owner records, which will supplement the documents already in your possession. If relationships are good, the owner may provide information that is helpful in preparing a claim, such as inspectors' daily logs. For more information on the presentation and settlement of claims, see Chap. 14.

If attempts to settle the claim are unsuccessful, the next step is to obtain all of the pertinent documents through discovery as explained in Chap. 8.

## D.  Dispute Review Boards and Neutral Experts

Knowing how to work with a project neutral, either a dispute review board (DRB) or a neutral expert, will help in resolving disputes faster, with less effort, and more equitably. See Sec. F of Chap. 4 for guidelines on why you should use project neutrals and how to select them.

### 1. Dispute review board presentations

*The secret to a successful DRB presentation is to be well prepared.* Halfway efforts will not suffice. Although it is an informal process, without participation of legal counsel, it should be accorded as much importance as an arbitration hearing. You *must* be prepared. Few DRB recommendations are taken to arbitration or litigation by a dissatisfied party. Although not binding, the board's recommendations are given great weight by arbitration panels and the courts because they were made by experienced, neutral individuals who were familiar with the facts. This will probably be the only chance you have for recovery. It would be false economy to skimp on preparation.

Although DRB members are seldom attorneys and they give less weight to case law than an arbitration panel or court, you should carefully research the law if the contract is not clear. In some cases, consider including an analysis of the pertinent case law by an attorney. It must be clearly written and directed to nonattorneys.

Contractors generally make their presentation first, followed by the owner's representative. The board may ask one or both parties for more information and then caucus for a decision. The decision is usually rendered within a few days and may address all the issues. Or it may address some issues and recommend how the parties can proceed to resolve the elements of the dispute.

When presenting to a DRB, remember they are seasoned construction experts who are familiar with the project and the parties involved and may already have an opinion on the dispute. Use jobsite personnel to "tell the story," even if the claim was prepared by an outside expert. Stick to the facts and don't try to educate them on means and methods or other issues they already know.

### 2. Neutral expert

A neutral expert may have a greater role in directly determining the facts or may act more as a DRB. In either case, full cooperation is essential and an active effort should be made to convince the neutral expert of the validity of your position.

## E.  How to Prepare a Simplified Claim

Too often, claim preparers tend to overdo fact finding and analysis, attempting to prove everything and ending up not being able to *persuade* the claim reviewer. They also run over the claim preparation budget. Although the facts are important and misrepresentation is unacceptable, settlement depends more on perceptions, timing, and the expectations of the other side. This is especially true for complex claims with inadequate records, disputed entitlement, ill-defined causation, multiple issues, and hard-to-prove impact and inefficiency. It also applies to smaller claims that cannot justify the expense of a thorough investigation and claim preparation effort, or when the contractor's records, lack of notice, or weak entitlement argument make recovery unlikely.

The processes described throughout this book are structured, rational checklists that should lead to resolution. Unfortunately, that doesn't always happen—because of insufficient data, inadequate time or budget for a thorough analysis, or the misperceptions and biases of the reviewer. More creativity, better communication skills, and a different approach may be necessary.

The generally recommended approach to preparing a simplified claim is as follows.

## 1.  Determine the magnitude of the problem

Use a quick, modified-total-cost claim analysis of the entire project. Verify that the actual total cost from the latest accounting report includes all costs. Many contractors, for example, don't charge all of their equipment and small tools costs against projects, or stop charging when projects go over budget. Few contractors charge project management time to the project unless the project manager works at the jobsite. Identify these costs and add them to the total. Also include the estimated cost of incomplete work, warranty costs, pending invoices, and potential claims by subcontractors and vendors.

Verify the budget for the work accomplished, using either the budget entered into the accounting system or the original estimate. Remove any markup, acknowledged bid errors, and construction mistakes (that are clearly the responsibility of the contractor and subcontractors), and add the cost (without markup) of all authorized change orders and pending change order requests.

Subtract the budgeted cost from actual. The difference (after adding markup for overhead and profit) is a modified total cost claim and normally indicates maximum recovery possible on all unfiled claims. If in arbitration or final settlement negotiations, don't forget unpaid contract balances, approved change orders, and retainage.

Failure to check the claimed damages against a modified total cost analysis can lead to disaster. For example, I have watched an inflated claim completely fail in arbitration, much to the embarrassment of the contractor's expert, when the amount claimed exceeded our analysis of the contractor's total overrun. The loss of credibility carried over to the arbitrator's perception of entitlement for valid issues.

## 2. Check the overrun by cost category and major cost code

Before proceeding, compare budgeted costs with actual costs by cost category, which may reveal the true source of the problem. For example, an overrun in both material costs and labor costs implies that the quantity estimate was too low. An overrun primarily in equipment costs may indicate either a bid error or rental instead of owned equipment. Or an overrun in one major cost code may be due to a bid error or field mistakes in that work only.

For example, a $700,000 claim on a $2.1 million commercial project was allegedly due to differing site conditions and multiple design problems. Our initial analysis, however, revealed that one cost code, rough framing, overran the estimate by $400,000 and extended overhead from delays accounted for another $100,000. Valid extra work claims totaled only $50,000, of which less than $20,000 were in rough carpentry. As we were defending the owner, our continued analysis found that the framing was grossly underbid by an inexperienced framer, who made serious mistakes, defaulted, and then continued to understaff and mismanage the project as an employee.

## 3. If necessary, submit a modified total cost claim

If you must submit a claim as soon as possible and lack the time, budget, or sufficient documentation for a more thorough analysis, submit a modified total cost claim. Modify it later with a more detailed analysis, and it may start the clock ticking on interest charges. After all, it is difficult to justify charging the owner interest when they don't know the amount due. An early indication of the cost magnitude may also prompt owner efforts to mitigate the problem. An early modified total cost claim starts the process of negotiation and establishes credibility on the extent of loss, even if entitlement and causation are still questioned.

For example, we prepared a quick, simplified delay claim on a radio-TV station renovation that forecasted $100,000 in delay damages due to haggling over a $12,000 electrical change. It got the owner's attention, the designer was on an airplane the next day, and we jointly resolved the problem a day later which capped the costs at $30,000. Had we taken the time to prepare a detailed analysis before submitting a claim, the damages would have exceeded our original forecast.

## 4. Develop a simplified, multiple-issue claim, based on a few fundamental elements

Most large, complex claims have a few major issues that constitute the bulk of the cost. There may be one or two major direct-cost issues that can be approximately quantified. If there have been significant delays, extended overhead (both jobsite and home office) can be easily quantified and the delay is simply the time from the original scheduled completion date to the actual date. Proving inefficiency will be the most difficult, and you may want to submit a claim for overtime premium costs plus a percentage of the impacted labor instead of performing a detailed analysis as described in Chap. 12.

When computing labor inefficiency, segregate the labor hours or costs by major cost code or time frame impacted, and apply different inefficiency factors to each depending upon the extent of impact. This is more realistic than a single percentage for all labor, and more likely to be accepted. To develop the percentage, use the standard charts for overtime inefficiency, rules of thumb for other impact, or the "judgment" of the claim preparer. If possible, provide quantified or anecdotal data to support your assertions. The "measured mile" and other approaches, as described in Chap. 12, may be applicable.

### 5. Identify the issues for which entitlement is acknowledged

Before preparing a simplified claim, agree to the facts that you both accept (the stipulated facts). Then determine what claim issues the claim reviewer believes may have entitlement. Focus your claim on those issues.

When preparing a detailed claim, there is little choice of what theory of recovery applies to each issue, as the facts will speak for themselves. Although sustainable in court or arbitration, that theory may be difficult to sell to the claim reviewer in negotiation (e.g., defective plans by the designer, interference by the inspector, or general contractor delay to the subcontractor) in negotiation. However, causation is less well defined for a simplified claim and several alternative theories of recovery may appear applicable.

As an example, inadequate records and a pending deadline for settlement of all issues forced us to submit a simplified claim for our client, the earthwork subcontractor on a major industrial project. Our preliminary analysis of the subcontractor's limited records indicated that the general contractor or other subcontractors may have been responsible for some (or all) of the delays, which forced our client to work through the winter. However, our simplified claim focused on damages, avoided entitlement, and described the issues as "wet weather," since the ultimate impacts were due to wet weather. The general contractor assumed responsibility for proving entitlement for this and other issues.

### 6. Build on established facts

Use perceived facts to prove a point, as long as they not misleading. Overlook minor discrepancies and don't attempt to change unimportant perceptions. If necessary, adjust for easily demonstrated changes. This avoids introducing too many new facts, which may confuse or provoke challenges.

### 7. Focus on your strengths

If entitlement is strong but damages are poorly documented, focus the presentation on entitlement. Conversely, emphasize damages if entitlement is weak. Obtain full acknowledgment of your strongest position, to the point of overkill, before moving on to the weaker argument. Then cover the weaker points quickly, returning to your strongest arguments as soon as possible.

For example, a mechanical subcontractor lost a great sum of money on an institutional project, owing to delays by another subcontractor. They had been unable to collect damages from the general contractor and were considering a $50,000 settlement. We obtained detailed scheduling data from the general contractor, which allowed us to prove that the delays were clearly the fault of other subcontractors and that the general contractor made inadequate efforts to accelerate them. Once entitlement and the amount of delay were established, proof of extended overhead costs was easy. By the time we had finished presenting the data, our rather weak proof of the amount of inefficiency damages was overlooked, and our client collected $1.4 million.

## 8. Use graphics and illustrative examples to convince

Nothing is more powerful than visualization and recollection of personal experience to give an appreciation of the facts. Describe framing on a wet, slippery roof, in the rain and wind, with a heavy rain slicker impending movement, fingers numb from the cold, and water running down your arm. Create an image in the reviewer's mind and a 20 percent inefficiency factor becomes not only believable but conservative.

Graphics can be equally persuasive. A banded bar chart, described in Chap. 10, was the deciding factor in more than one major dispute. Histograms of planned versus actual crew size are another effective graphic for labor inefficiency from crew stacking and weather impact. Percent complete curves clearly show reduced productivity, delay, and recovery. Comparison bar charts demonstrate delays or acceleration and their impact on the work.

## 9. Submit a professional-looking claim document

Package the claim in a well-designed, cardstock cover, either comb-bound or in a three-ring binder. Use an attractive large-sized font and laser printing. Draft a carefully worded document that clearly explains the claim, with headers, lots of space, etc. Use a larger font and lots of space (but not excessively) to expand a limited amount of text. Include prepared exhibits and pertinent source documents. Page number the document. Assign exhibit numbers and page number the exhibits in the appendix to facilitate access. Provide a sufficient number of copies.

In some cases, a brief, simple document can be more effective than pages of text—if it strikes to the heart of the matter and makes the essential points. A poorly written, typed document with hard-to-read copies of exhibits that is stapled in the corner will not engender respect. Obvious errors will detract from your credibility.

## 10. Be willing to compromise

It is better to compromise and reach a reasonable agreement than fail and be forced to prepare a detailed analysis or submit a claim through the disputes process. One exception is when you are forced to create a precedent to show that you will not be bullied into unfair settlements. Be aware, however, of the costs involved.

Contractors submitting a simplified claim often accept a less than optimal settlement. That may well be the best choice given the investment required for a detailed analysis, the risk of ultimate rejection, and the benefits of moving on to new projects.

### 11. If unsuccessful, prepare a detailed claim

When presenting a simplified claim, verbally identify it as such (but not on the document). This will make it easier to replace with a more detailed claim using a different approach if the simplified claim is rejected.

One of the most contentious issues on a claim submitted to arbitration several years ago was the original claim prepared by another expert. It inadvertently supported one of the owner's major defenses and had to be refuted before our detailed analysis could be accepted. Fortunately, we were able to convince the arbitrators that it was only a preliminary effort, sufficient for negotiation but without benefit of a detailed analysis which would have established all of the facts.

See Chap. 9 for details on preparing a detailed claim.

# 8

# Phase 3—Obtaining and Organizing Documents for Review

Obtaining and organizing the facts is an important task, whether preparing a change order request while construction is in progress or a claim long after the work is completed. All pertinent documents and other information must be reviewed in a logical order. Nonessential documents should be excluded, in order to limit the amount of information to be evaluated. This chapter describes how to do this in the following sections:

The process of obtaining and organizing documents is as described in Fig. 8.1.

## A. Obtaining and Copying Documents

Before reviewing the documents, understand the project scope and the key issues in dispute, develop a strategy, and know what to look for in your review. Otherwise you may overlook essential information and end up repeating the document review or failing to identify important facts.

This section describes all of the documents needed for a major claim. In most cases, many of these documents are not relevant and need not be reviewed. Whenever possible, claim preparers should limit their document collection and review to only those documents needed.

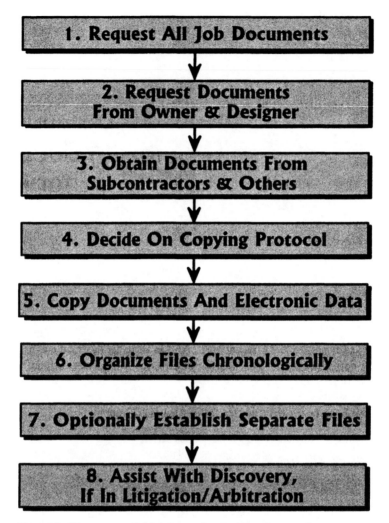

**Figure 8.1** The process of obtaining and organizing documents.

### 1.  Obtaining all job documents

The project team or contractor's home office staff should organize the contractor's documents for review, verify their completeness, put correspondence in chronological order, and do anything else to facilitate the review. *All* relevant documents should be copied or made available for review. Too frequently, individuals within the contractor's organization don't provide essential information—such as their "personal" diary, computer files, photographs, or stored documents.

Items such as vendor invoices and delivery tickets generally are not provided, except for special cases identified by the expert. For example, delivery tickets from a certain date may be needed to confirm that the structural steel was delivered that day. Or a specific supplier's invoice may be needed to determine when materials were available.

When obtaining project records, check for and remove any privileged documents (e.g., communication with legal counsel). Otherwise they may become discoverable.

**Track and verify all documents are provided.** *Insist on receiving all documents.* Do not let the project team or home office staff decide what is important. If working with legal counsel, do not let a paralegal or attorney select the documents to be reviewed and do not limit review to those documents that opposing counsel has copied in discovery. Neither they nor you know what is important until the documents are reviewed. In some cases, you won't realize a document is important until later during the analysis. However, you will probably remember reading something about the issue and the approximate date of the document so that it can be found.

Missing documents (when the project team doesn't provide everything) is one of the biggest constraints on efficient claim preparation. Either the analysis is completed without all the information or the material must be re-reviewed and the analysis changed after the additional information is received. To avoid this, do the following.

- Confirm the request for documents in writing. Provide a list of the documents requested, highlighting the most important or those needed first, and specifying the date needed.

- Follow up if documents are not provided as promised. Obtain another promised date and track the new date.

- Inventory the documents upon receipt to confirm that all documents requested were provided. If any are missing, request them immediately.

- Clearly inform the claim support team, in writing, if a lack of documents hinders the review and analysis. Note the delay and additional cost and request early resolution or alternate instructions on how to proceed.

**List of documents to request.** Unless certain they won't be needed, request all of the following categories of documents:

---

**DOCUMENT CATEGORIES**
1. Pre-Bid Documents
2. Estimate and Bidding Documents
3. Other Pre-Construction Documents
4. Contract Documents
5. Clarifications and Changes Documents
6. Payment
7. Correspondence and Administrative Documents
8. Meeting Records
9. Shop Drawings and Other Submittals
10. Schedules

---

*(Continued)*

11. Cost Accounting Records
12. Field Records
13. Other Contractor Records
14. Subcontractor Records
15. Vendor Records
16. Records Received from the Owner or Designer
17. Computer Records

**1. Prebid documents.** These include any documents prepared prior to bid preparation.

- Newspaper clippings and correspondence about the pending project.
- Documents regarding its funding and design.
- Engineering and financial study reports—probably obtained from the project owner.

**2. Estimate and bidding documents.** The bidding documents describe the contractor's planned jobsite overhead and project duration, as well as the quantities of work and detailed estimate of direct costs for each cost code item.

- Request for proposals or qualifications, if a negotiated contract, or invitation to bidders, if publicly bid.
- Dodge reports, Brown's letters, etc.
- Minutes and contractor notes of prebid meetings.
- Prebid site inspection and findings report.
- Geotechnical reports
- Bid estimate documents including requests for subcontractor quotes, subcontractor and vendor quotes, quantity takeoffs, detailed pricing sheets, summaries, the bid summary, markup computations, etc. (On some contracts, these are placed in escrow for possible use by the owner in disputes.)
- Record of M/WBE contacts, if required by the bid documents.
- Bid form if publicly bid, or proposal if negotiated contract.
- The bid schedule, if prepared, with computations of jobsite overhead costs.
- Bid tabulations by the owner listing all contractor bids, if publicly bid.

**3. Other preconstruction documents.** Always request the preconstruction meeting information, including the minutes and handwritten notes by the participants.

- Correspondence and documents related to bond, insurance, award, and other preaward issues.
- Correspondence and other documents regarding mobilization.

- Minutes and notes of preconstruction meetings.
- Permit applications, permits, and related correspondence.

**4. Contract documents.** The plans, specifications, and contract are always needed, along with subcontracts for the disputed work.

- Plans, specifications, and the contract, with all addenda and approved change orders.
- Superseded contract documents—drawing revisions, etc.
- Referenced industry standard specifications and codes.
- All subcontracts and purchase orders for issues in dispute.
- Copies of approved change orders (see item 5).
- Design clarifications, sketches, and supplementary information from the designer.
- Drawing issuance logs, if provided by the designer.

**5. Clarifications, notices of change, change order requests, claims, and change orders.** If changes are an issue, obtain all related information including sketches, revised drawings, etc.

- All requests for information (RFIs) with attached documents and the RFI log.
- Notices of change, requests for proposals, work orders (force account work), etc. (one file folder each), with logs.
- Change order requests (one file folder each), with all referenced documents and owner responses and a COR log.
- Claim files with all documents related to each claim (one file folder each) and a claim log.
- Change order files with all documents related to each change order (one file folder each) and a change order log. Include estimates, record of negotiation, agreements, etc.

**6. Payment records.** The progress payment requests describe the work accomplished at the end of each month and can be used to determine progress after adjusting for overbilling and disputes over work installed.

- Schedule of values for lump sum contracts or bid schedule for unit price contracts.
- Progress payment requests, with supporting calculations and records including lien release forms.
- Records of when payments were actually requested and received, or a log.
- Cash flow forecasts and actual cash flow chart.
- Reconciliation of revenues and invoices and payments to subcontractors and vendors with retainage.

- Computed approved rates for labor, labor fringes, equipment use, jobsite and home office overhead markups, and extended jobsite and home office overhead rates.

**7.  Correspondence and administrative documents.**  Request all correspondence and administrative documents. They should be placed in a single chronological file with the oldest on top, in order to facilitate review.

- Correspondence and other records of communication (i.e., faxes, speed memos, memorandums, etc.) with correspondence logs.
- Telephone conversation notes.
- E-mail transcripts and backups from job computers.
- Transmittal forms, if not included in submittal files.
- Weekly and monthly reports to company management.

**8.  Meeting records.**  Request records of all meetings with the owner, subcontractors, or others.

- Weekly progress meeting agendas, minutes, handwritten notes, tape recordings of meetings, etc.
- Periodic meetings with the owner's representative.
- Special meetings with any party.

**9.  Shop drawings and other submittals.**  Submittals are needed only when specific issues are identified.

- Submittal logs.
- Copies of submittals, including transmittal forms and the attachments, if requested. Copy only the shop drawings and equipment catalogs specifically requested. Or make the originals available for review when needed. (Obtain the copy reviewed, stamped and returned by the designer, not the copy retained by the contractor when the drawings were submitted.)

**10.  Schedules and progress reports.**  The most critical documents for a scheduling claim are the original (as-planned) schedule and the last update. Interim schedule updates, short interval schedules, and progress reports are important if needed to ascertain status at a particular time. Draft schedules that were prepared but not submitted might be helpful in proving intent.

- The original schedule and all updates.
- Short interval (look ahead) schedules and any schedule to completion.
- Weekly or monthly progress reports from the project team, the owner, or designer.

- Subcontractor schedules, if available.
- All fragnets and what-if analyses.

**11.  Cost accounting records.**  The most important cost document is the most recent ledger report showing the budgeted and actual cost for each cost code, by category (labor, equipment, materials, subcontract, etc.).

Before requesting specific data, determine the reports the accounting software generates and how to request specific reports with the desired sorts and other features. Have reports sent to disk, in addition to being printed. The resulting ASCII file should be provided on diskette so that it can be imported to other software for analysis.

- Budgeting information on how the estimate was converted into a chart of accounts for job cost accounting.
- Weekly labor cost reports, especially if doing a measured mile inefficiency estimate.
- Monthly job cost summaries and other reports, including the final report.
- General ledger reports—with the primary sort and subtotal three different ways: by vendor, by cost code, and by cost category.
- Cost ledgers.
- Balance sheets.
- Annual financial reports, monthly and quarterly profit and loss statements, cost-to-complete analyses, etc.
- Overhead information necessary to develop home office overhead markup and extended home office overhead costs for delays—detailed G&A (general and administrative) expenses for the corporation and subsidiary, if involved, plus direct job costs and revenues for the year or years of the dispute.
- Payroll registers
- Certified payroll reports, if needed.
- Subcontractor and vendor invoices, if needed.
- Time sheets, especially if annotated.
- Special audits or cost studies.
- Cost-to-complete analyses.
- Internal auditor's reports and work papers.
- Financial statements and tax returns, if needed.
- Income analyses.
- Materials and equipment logs.
- Job financial reports.
- Invoices.
- All other accounting reports relevant to the project.

**12. Field records.** All daily reports by field personnel should be provided.

- Daily diaries by all project personnel.
- Supervisor's and superintendent's reports.
- Subcontractor reports.
- Designer and owner representative reports, if available.
- Inspection and test reports—concrete placement inspections or cylinder breaks (to determine pour dates and when steel can be set or the next floor placed), soil tests (for when ready to form and pour footings), etc.
- Field survey notes, if needed.
- Delivery tickets, if needed.

**13. Other contractor records.** All other available contractor records should be requested.

- Photographs or videotapes. (Ask for color xerox copies if the originals aren't available.)
- News clippings about the project.
- Weather data.
- Safety plan, safety reports, accident reports, etc. (They may indicate what the crews were doing, when work was complete, owner occupancy or startup impacts, etc.)
- Union agreement and labor relations files.
- Quality assurance and quality control plans and records.

**14. Subcontract files and subcontractors' records.** Contractors may have records from their subcontractors and vendors. These should be requested.

- Subcontracts and all records of quotes and negotiations with all proposers.
- Correspondence.
- Other files, similar to those above.

**15. Purchase orders and material supply records.** This would include all documents related to each purchase order—in alphabetical order by vendor plus either a separate file or cross reference list by purchase order number.

- Purchase orders.
- Release orders.
- Change orders.
- Submittals and approvals.

- Correspondence and other communication.
- Packing slips.
- Delivery tickets and receiving reports.
- Invoices, statements, and records of payment.
- All other documents related to the material, including those from the designer and owner's representative.

**16.  Records received from the owner or designer.**  In many cases, the contractor will have copies of some owner records.

- Clerk of the work's, resident engineer's, and inspectors' diaries and daily reports.
- Inspection and test reports.
- Correspondence between the owner and the designer or third parties.
- Memorandums and internal documents of the owner and designer.
- For a subcontractor, correspondence between the owner and general contractor.

**17.  Computer records.**  Request, as a minimum, a printout of all computer directories. Review the file names with the user of the computer to determine what the documents are, and copy those that appear to be important.

## 2.  Requesting documents from the owner and designer

If relations with the owner and designer are good, request pertinent files such as inspectors' reports and test reports. Do not be overwhelming, and avoid triggering a negative reaction or a refusal to provide other documents by asking for sensitive documents such as daily diaries.

## 3.  Obtaining documents from subcontractors and others

Check with subcontractors for additional documents to help establish the facts of the claim. They will usually cooperate, especially if they are preparing a pass-through claim. Unfortunately, their records are often incomplete.

Neighborhood residents adjacent or near the project site may have important information. On a claim in southern Oregon, a neighbor videotaped nearly all sitework on a daily basis, which helped establish what happened. Inspections and regulatory agencies reports can also provide needed information.

## 4.  Copying all, or only selected, documents

Strive to obtain and review all *pertinent* client records of the project, but avoid the large mass of irrelevant data. The two approaches to achieving that goal are (1) copying all files for later review (except obvious duplicates, vendor invoices, shop drawings, etc.) or (2) reviewing the files and marking selected pages or sections for copying.

**Approach 1—copy all documents.**  If copying *all* documents (except obviously unneeded items) before starting the analysis, clerical staff can reorganize the files chronologically. Then review, mark up documents, and identify exhibits in one pass through the files.

Generally purchase orders, catalog cuts, vendor invoices, and nonessential submittal files will not be copied. Record the general extent of such files, obtain any logs or summaries of those files, and request them later if they are needed. If in discovery, ask counsel to obtain an agreement with the opposing attorney to allow reviewing and copying of additional owner files not reviewed earlier, if you need specific documents.

**Approach 2—copy selected documents.**  If copying only selected documents, skim the files and mark only selected documents for copying. This saves copying costs and reduces the number of documents to be examined but prevents highlighting sections of interest on the original while skimming the documents. Documents cannot be tabbed for making a second copy for an issue file, and summary notes cannot be entered in the narrative text. Consequently you will have to wait until copies are made to re-review the pages tabbed, identify which file they belonged to and why they were tabbed, highlight the text of interest, enter summary notes in the narrative text file, and tab those pages to be copied for the issue files.

Copy all the relevant files the first time around because information reviewed later may force a change in your approach. Although documents may need to be added after obtaining the owner's records in discovery, this shouldn't happen with the contractor's files and files from subcontractors.

**If additional documents are found later.**  If additional documents are found after reviewing the files, have a clerk determine which are new, and either tab and insert them at the proper location in the chronological file or put them in a separate file for review. Then you need only review the new documents, highlight important text, mark those to be copied, update the narrative text, etc.

**Recommendation—copy all documents.**  It is usually more economical to copy the entire file—except for obvious duplicates and specified files. This is especially true if the documents are not available for a second review, if they are badly organized, or if there are potential issues that are not yet identified. To help compensate for the *apparent* waste of resources (but saving of labor), segregate and use duplicate documents for the issue files and recycle all unneeded papers.

## 5.  Marking documents to copy

When marking pages for copying, mark single pages with Post-it note tabs at the top or side, using a different color than existing tabs. Paper clip multiple pages to be copied and tab the first page of the group. Tab the front of folders which are to be entirely copied. Be liberal. Mark anything that might be useful. Copying a thousand pages at $0.08 per page is less costly than spending $150 an hour for an expert to return and search the files for something that wasn't copied during the first review.

## 6.  Copying documents

The following copying protocols will make your document review easier and more effective:

- Check for clarity and make documents darker or lighter if necessary.
- Reduce legal size documents to $8\frac{1}{2} \times 11$ inches. Reduce computer paper ($11 \times 14$ inches) and $11 \times 17$ inch paper to $8\frac{1}{2} \times 11$ inches if the result is still readable.
- Reduce sheets with notes close to the edge to 95 percent, to avoid losing information on the original.
- Paper clip and/or staple as per the original, use colored divider sheets to signify folders, and write the name from the folder label on the divider.
- Keep files in original order as much as possible, so undated material stays in approximate chronological sequence.
- Copy large drawings onto paper with dimensions in multiples of $8\frac{1}{2} \times 11$ inches for easier folding to that size. If the original is slightly too large, trim if possible without losing information. If slightly too small, copy onto larger paper.
- If a fax transmittal sheet is attached to a letter or other transmittal, staple it behind the letter so you can read the document without pulling back the transmittal sheet. The fax transmittal sheet is needed only to confirm the date, time, and to whom it was faxed, along with any comments. Only if the comment on the transmittal is the important information should the transmittal be on top. The expert can determine that and restaple it.
- When copying several photographs on one sheet, ensure that they are all orientated correctly. Try grouping horizontal shots with horizontal shots and vertical shots with vertical shots when possible, as it makes them easier to read. Sort photos by chronological order, issue, or work area.

Files from discovery will be stamped with a "Bates number," which is a letter identifying the party providing the document and a sequential number for reference. Bates numbers ensure that both sides are discussing the same document, since minor variations of a document or different annotations may alter its significance and meaning. If possible, use a Bates stamped document from the owner in lieu of an unstamped copy from the contractor.

When having copies made, specifically request that a named supervisor of the copy person confirm the copy quality for legibility and completeness. Otherwise some of the copies will be illegible.

### 7. Recording and processing printed information

In addition to using source documents and files generated by others, you will be generating and modifying information in printed or computerized form. Be efficient to avoid wasting time and money, and be effective to ensure the information is discovered and fully utilized in the analysis.

**Handwriting.** Handwriting on paper is the common means of recording information, but it is slow, difficult to read, hard to modify, and must be input to a computer for processing. The most effective means of recording information is typing it into a computer or using voice-recognition software. There are exceptions, including:

- Pencil and paper (plus a camera, tape recorder, or videotape) are more convenient for jobsite visits.

- Photographs are good for recording the status of the project or work area and the physical appearance of items in dispute. Basic photography skills; a flash and wide-angle lens are useful.

- Tape recording can be useful if a large amount of spoken information must be recorded and an exact transcript is needed. However, it must be transcribed and is more usable if you insert edit comments such as a change in topics, an explanation of who is speaking, etc. Summary notes by a person familiar with the topics discussed, who is able to condense the discussion, is more effective and less expensive.

- Videotaping is the best means of recording ongoing action and visual/audio information. Although inconvenient to review, videotapes can be very effective exhibits for understanding spatial relationships, crowded or other difficult working conditions, work flow, and patterns of activity.

**Scanners.** Scanners can save a great deal of time when importing data. The critical function of scanners is the optical character reading (OCR) software that converts images of letters and numbers to computer-recognizable characters. On very large claims, it may be economical to scan all relevant documents and "verify" them so that computer-generated copies would be acceptable as evidence. Be alert for what is practical and economical.

### 8. Copying electronic data

Obtain pertinent files from the jobsite computers and the "personal" computer files from project team members. Ask for a printout of all file directories, to determine what might be available but forgotten. If on good terms with the owner and designer (i.e., partnering the project), or if in discovery, obtain needed files from the owner and designer on diskettes.

**What types of data and what format.** During review of the documents, note which are computer-generated. Request them and all similar files and ask for others that might be available. Insist that every project team member be asked what files they have.

Ask for all data normally on computers such as project schedules on scheduling software, change logs on spreadsheet or word processing software, cost analysis data on spreadsheets, cost accounting data on the company's accounting software, correspondence and internal memorandums or reports on word processing software, and a variety of other data on database management software.

Request computer data in its original format, if you have or can obtain the application software. If not, obtain it as an ASCII file, either in an export format or as a report sent to a disk file, so that it can be imported into other software for review and processing. Reports sent to a disk file should be one line per record with fixed column widths, as multiline reports are difficult to edit. Single-line reports can be edited with a word processor to remove page breaks, column headers, and blank lines before importing to a spreadsheet.

**E-mail.** Most E-mail is read and discarded, with no physical copy maintained. In many cases, however, a printed transcript is made and copies may be present on individual's computers or the network server. Request all transcripts and copies that are available on the hard disks.

Many public agencies are required by law or ordinance to create and maintain transcripts of all E-mail, that can be obtained through discovery or a Freedom of Information Act request. If transcripts aren't made, copies of the E-mail may reside on the server backup tapes.

**Scheduling data.** The three most commonly used construction scheduling programs are Primavera or Suretrak (for larger projects) and Microsoft Project (for smaller projects). Request the original schedule and all updates along with any supporting files. Determine what version of the software was used to generate the data, as different versions of the software may not be able to read data generated by later versions of the same software.

Primavera and other major scheduling software packages can generate an ASCII file in a special format specified by the U.S. Army Corps of Engineers, which can be imported to other scheduling software for processing. For more information on the Standard Data Exchange Format (SDEF) for CPM scheduling, contact the Corps' Construction Engineering Research Laboratory (CERL) in Champaign, IL (call 1-800-USA-CERL). Ask about the Guide Specification 01310 (Dec94), "Project Schedule" and the SDEF Technical Specifications. In addition, several programs can interface with Microsoft Project via their export file structure. Be aware, however, that different scheduling programs have minor variations in the way they process data and under certain circumstances will produce slightly different results.

Pinnell/Busch has developed scheduling software, PMS80, with unique features for analyzing claims. It is the only program with a true timescale arrow diagram and has the ELIPSE schedule plus a number of other features described in this book.

**Accounting data.** Cost accounting data will be available on either spreadsheets or through the company's accounting software. If available on diskette, it can be imported into scheduling software to graphically display daily labor hours by cost code, or into spreadsheets for analysis, reorganization, and graphing. Chapter 10 describes that scheduling software feature and how it functions.

**Importing data from other software.** You may need to transfer information between other software programs. Fortunately, most programs have import/export routines for transfer of files in either ASCII or dBase format. If not, you can send a report to disk, which generates an ASCII file that can be imported by most programs. Alternately, use special ASCII import/export programs, or ask a computer expert to transfer files using database management software.

## B. Assisting with Discovery

Discovery is the process of obtaining relevant documents from the other party in a dispute. It is a legal process, dictated by law and court procedures, and is available only when in arbitration or litigation.

### 1. Involvement of expert in discovery

The expert should assist the contractor's staff, attorneys, and paralegals with discovery, to identify those documents that may be overlooked by a nonexpert. For example, penciled notations by the reviewing architect/engineer on the file copy of shop drawings may be crucial to proving claims of slow review or the unreasonable rejection of substitutions. Document discovery should not proceed until the expert has conducted an initial review of the contractor's files and prepared a preliminary analysis with conclusions, hypotheses, questions, and a list of issues to be analyzed.

### 2. Importance of document discovery and difficulty of obtaining all documents

In arbitration, discovery is limited to whatever the parties agree to share, unless clearly specified in the rules under which the arbitration is held or as governed by state law.

Document discovery is always important and is in some cases vital to determining what happened on the project. This is especially true if the contractor has inadequate records. Unfortunately, the owner may not have adequate records or may refuse to share them. If legal counsel is not aware of the importance of all documents, inform them before an agreement is reached on more limited discovery.

To help "shake out" those documents that "can't be found," ask for them in very specific terms. In some cases, you may ask legal counsel to request a subpoena from the arbitrator to force the other party to provide requested documents. When requesting documents without a subpoena, request them through the opposing attorney, who may be less likely to "overlook" something. Check the other party's practice on other projects to determine what records they normally keep and request them by name. Also, identify and specifically request documents missing from discovery but mentioned in the documents provided. If documents are still missing, prepare detailed, specific questions for depositions to determine the party's recordkeeping practices and to identify what documents may exist but have not been provided.

**3.   What documents to request**

The normal practice is to ask for all project documents. Often the opposing party will label this a "fishing expedition" and will not provide many documents. More frequently, they "forget" to provide some documentation. Therefore, specifically enumerate the most important documents needed and use generally accepted terminology. When requesting voluminous files of which only certain documents may be needed, request that the files be made available at a mutually agreeable time and place.

Request the same items from the owner as requested from the contractor (listed in Sec. A.1 above). In addition, request the following:

- Predesign studies and other documents related to the design.
- Reports from the owner's facility manager to the president or others regarding the project.
- All communication to and from regulatory authorities and financial institutions.
- Permits, easements, and other governmental approvals.
- Correspondence and other written communication with the designer and their subconsultants.
- All jobsite records, including inspection reports, concrete form approvals, test reports, etc.
- Diaries of onsite personnel and other personnel involved in the project.
- Copies of written field directives or clarifications.
- Shop drawings and submittals with logs.
- Change order and notice of change files and logs.
- Progress payment calculations and records of negotiation.
- Safety observations and directions.
- The project manager's after action report.
- If the owner counterclaims for lost profits, request profit and loss statements for the period allegedly impacted, revenues and expenses, sales information, etc.

Request discovery of the designer and their key subconsultants, plus the construction manager if involved.

### 4.  Freedom of Information Act requests

On public projects, the federal or state Freedom of Information Act (FOIA) allows access to most of the owner's records without going through discovery. It is especially helpful for a subcontractor wishing to obtain communications between the owner and the general contractor. However, the public agency may balk at providing the documents. Ask your attorney about making FOIA requests.

### 5.  Depositions and interrogatories

An expert should prepare deposition and interrogatory questions for legal counsel to answer the questions arising from the review and analysis. The responses may resolve issues not clearly documented or help identify the owner's defense and anticipated legal strategies.

The expert should attend depositions of important opposing expert and material witnesses, in order to suggest follow-up questions to unanticipated responses. Use a laptop computer to make a summary unofficial transcript, with notes for review by counsel and additional questions to ask opposing witnesses.

### 6.  Deposition transcripts and exhibits

To help in understanding the testimony from deposition transcripts, ask for copies of the exhibits used during the deposition. Otherwise you can't evaluate the responses.

### 7.  Integrating discovery files into your files

It is time-consuming to review the new set of files received through discovery, separate new material from material already reviewed, review the new material, and integrate your findings into the previous analysis. To utilize time more effectively:

- Have staff organize the new files chronologically.
- Have staff review both files and remove duplicates from the new file.
- Review only the new material; make notes, and revise the analysis as needed.
- Tab any new documents to be copied for the separate issue files.
- Have staff copy and post all new documents to the chronological file and issue files.

## C.  Organizing the Files

Organizing the files for more efficient review is simple but extremely important.

### 1.  Organize the files chronologically

For review, group all files by general category and sort them chronologically with the oldest on top.

**General file categories.** On small projects, a dated source document can be merged into a single chronological file. In most cases, however, the best organization is a main chronological file plus several subsidiary chronological files. The main chronological file should include the following documents:

- Letters to and from all parties.

- Telephone conversation notes.

- Handwritten notes that are dated.

- Memorandums.

- Agendas and minutes of meetings.

- Miscellaneous dated documents and documents without dates, if kept in their original sequential order, which would be roughly chronological.

On large projects, consider creating a separate chronological file for separate project phases. For example, a design file might be kept separate from the construction file. Other documents are accessed more efficiently if kept in separate chronological files, depending on the size of the single file and how frequently it is used. These documents include:

- Progress payment requests, which help determine the project status at the end of each month (when the as-built schedule is absent or unreliable). Do not rely too much on progress payments, as contractors sometimes get paid for more than they have accomplished and owner's representatives sometimes refuse to pay for some items until months after they are largely complete.

- Daily reports, which should be organized chronologically, with all reports for each day stapled together in a consistent order by field supervisor or subcontractor.

- Timecards.

- Cost accounting reports.

- Photographs.

- Schedules, including the original schedule, all updates, and short-interval schedules.

Leave invoices from subcontractors and vendors in alphabetical order by vendor/subcontractor name. Organize each invoice chronologically if you plan to review them. Copy those that establish an event date that can't otherwise be substantiated, and place them in the chronological file. Consider keeping the vendor's correspondence files separate from the main chronological file if there are special issues related to that vendor.

**Which date to use.** Many documents will be undated or indirectly dated. If faxed, use the date of the fax, unless the date of origin is most important. If handwritten information is important and dated differently than the original document, file the document under both dates. If the date of receipt is important, file it under both the date received and the date issued.

If uncertain of the date, add an approximate date (based on its location in the original files) such as Jan98, May96, Summer97, etc. Write the date on the document with a colored, erasable pencil and a distinctive style in a standard location (upper right corner).

If two documents related to the same issue have the same date, put the one that was generated first (i.e., a request for information) before the later one (the response). Place lengthy typed documents that took considerable time to write before those that were handwritten or obviously generated the same day.

**Documents with annotations.**  If some copies of a document are marked, use a marked copy for the chronological file. If multiple copies of a document exist, but with different markings, keep all that are possibly significant. Paper clip them together in the chronological file with the most important on top.

**Attachments.**  If a letter has another letter attached as a transmittal, leave it attached as it may be important to review what was transmitted at that time. Insert a copy of the attached letter in its original chronological order.

**Source.**  In some cases, knowing which party provided a specific document may be important. If a document is Bates stamped, the letter(s) preceding the number identifies the source. If not (e.g., documents obtained from the contractor or through informal discovery from the owner or designer), identify the source if important but not apparent.

**Draft documents.**  Paper clip draft copies of letters or other documents behind the final transmitted copy.

**Who should organize the files and how.**  Use an experienced secretary to organize the files chronologically, as that is a considerable cost savings over using an expert to perform a clerical task. However, if files are poorly organized and confusing, the expert should review them in the order in which they were stored by the opposing party, to determine which fragmentary and unidentified documents are related, their approximate date (if undated), and who prepared them.

When sorting documents by date, lightly mark the date with yellow highlighter to aid in inserting the next document before or after it. Or use a general file sorter (available from office supply catalogs for about $20 with 31 divisions indexed by letter, number, and month).

**2.  Optionally on large claims, set up separate files by major issue or subcontractor**

On large, complex claims, retaining continuity is difficult when reviewing a single, large chronological file. It is also time-consuming to locate a specific document in a single large chronological file. Therefore, it may be advisable to create a separate file of chronological documents for each issue or for each sub-contractor, in addition to or in lieu of the main chronological file. Do this during document review, as described in Chap. 9, instead of during file organization—unless the files have already been organized by issue or subcontractor.

**Place only critical documents in the issue files.**   Two sets of files require a lot of copying, especially when some documents refer to multiple issues. A solution is to copy only the most critical documents for the issue files and use the summary notes in your narrative text to chronologically review the documents related to that issue. This way you can refer to the actual documents in the issue file for the specifics of those critical documents.

**Or, place all documents in issue files (except for documents not relevant to any issue).**   Alternatively during the document review described in Chap. 9, place all relevant documents in separate issue files, leaving only nonessential and duplicate documents in the main chronological file. This avoids making extra copies of files, except for those documents relevant to two or more issues.

If a document refers to multiple issues, place it in one of the issue files and tab it with a note *"COPY for ____ file."* The copy person can return the first copy to the first issue file and the second copy to the second issue file, after highlighting it in the same way. During document review, the expert tabs documents for posting or copying to an issue file and places irrelevant documents in the main chronological file.

If files are organized by issue only, and you want to retrieve a document by date and who from/to, refer to the main chronology of summary notes in the narrative text to identify the issue. Then search the issue file to find the document.

## 3.  Save master copies of source documents

If in discovery and attorneys are involved, files will be kept in Bates number order and documents may sometimes be accessed by Bates number. If there are duplicate copies of a document, keep those with Bates numbers.

## 4.  Identifying documents

The recommended standard method of identifying a source document is:
TypeOfFile-WhoFrom/WhoTo-Date e.g., [L-GC/ENGR-14Aug98].
The reference is used to retrieve documents from the chronological files. Other means of identifying documents are:

- Bates stamp number—added by the attorneys in discovery.
- Exhibit number—added for hearings by each party to their set of documents, or as referenced by and attached to a document.
- Author's coding—letter number, pay request number, RFI number, etc.
- Description on the document—General Ledger Report for 18Sep94, Speed-memo reference the blockout on pour 4, etc.

## 5.  Digitizing Documents

Rapidly advancing technology has made digitizing all job records more economical, which greatly facilitates chronological sorting, coding, and review. If the records are digitized, the recommendations described above are applicable but even easier to implement.

## D.  Interviewing Project Participants and Technical Experts

Obtaining valid useful information from project personnel requires a well-grounded knowledge of the project, good people skills, and a checklist of what to do and what questions to ask.

## 1.  When to interview

Talk with project team members and key subcontractors as soon as possible to get their input and comments on key issues. You may want to wait until you are familiar with the project before interviewing some individuals, especially if they may be uncooperative or hostile. A two-step process is often best: a brief call early to introduce yourself and obtain initial input and an in-depth interview after you've done your research to ask detailed questions. If delaying too long, however, especially when arbitration or litigation hearings are imminent, the opposing attorney may contact employees of other firms. If so, they may become cautious or unwilling to talk.

Start with the project manager and project superintendent, but also speak with the project engineer, project scheduler, key foremen, and key subcontractors. If available and willing to talk, speak with the owner's representative, designer, and others—but only after you are more familiar with the dispute.

If the claim is in arbitration or litigation, check with counsel before attempting to interview a potentially hostile witness or ex-employee of a party to the dispute. Do not contact employees of the owner or designer unless authorized by legal counsel.

## 2.  The process

The interview process varies depending on whether the witness is friendly, cautiously neutral, or unfriendly. Most unfriendly witnesses won't consent to be interviewed outside of the deposition process. In general, the recommended process of interviewing witnesses is as follows.

**Preparation.** Being well prepared saves time, avoids forgetting important questions, and encourages a more cooperative effort.

- Plan for the interview. Research the witness's experience, attitude, position, interests, and sensitivities. Develop your questions based on your prior research of the claim.
- Develop questions ahead of time and organize them in a logical, useful sequence. They should be:

- Nonthreatening.
- Based on adequate prior research.
- Organized for a logical flow of questions.
- Pertinent, with a need for a response, even if it confirms an unfavorable opinion or fact.
- Within the witness's personal knowledge or expertise.
- Not likely to tip off the opposing party to your strategy or to an important fact they may not know or may not know that you know.

- Schedule the interview at a mutually convenient time, at a comfortable location for the interviewee.

**Initiation.**   How the interview begins sets the tone for the entire session.

- Introduce yourself and briefly explain your role in the dispute and the need to obtain information. Asking for help while expressing your sincere appreciation will go a long way toward obtaining cooperation.

- Confirm their full name, address, phone number, how to contact them again if needed, role on the job and title, dates on job, current employment situation, etc. Such mundane routine will put the witness at ease and provide background information that may be needed.

- Minimize extensive note taking at the beginning of the interview, especially if it makes the witness uncomfortable. Do not ask to tape the conversation unless certain it won't cause concern. If the witness is comfortable with it, use a laptop computer to take notes.

- Listen carefully and use listening, paraphrasing, and affirmation skills to encourage further discussion beyond the specific questions.

- Establish rapport and encourage them to relax. Find some mutual interest, regarding either work or personal life.

- Try to convince the witness that your goal is to simply identify the truth, which should help all parties, including the witness and his or her company.

**Questioning.**   It takes skill and practice to successfully question a witness. The following guidelines will help you gain this skill.

- Start with fairly simple noncontroversial questions, whose answers you generally know, to establish a pattern of cooperation and to confirm their veracity.

- Carefully frame questions to obtain the needed information. If the answer doesn't fully respond to the question, ask amplifying questions.

- Encourage the witness to continue with a monolog, if pertinent. Gently bring them back on subject if they stray too far.

- When asking critical questions that may elicit self-serving answers, ask test questions whose answers you know. Develop verification/cross-check questions to confirm questionable statements.

- If side issues develop, either pursue them immediately or make a note and return to them later.

- Emphasize the importance of being objective. Don't lead the witness so as to influence his or her recollection of the facts, but guide the discussion to the issues you want discussed. In some cases you may suggest a response. If they agree with you, carefully challenge them to confirm that they are secure in that position.

- If necessary, agree to confidentiality regarding certain issues, but only if you can obtain the information independently once you are aware of its existence. If there is no other way to get needed background information, if you are certain you won't need their testimony, and if your attorney agrees, assure them they won't have to testify.

- Avoid arguments. If necessary, try to convince them of a contrary opinion or fact, but use caution and be certain you want to change their mind. It may be advantageous to prove an unfriendly witness wrong in hearings.

- Don't disclose sensitive information, as the witness may reveal it to the other side. If you need to divulge such information, explain why it is confidential and obtain a commitment not to reveal it. Give friendly witnesses enough information to make them feel a part of the team. Be careful in answering their questions; remember who is being interviewed.

**Wrap up the interview.**   How the interview is closed often determines how the witness will respond to future requests for information and may affect their testimony. Conclude your interview by following these guidelines:

- Summarize the key points to confirm their statements, especially if they will testify.

- Encourage the witness to add helpful information.

- Thank them, acknowledge their assistance (even if not especially helpful), and request that they call you if they remember other useful information. Give them your number, and ask if you can call them with brief questions later.

- If in litigation or arbitration, tell them the other party may contact them and ask them how they will respond.

- If further interviews are needed, schedule them.

- If they are expected to testify, notify them of the planned hearing dates so that they can schedule the time.

**Follow up questions.**   Once a rapport is established, it is easy to call and ask another question about the project. But always prepare your questions in advance to avoid wasting time.

### 3.   Project team review and brainstorming sessions

In some cases, it is more effective to discuss some claim issues with all project team members who are familiar with the issue—the project manager, superintendent, project engineer, office manager, foremen, and subcontractors' supervisors. This is especially true if needing to reconstruct the origin and impacts of multiple small changes.

**Identify and confirm who is knowledgeable about an issue.**   Find out who may have knowledge about a specific issue. This normally includes the project manager, field supervisors or the subcontractor's superintendent. Briefly talk with them to confirm the extent of their knowledge about the issue, reassure them there will be no negative consequences, and motivate them to remember what happened.

**Prepare for the review session.**   Schedule the review session at a convenient time, when the team members aren't too busy. This can be on a Saturday or evening, or during the day if work is going well. Provide a pleasant work space with enough space, comfortable chairs, a table that everyone can use, a whiteboard, coffee, cold drinks, etc.

Before the meeting, prepare and distribute an agenda and a description of each issue that identifies the key facts and items in question. Arrange for a good tape recorder and someone to take notes and to transcribe the tape recording supplemented with their notes.

**Conduct the session.**   Open the session by expressing your appreciation for their attendance and understanding of the inconvenience. Ask what can be done to facilitate their attendance. Assure them the information is sought only to prepare a claim and will not reflect on anyone's capabilities. Emphasize that they must be accurate, as exaggeration will end up causing you to waste time and pursue an issue that will eventually be rejected. Then follow the agenda, addressing each issue in order.

**Write up the record and confirm it.**   When done, have a transcript prepared and prepare an issue analysis based on it. Distribute the written analysis to everyone who attended and ask them to correct any errors and add any additional information. Briefly meet with them and discuss it to confirm what was recorded, to expand it with additional facts, and to identify any additional issues. The revised issue analysis can then be incorporated in the claim analysis.

## 4. Other experts

No matter how experienced, you will encounter unfamiliar types of construction and will need to rely on others for knowledge of special techniques and materials. Asking advice from an expert in the area will avoid overlooking significant issues or making wrong assumptions. Engage a highly qualified expert to do research and to testify on critical issues outside your expertise.

## E.  Visiting the Site

Site visits are always recommended and should be made as soon as possible. You get a faster and better understanding of the project and the work in dispute.

It is vital to visit the site and view the project before testifying in arbitration or litigation. Failure to do so will damage your credibility if opposing counsel asks whether you have visited the site.

## 1. Preparation

To be most effective, become familiar with the project and key issues before the site visit. Don't delay too long in visiting the site, as you may discover items that should be considered earlier in the analysis. Prepare a checklist of items to examine, based on the narrative text questions and comments.

Ask to be accompanied by one or more project team members to point out areas of interest and answer questions. Identify onsite personnel to interview (either construction personnel, if work is ongoing, or operations and maintenance, if not). An owner's maintenance supervisor may have vital information about continuing problems and their cause, or how well the project was constructed. Often they were on the project for part of construction and know the progress and problems, even if only by hearsay. Ask the facility manager to arrange access to nonpublic areas.

Take along reduced size plans or a few key drawings and a scale. Carry a clipboard and pad for notes, a tape recorder, a flashlight, a tape measure, and a camera. Arrange for a ladder or the use of other equipment if needed. If unable to mark up the drawings, take along translucent tracing paper to overlay the plans and mark up.

## 2. Collecting additional data if work is ongoing

If the project is ongoing, you can obtain a great deal of valuable data, even if much of the work in dispute is complete.

First, review existing recordkeeping practices to confirm their adequacy and completeness in recording critical information, as described in Secs. B.3 and C of Chap. 4.

Second, have the jobsite personnel increase recordkeeping of work in dispute. Record weekly labor hours and production per cost code, so that you can compute productivity under all conditions (impacted and nonimpacted). Consider recording daily labor productivity and making time and motion studies of current production for use in analyzing inefficiency. Take more photographs, increase the level of detail in diaries, and ensure that important information is being recorded.

Third, verify that notice of change has been given and that other contractual requirements have been met.

Fourth, consider implementing a change order management program, as described in Chap. 4. It is never too late to begin.

# Phase 4a—Determining the Facts and Analyzing Entitlement

Chapter 9 is addressed to a project team member already familiar with the dispute, an in-house expert with limited prior involvement, or a consultant with no prior knowledge of the dispute. It explains the recommended process for determining and documenting the facts, analyzing entitlement, and preparing narrative text containing the results of the investigation. Scheduling analysis is covered in Chap. 10 and costs in Chaps. 11 and 12.

This chapter is divided into the following sections:

If you are reading this chapter from start to finish, break your review into related sections (A–B, C, and D–E). After completing your analysis of entitlement, you will need to finish computing damages, which partially overlaps analyzing entitlement. Computation of damages is covered in Chaps. 11 and 12. After computing damages, see Chap. 13 for suggestions on assembling the claim document and preparing final exhibits.

## A. Strategy and Perspective

Before proceeding, verify your objectives, strategy, and recovery theories as documented in the preliminary analysis. Adjust them based on the completeness and extent of the documents obtained and for any other information received since completing the preliminary analysis.

## 1.  Establish objectives for the detailed analysis

Establish clearly defined, measurable objectives and use them to guide your detailed analysis:

- Determine the facts—what actually happened.
- Briefly examine damages—to help prove entitlement and to prioritize your efforts.
- Establish entitlement—that you are entitled to payment of the extra cost (damages).
- Document your analysis—to convince the reviewer of the validity of your claim.
- Work in an economical, timely, and effective manner—to produce the best possible product.

## 2.  Vary the analysis depending on its purpose

If expecting to negotiate a settlement and avoid arbitration or a trial, the claim preparation effort can be simpler, faster, and less expensive than if expecting to arbitrate or litigate. Preparation for arbitration or litigation should be more detailed and the content should be more aggressive.

Don't skimp. It would be false economy to fail to settle in negotiations because of an incomplete analysis, and go through arbitration or litigation to get paid. Compromising with a less-than-fair settlement is also not desirable. Knowledge is power. The contractor's superior knowledge and the serious intent demonstrated by thorough preparation will go far in helping settle a claim.

## 3.  Identify the issues and conform to your legal theories of recovery

Identify all the issues involved in the dispute, group them in logical categories, and prioritize them in terms of importance, ease of investigation and analysis, probable recovery, and sequence of investigation.

After identifying the issues, identify at least one theory of recovery for each claim issue. Identifying realistic recovery theories is one of the most important steps to a successful claim. If involved, an attorney will have established tentative theories of recovery. Discuss the recovery theory with the claim reviewer to get his or her input, and concurrence, if possible. Conform the analysis to the theory the reviewer prefers, if it is valid.

When selecting one of several alternative theories of recovery for an issue, select the one that will be best received. For example, a differing site condition claim will generally be better received by a designer than defective plans and specifications. Likewise, characterizing a claim as being a result of incorrect as-builts will be preferable to characterizing it as defective plans and specifications.

When presenting a number of major claims, it helps to "package" them with the easiest-to-sell claims being included in the title, presented first, and receiving larger-than-proportionate attention in the claim. The more difficult to sell or controversial issues should be listed last and deemphasized.

### 4. Focus on your primary objective—mutual understanding

First, understand the issues. Then, clearly communicate the facts and your position to the claim reviewer. The reviewer needs to understand the work performed and the issues in dispute. Otherwise it is easy to reject your claim or demand a compromise. Once the reviewer understands the facts, present your analysis of those facts.

### 5. Document everything

One of the most important steps to analyzing a claim is to document the facts supporting your analysis and conclusions, and reference them to source documents. That allows a reviewer to check the basis of your opinions and to confirm what the documents really say. If claim preparation is put on hold and then resumed, you can pick up where you left off with little loss of effort.

## B. Using Narrative Text to Organize and Analyze Facts and Entitlement

Written documents are essential for recording, organizing, reviewing, analyzing, displaying, understanding, and communicating claims. *The recommended form of documentation is narrative text,* using a combination of software, procedures, and techniques. Narrative text is created first for the preliminary analysis, explained in Chap. 7. It is continuously revised and expanded during the analysis explained here in Chap. 9, and eventually converted into the claim document, as described in Chap. 13.

### 1. Features and benefits of narrative text

Narrative text describes the project, the claim, the documented facts of the issues in dispute, the pertinent contract clauses, and your analysis of entitlement. *Narrative text is a continuously evolving document that helps to organize the facts.* It references significant factual statements and your analysis and opinions to source documents. It documents the analysis process and communicates the facts, the analysis, and administrative items to the claim support team. Significantly, *narrative text differs from typical claim documents and procedures because it is organized differently and focuses on issue analyses.*

The use of narrative text facilitates managing support staff, supervising less experienced claim preparers, and training new claim preparers. The senior expert either prepares the narrative text for the preliminary analysis or supervises its preparation. The junior expert then completes the claim. The senior expert tracks progress and maintains control by regularly reviewing the narrative text, which is being continuously updated, and performs those portions of the analysis requiring greater expertise. Everyone works off the same document, contributing their portion.

Upon completion of the analysis, portions of the narrative text become the draft of the claim document. This avoids having to write up the findings from scratch at the last minute, as sometimes happens in a conventional claim analysis.

## 2. Organizational structure of the narrative text file

For a medium-sized to large claim, the narrative text will normally have the following outline, depending upon the results of the preliminary analysis. The text simply expands on the preliminary analysis.

---

**TABLE OF CONTENTS FOR NARRATIVE TEXT**

A. ADMINISTRATIVE MATTERS
1. Summary of Status and Direction
2. Scope of Work, Deliverables, Budget, and Schedule
3. Work Accomplished Last Period (or to date)
4. Work Planned Next Period (or to completion) with a brief work plan, budget, and schedule
5. Questions and Comments (reference those inserted in the issue analyses)
6. Recommendations
7. Action Needed, by Whom, and When
8. How to Use the Narrative Text, Definitions, Abbreviations, Miscellaneous Information

B. DRAFT FINDINGS AND CONCLUSIONS
1. Executive Summary
2. Factual Findings (including project background and history)
3. Contractual Findings (with legal questions for response or action by counsel)
4. Conclusions

C. ISSUE ANALYSES—with a subsection for each issue, each having five sub-subsections:
1. Issue 1
   a. Summary
   b. Description
   c. Analysis
   d. Conclusions
   e. Chronology of Summary Notes
2. Issue 2, etc.

D. SCHEDULE ANALYSIS

E. COST ANALYSIS

F. REVIEW NOTES—from document review, telephone conversations, meetings, and interviews

G. CHRONOLOGY OF ALL SUMMARY NOTES—optional, in addition to chronology for each issue

H. APPENDIXES—with exhibits, detailed calculations, and subcontractor pass-through claims

---

## 3. Section A of narrative text—administrative matters, questions, comments, and action lists

Section A is started at the kickoff meeting for the preliminary analysis. It is updated periodically to reflect the current status and plans, before being distributed to the claim support team. Section A contains recommendations and notes for action by the claim preparer or others, a record of work accomplished and planned, information for communication with the claim support team, and other information for convenient access. This section will be deleted when the narrative text is converted into a claim document.

Create a subsection listing the parties involved in the project. List the firms, including all relevant subcontractors and vendors, their trade or the materials they provided, telephone numbers, and key people identified by title and role, etc. This becomes very useful when calling subcontractors and others for interviews. It also helps identify the parties mentioned in the correspondence.

### 4. Section B of narrative text—draft findings and conclusions

This section contains the most important findings, reworded and summarized for easy reading. Initially, the findings are preliminary. As the analysis proceeds, they become the draft findings. Upon completion of the analysis and after editing, the findings and conclusions become the body of the claim document.

### 5. Section C of narrative text—issue analyses

One of the primary elements of narrative text is the analysis of individual issues. These analyses are discrete tasks done independently and analyze either identified claims (e.g., delay, constructive acceleration, extra work) or other issues (e.g., whether an event actually happened, possible concurrent delay by the contractor, critical clauses in the contract, etc.).

The issue analyses are an interim step in developing the claim and will be incorporated into the claim as part of the findings. The most important issue analyses may also be included as separate sections of the claim. If a particular issue has no relevance to the claim, do not include it. However, preserve unused material for possible use in rebuttal of owner defenses or counterclaims.

**Contents of issue analysis section.** Section C contains the detailed analysis of each individual issue in a separate subsection. Each subsection starts with a summary of the findings on that issue. The summary is followed by a description of the issue and why it is important, an analysis of the pertinent facts and contract terms, and factual or contractual conclusions. A chronology of the summary notes created during the document review ends the subsection.

**Issue listing, organization, and coding.** A list of disputed issues and uncertain facts is developed during the preliminary analysis and then becomes the table of contents for the Issue Analyses section of the narrative text. During the detailed analysis, expand the list of issues to include newly discovered issues. Each new issue will be addressed by a new subsection of the Issue Analysis section of the narrative text.

Develop abbreviated descriptions or codes for the issues and enter them in the chronological summary notes, if entering summary notes in section G of the narrative text. You can then selectively print the summary notes for each issue.

**Written form.** Since issue analyses are in written form and included in the narrative text, they can be shared with the claim support team. This helps ensure that everyone understands the key topics. Unlike a verbal briefing, the written issue analyses can be referred to later to refresh memories.

Your ultimate goal is to incorporate the issue analyses into the findings and the claim document. The analyses describe the issues, the recovery theories, the project chronology, how the contract and contract law support entitlement, what the linkage is between the events and damages, and how the damages are computed. They are essential to the claim and the development of your findings.

The issue analyses extensively reference source documents, statements of opinion by the expert, and comments by project team members. These references provide an audit trail from every conclusion back to documented facts in the source documents. When referencing source documents, keep in mind that it is most effective to reference the opposing party's records, if available, since they are less likely to be contested than the contractor's records.

## 6. Subsection C.xx.e of narrative text—chronological summary notes from document review

When reviewing the chronological document file, enter summary notes of the dated information in either a chronology sub-subsection for that issue in section C of the narrative text or in a chronology for all summary notes in section G. The notes include a summary or paraphrasing of the information in the source documents (and sometimes direct quotes of key items). They also include the date and a source document reference for retrieval of the document, and a code for the issue if you enter the note into section G.

An example chronological summary note is:

> 14Aug98  Formed and poured E footing slab. Had problem with last minute changes by owner's rep and had to shut down for 4 hours to make adjustments. CONC DELAY [DailyLog]

Reviewing the summary notes provides a clear understanding of the issue, as it developed, from the viewpoint of all parties. This is essential to understanding and analysis. References to source documents allow easy retrieval of the original documents for further examination or to confirm the summary. The notes include the following information:

- *Date of the event or activity.* Enter the date of the event or document at the left margin; then tab over for the body of the note. The day/month/year format (14Jan98) is easier to read, but you may use any format your software can sort. The date enables the file to be sorted chronologically and allows retrieval of the referenced source document. Use hanging indent paragraphs for easier viewing.

- *Summary of the facts relevant to an issue.* Enter a brief summary of the facts in the balance of the note. The summary is indented so that the dates appear in a separate column. Use abbreviations and paraphrase the summary for conciseness and clarity, except when a full quote is necessary. Conciseness saves space, speeds reading, and facilitates understanding.

- *Issue codes.* Optionally, if entering the notes in the common chronology (section G), establish and enter one or more codes in the summary notes for the issues referenced. The recommended method is to truncate and capitalize the first three or four letters of the key word describing the issue, so that the issue stands out when reading and the space used is minimized (e.g., *"CONC DELAY"* above). If the summary notes are entered directly into the chronology of narrative text sub-subsection for the issue it describes, you won't need to enter the issue code.

- *Source document code.* Lastly, enter a source document code in square brackets at the end of the note. Use the first one or two letters of the type of document (Letter, Fax, Memo, Telephone conversation note, Transmittal, MM for meeting minutes, etc.), followed by a hyphen, a three-character code for the originator, a slash, and a three-character code for the recipient (e.g., [L-GC/OWN]). You can readily retrieve the document from the chronological files using the code and the date of the event or using the source document date within the source document reference, if the document refers to something that happened at a different date. Except when quotes are helpful, paraphrase and clarify the original text.

### 7.   Sections D and E of narrative text—schedule analysis and cost analysis

Owing to its importance, complexity, and length, the schedule analysis is normally placed in a separate section, section D. The cost analyses can also be accomplished in a separate section (section E) or included in a subsection for each issue.

### 8.   Section F of narrative text—review notes of nondated information

Take extensive notes of meetings, telephone conversations, interviews of witnesses, while reviewing the documents. Most of this information is nondated (e.g., the plans, specifications, and contract). It doesn't fit in a chronology, but it needs to be analyzed. Enter these as review notes into section F of the narrative text. Portions of the notes that are time-related can be transferred to a chronology. The balance will be left in section F and reviewed when you analyze the issues.

An example of review notes is as follows. It references Volume I (Architectural) of the contract documents, Division G-F.1.5, which refers to markup on changes. The summary note lists the allowable percentages of markup on different cost categories. Following it is a comment in all caps by the claim preparer noting that the claim will propose actual overhead and a reasonable profit, in lieu of the specified markup.

> 3. **Specifications—Volume I—Architectural**
>
> *Division G - F.1.5—Markup*
>
> Specifies 15% on labor, 10% on equipment and materials, and 5%-10% on subcontract work under/over $2,000. C: PROPOSE ACTUAL OVERHEAD & REASONABLE PROFIT

## 9. Section G of narrative text—chronology of all summary notes (optional)

This section is needed only if you enter all summary notes in a common chronology.

## 10. Section H of narrative text—appendixes

Exhibits of important source documents and generated documents, or additional information too voluminous for the body of the text, are included in the appendixes. The appendixes also contain analyses and calculations too extensive or detailed for inclusion in the body of the claim.

## 11. Selectively printing of chronological summary notes from section G to section C

If the chronological summary notes are entered in section G of the narrative text, you can selectively print the file for each issue (using the embedded issue code), or you can post the summary notes for each issue to the chronology sub-subsection of each issue subsection in section C. The latter requires special software.

## 12. Tips on how to prepare narrative text

**Use descriptive headers and subheaders.** The standard legal recitation method of page after page of text, unrelieved by figures and headers, is too hard to read. Organize the text in hierarchical sections and subsections with headers and subheaders clearly describing the contents, using at least one header per page. Create headers that clearly describe the contents and purpose of the section.

**References to source documents.** Reference the narrative text to source documents so that the reader can verify all factual statements.

**Dates and chronologies.** The recommended method of specifying dates is DaMonYr (e.g., 14Aug95). It takes the least possible space (7 characters), doesn't require translating the number of the month, and avoids the confusion of the military/European versus the U.S. system (i.e., 14/8/95 vs. 8/14/95). Hyphens between day, month, and year (14-Aug-95) makes the date longer and harder to read.

Use a type size (10), which is easily readable by most people. Put a tab at 0.50 inch and a hanging indent at 0.50 inch. Enter the date, tab over to enter the text, and the hanging indent will keep subsequent lines of text indented, leaving the dates clearly readable in their own column, as you see in the example above. If the dispute is limited to a single year, leave the year off the date if your word processing software can sort by day/month.

**Direct quotes from documents and individuals.**    Direct quotes of critical information from source documents or interviews are very convincing. Embedding quotes in the body of the text saves time and improves understanding. That way, reviewers do not have to refer back to an attached exhibit. If the quote is no more than one or two lines, important but not essential, place it in quotes within the body of a sentence. If, however, the quote is several lines long and essential to understanding, a separate subparagraph indented at both margins is needed. Delete nonessential text and show your deletion with a series of ellipsis points (periods and spaces). Insert words that are not part of the direct quote but are needed for understanding within [square brackets]. For example,

> "...stated that the contractor would have to construct the footing as he wanted it done...[regardless] of what the specifications said..."

Careful selection of the best portion of the quote, elimination of nonessential wording, and added (but clearly identified) words for clarity greatly facilitate understanding. Identify the source document so the reviewers can verify the quote. To aid review of the source document, highlight the quote on the source document as described in Sec. C.7.

**Bullets.**    When listing multiple and nonsequential items of roughly equal weight, use bullets. This leads to a much more succinct presentation than paragraph-style text. If the items are sequential, use numbers. For example:

> **DAMAGES.**
>
> The damages are computed in 5 categories:.
> - Contract balance due, held as retainage.
> - Unpaid CCDs, held due to lack of agreement on value.
> - Direct costs of constructing the attic sprinkler chase.
> - Extended jobsite and home office overhead.
> - Subcontractor claims.

**Figures: charts, graphs, and tables.**    Figures are extremely helpful in conveying a point. Keep them simple, possibly summarizing a more detailed figure from the appendixes. Graphs are preferred to tables for most items, as exact numbers often obscure the important issues and are usually less important than relative values.

Insert figures in the narrative text instead of in the appendix. If the figure is a full page, place it on the page facing the narrative. This livens up the text and communicates information in an interesting and understandable format. Figures within the body of the text should not exceed one-third of a page. They are particularly good for comparisons, as in this example:

| Original versus Revised Scheduled Dates | | | | |
|---|---|---|---|---|
| Critical Path Activities | Orig Sch | Rev'd Sch | Delay | Comment |
| Concrete footings | 15Nov | 1Dec | 15 | No delay |
| Masonry | 2Dec | 6Jan | 26 | AS=2Jan |
| Erect structural steel | 5Jan | 1Feb | 25 | Late delivery |
| Glass framing | 25Jan | 4Mar | 32 | Understaffed |

| ID | Activity Description | Duration | Start | Finish | Comments | Nov '96 | | | Dec '96 | | | | Jan '97 | | | | Feb '97 | | | | Mar '97 | | |
|---|---|---|---|---|---|---|---|---|---|---|---|---|---|---|---|---|---|---|---|---|---|---|---|
| | | | | | | 10 | 17 | 24 | 1 | 8 | 15 | 22 | 29 | 5 | 12 | 19 | 26 | 2 | 9 | 16 | 23 | 2 | 9 | 16 |
| 1 | Concrete Footings | 30d | 12/2/96 | 1/10/97 | Weather Delay | | | | | | | | | | | | | | | | | | | |
| 2 | Masonry | 30d | 1/6/97 | 2/14/97 | Actual Start=2JAN | | | | | | | | | | | | | | | | | | | |
| 3 | Erect Structural Steel | 21d | 2/3/97 | 3/3/97 | Late delivery | | | | | | | | | | | | | | | | | | | |
| 4 | Glass Framing | 10d | 3/4/97 | 3/17/97 | Understaffed | | | | | | | | | | | | | | | | | | | |

Project:
Date: 9/10/97    Revised Schedule ▭    Original Schedule ▬

Page 1

**Figure 9.1.**   Comparison bar chart of original versus revised schedule dates.

**Photographs.**   Photographs can communicate conditions and work areas better than text, especially when labels and symbols are used to highlight and explain what is shown. Although generally documenting static conditions, multiple photographs (or video clips) can demonstrate construction operations and impacts. However, they *must* be clear, as indistinct photographs provoke frustration.

**Emphasizing with bold text.**   To emphasize critical text in a chronology, make it **bold.** This helps the reviewer identify the most important commentary.

**Standardized terminology.**   To facilitate easy searching with the word processing software and to avoid confusion, establish standard descriptions, terminology, and coding. If multiple terms or codes were using during construction, select a standard and create a cross-index reference. Also create a table of individuals with their names, nicknames, company, and roles.

**Questions and comments.**   Insert questions and comments in the body of the text, as a reminder to ask a question or as a note, for instance, on how to analyze or present an item. Use all capital letters and start with a special code (e.g., "C:" or "Q:"), so that it is noticed during reviewing and can be accessed with word processing software search commands.

## 13.  Risk of discovery

One drawback of narrative text is that it may be discoverable in arbitration or litigation. Discovery is limited in arbitration and many state courts. However, if the attorney is concerned about potential discovery, consider the procedures in this section.

If you are a consultant and concerned about possible discovery, work directly for the attorney and provide your work product exclusively to the attorney to limit, or at least delay, discovery.

Before creating any written document, evaluate the risk of discovery by opposing counsel. Ask your attorney to define limits on preparing and distributing documents and mark every page of every work product "Confidential—Attorney Work Product."

There is a risk in that you may discover and record unfavorable information that could be obtained by the opposing side in discovery. Careful wording will minimize this problem.

Another risk is that earlier, incorrect opinions can be used in cross-examination to discredit your final conclusions. To minimize this risk, date all versions and clearly label them as a draft or preliminary document. When revising, update the current version, keeping only a single copy of the version on disk as a backup. Limit the distribution of draft versions of the work product and ensure that they are destroyed or returned. Discard all earlier versions in your possession, including files on computer backup disks. Communicate your findings without providing damning statements that opposing counsel could use to impeach your testimony if the document is provided in discovery.

Do not include privileged information in files taken to deposition or hearings, as opposing counsel may ask to see your files. Check the files of inexperienced material witnesses and technical experts and remove any privileged information.

Narrative text prepared after discovery won't be disclosed, which has some benefits. However, there is the risk, in court, that the judge will disallow any late-developed facts.

In some jurisdictions, communications between an expert witness and an attorney are discoverable in litigation, as are the documents the expert relies upon to form his or her opinions. This may jeopardize the confidentiality of privileged documents reviewed by the expert. Many attorneys insist that their expert not prepare written reports, and will read privileged documents to the expert instead of allowing them to personally read those documents. This complicates analysis and communication within the claim support team. Some attorneys will not provide their expert with privileged documents that contain adverse information, which often leaves the expert unprepared. The result is loss of credibility if questioned on those issues. Nor can the expert help develop a defense for those issues. In some jurisdictions on larger projects, attorneys sometimes hire two experts—a consultant whose work product is protected and an expert witness who will testify.

This is expensive and awkward and may result in misunderstanding. There are no easy solutions.

I strongly recommend developing narrative text, even with the risk of discovery and the possibility of adverse information being revealed to the opposing party in arbitration or litigation. In most cases, the use of narrative text and the procedures described will enable you to settle the dispute without arbitration and discovery. Even with discovery, however, the advantages far outweigh the risks. Although the other party may gain access to your information, you can do a better job of developing your position and are likely to recover the full cost of the changes.

## C.  Determining the Facts and Analyzing Entitlement

Analyzing a dispute and preparing a claim is a complex mental process. Thousands of facts and conclusions must be evaluated to develop a coherent explanation of what happened and why. To assimilate and understand this volume of data, break it into phases and study it one step at a time. This limits the amount of information to consider simultaneously and keeps you focused and within budget. In addition, at the end of each phase, identify dead-end approaches and redirect your efforts.

The detailed process of analysis is described as orderly and logical, but the actual process of preparing a claim may sometimes feel disorderly and unstructured. This is because the sequential steps are frequently iterative, with the most recent results changing prior findings and requiring additional review and analysis.

The steps for claim analysis are as follows:

---

### 16 STEPS TO CLAIM ANALYSIS

1. Conduct a Kickoff Meeting.
2. Use Narrative Text for Recordkeeping, Analysis, and Communication.
3. Revise the Goals and Objectives, Deliverables, and Strategy.
4. Prepare a Realistic Work Plan, Budget, and Schedule for Detailed Analysis.
5. Understand the Project, the Issues in Dispute, and Pertinent Contract Law.
6. Examine the As-Planned Schedule and Schedule Updates.
7. Review the Chronological Document Files and Summarize.
8. Establish Entitlement.
   a. Determine the Facts.
   b. Develop a Legal Theory of Recovery and Determine Liability.
   c. Prove Causation—the Linkage from Initiating Fact to Damages.
9. Complete the Analysis of Each Issue.
10. Analyze the Schedule.
11. Identify Additional Claims and Owner Counterclaims or Defenses.
12. Compute Damages.
13. Meet Periodically with the Claim Support Team to Review Status and Action.
14. Summarize Your Draft Findings and Conclusions.
15. Verify Your Results.
16. Obtain Authorization to Proceed with the Next Phase.

---

In many cases you may be able to skip some of the steps and only briefly perform others. Do only the steps needed and only to the level of detail required for a specific claim.

### 1.   Conduct a kickoff meeting for the detailed analysis

After obtaining and organizing the documents, hold another meeting. Review the current status of the dispute and the objectives, work plan, budget, and schedule to complete the work. This meeting is similar to the kickoff meeting for the preliminary analysis, which is described in Sec. B.1 of Chap. 7.

### 2.   Use narrative text to record notes, develop the analysis, and communicate with the team

Expand the narrative text created during the preliminary analysis by recording meeting notes, review comments, and chronological summary notes. Use the narrative text to communicate with the claim support team and to record miscellaneous information (such as phone numbers, subcontractors on the project, etc.) for easy retrieval.

### 3.   Revise the goals and objectives, deliverables, and strategy

Your goal is to settle the dispute for a fair price in a timely, efficient manner. The objectives should include a budget and a completion date for claim preparation, and a description of all deliverables.

The claim document is the most important deliverable. Other deliverables may include affidavits or issue analyses of specific topics, as requested by legal counsel or company management.

The strategy defines the approach for producing the deliverables and achieving the objectives. The strategy varies depending on the known facts, the contract terms and tentative recovery theories, relative strengths of the contractor and owner, personalities of the parties involved, and your assessment of the best means of achieving settlement. The strategy must address how to collect needed data, analyze the issues, develop and format the claim, present it, and negotiate resolution.

### 4.   Prepare a realistic work plan, budget, and schedule for the detailed analysis

The preliminary analysis was conducted according to a work plan, budget, and schedule. Now you must determine, and agree on, the work plan, budget, and schedule to complete the claim. Do this whether you're an employee of the construction company or a consultant.

The detailed work plan, in the form of a bar chart or timescale arrow diagram, guides you in accomplishing the objectives while maintaining the schedule and budget. If the work plan is resource-loaded with hourly labor costs and estimated expenses, it can generate the claim preparation budget and control the claims preparation effort. A typical work plan in the form of a timescale arrow diagram appears as in Fig. 9.2. The budget for preparing the claim must be realistic in terms of achieving the desired work product, yet ensure a high return on investment. Include all costs—both out-of-pocket expenses such as consultant fees and hidden costs such as management time and lost opportunities.

When budgeting time and cost for preparation of the claim document and exhibits keep in mind that most technical reports require considerable time for editing the text, generating figures, printing, and binding. The effort expended on finalizing the claim document and exhibits will be considerably less owing to the effort expended in generating the narrative text, but not too much less or the quality of the finished product may suffer.

## 5. Understand the project, the issues in dispute, and pertinent contract law

Next understand the project, the issues in dispute, and the contract. This requires reviewing additional project documents, interviewing project personnel, and visiting the site. As you do so, make summary notes in the narrative text—either nondated notes about items such as the contract in section F of the narrative text, or dated (chronological) summary notes of actual or planned progress in section C under the subsection for the issue discussed or in section G.

**Review any existing claim documents and ask questions.** Review the preliminary claim document and be prepared to defend it or to provide reasons for changing previously asserted facts, recovery theories, or damages. If that claim document has already been submitted to the owner, retain the same organization and language in your revised claim document where practical. Check your list of the issues in dispute and any unknown or uncertain facts that may affect resolution. Add to the list as new items are discovered.

**Prepare a preliminary estimate of damages and recovery for each issue.** Determine the possible damages for each issue, the probable recovery, and the difficulty of obtaining it. Prioritize your efforts and focus on those issues with the largest potential return. See Chap. 11 for details on determining damages.

**Review the contract documents.** Review the contract documents and schedule of values. Identify clauses affecting entitlement and damages, define possible recovery theories, and prepare questions for legal counsel. Review the plans. Examine the technical sections of the specifications for performance and construction methods, measurement for payment, and other requirements for the work in dispute. Examine the change orders to determine how the contract has been modified, and check the final progress payment for approved change orders and the current contract balance due.

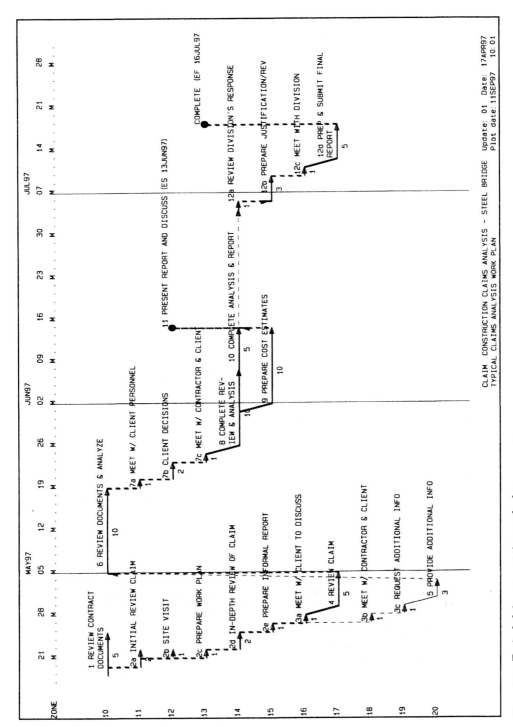

**Figure 9.2.** Typical claim preparation work plan.

229

**Update and organize the list of issues, theories of recovery, and strategy.**  Add additional issues identified during review of the contract documents. Draft an analysis of each issue along with questions on factual and contractual points. Ask the claim support team to review the list and comment. Identify the theory of recovery and damages for each issue and refine your strategy.

**Visit the site.**  Visit the site, if you have not already done so during the preliminary analysis. This provides a better understanding of the work than hours of studying the plans and specifications.

**Interview project personnel.**  Interview project personnel. They often have a very different view of the project and always have more specific information about what actually happened. By now, you should understand the dispute, have most of the necessary information for the analysis, and know what information is missing. The project personnel can often provide this information.

When interviewing field personnel, consider using them as witnesses, if in arbitration or litigation. Arbitrators and other fact finders see field personnel as being nonbiased and more knowledgeable of actual conditions and events. A down-to-earth appearance and demeanor can be an asset in some circumstances, as long as the individual is straightforward and cooperative.

**Identify technical experts who might be needed.**  Identify design experts, specialty trade experts, computer analysts, etc., who may be needed to supplement your investigation and testimony. Involve them in the investigation as required and coordinate their work.

**Ask legal counsel to review contract issues, recovery theories, and strategy.**  Ask counsel to review your questions and to comment on legal issues. If an attorney is not yet involved and there are significant legal issues, ask that one be retained.

## 6.  Examine the as-planned schedule and schedule updates

A basic understanding of the planned schedule and actual progress, as documented by the schedule updates, is crucial to understanding the project. Study and understand the planned schedule before reviewing the chronological document file. Otherwise you are likely to miss important issues. The objective at this time is not a full schedule analysis but rather a general understanding of what was planned and what was done. Later you will make a more detailed analysis as explained in Chap. 10. If it is difficult to understand the as-planned schedule, convert it to a timescale arrow diagram, as described in Chap. 10, to facilitate your review and analysis.

**Optionally, create a modified as-planned schedule.**  In some cases, the as-planned schedule will have obvious errors that would have been corrected as construction proceeded. If so, prepare a modified as-planned schedule to use in the schedule analysis described in Chap. 10 and to facilitate a computerized comparison with the as-built schedule.

**Break the project into its major phases for analysis.**    To minimize the amount of information being analyzed concurrently and to provide a convenient point for summarizing progress, *focus on one phase of the project at a time.* Base the phases on major time-related claims or overall construction. For example, building construction phases could be excavation, structural, enclosure, rough-in, finish work, and punch list. To explain the project schedule, create a summary bar chart by phase. The phases can also be included in the complete schedule as hammock activities.

**Review the schedule updates.**    Review the final schedule update to determine when and how the project was completed and if the project team recorded actual start and finish dates. Unfortunately, the actual job logic will differ from the plan, and the updates seldom include logic revisions. In addition, many projects are controlled by short interval schedules, with different tasks and logic than the master schedule. Review the sequential schedule updates to determine interim progress and logic changes as made. This will reveal the intent of the project team and what would have happened if conditions had been different.

You now have a basic understanding of how the project team expected to build the project and what generally happened. You may have discovered additional issues that also need investigation. The next steps will be to detail what actually happened (Sec. C.7) and then to analyze who is responsible (in Sec. C.8). A more detailed analysis of the schedule will be accomplished later.

## 7.   Review the chronological document files, prepare summary notes and detailed as-built

After completing the preliminary tasks, determine the facts of the dispute. Review the chronological document files, starting with a concurrent review of the main chronological file, the daily field report file, and the photographs. The schedules and cost accounting reports may be reviewed concurrently with the main chronological file or separately. The files include:

- The main chronological file.
- Daily reports, inspection reports, and other daily field records.
- Photographs and videotapes.
- The bidding records and preconstruction documents regarding award, obtaining bonding, etc.
- The contract documents—plans, specifications, and contract (if not reviewed in detail earlier).
- Progress payment requests and records of payment.

- Change documents: notices of claims, change order requests, and the change order backup files.

- Requests for information (RFIs) and associated documents.

- Submittal logs (and specifically requested shop drawings and submittals).

**Decide whether to prepare chronological summary notes, a detailed as-built schedule, or both.** Before proceeding, decide whether to make only chronological summary notes of pertinent facts regarding the issues, or to prepare a detailed as-built schedule, or both. A detailed as-built schedule is recommended for most claims involving scheduling issues. However, the cost may be difficult to justify when the claim amount is small and the exact sequence of construction is not in dispute. The cost to prepare the detailed as-built schedule is usually far less than expected.

The chronological summary notes are entered in the narrative text using word processing software. They are based on the correspondence and other dated records, but normally not on the daily construction reports. The chronological summary notes generally focus on notice and discussion of disputes rather than on the details of field operation. Chronological summary notes should always be prepared as part of the issue analyses and referred to when writing up the analysis of each issue.

The detailed as-built schedule (Fig. 9.3) is either handwritten on large drawing paper or keyed into a computer spreadsheet program with special features for displaying a calendar and activities, as described in Sec. G of Chap. 10. The detailed as-built schedule is based primarily on daily construction reports and photographs, supplemented with other records as needed (progress meetings, correspondence, etc.). It focuses on construction operations and activities and contains an enormous quantity of data that is entered one day at a time. This mass of data becomes meaningful when organized by activity and plotted to timescale. When that is done, add major milestones and critical events from the chronological summary notes or from an analysis of the schedule.

The detailed as-built schedule is referred to when writing up the analysis of those issues related to the activities or events recorded on the schedule. It is also used to analyze the schedule, after being condensed into a normal as-built schedule. It should be prepared when the schedule updates are nonexistent or inadequate, or when more detailed information is needed on delays and impacts.

Since the chronological summary notes are primarily based on correspondence files and the detailed as-built schedule is primarily based on the daily field reports, they can be prepared separately. However, it may be more efficient to do them concurrently. In either case, there will be some information on the daily field reports that should be posted as chronological summary notes to the narrative text, and some information from the correspondence that should be posted to the detailed as-built schedule. Keep in mind that the information posted from the daily field reports to the detailed as-built schedule is so voluminous and detailed that little of it will be posted to the chronological summary notes. Much of the information in the chronological summary notes is not relevant to the construction field operations.

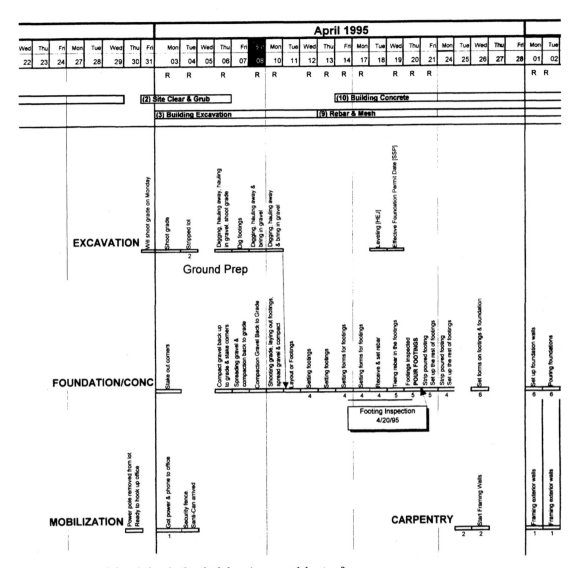

**Figure 9.3.** Typical detailed as-built schedule using spreadsheet software.

The chronological summary notes will always be prepared at this time, when reviewing the chronologically organized source files. The detailed as-built schedule, however, can be prepared either at the same time or immediately afterward. Or it can be prepared later, as described in Sec. C.10.

**Review the chronological files.** Center your review on the main chronological correspondence file, but concurrently review the other chronological files (i.e., daily field reports, payment requests, etc.) so that you have a complete picture of the job for each day, each week and month, and each major milestone. Review the daily field reports (for example) a week at a time, after reading the correspondence file for that week. Review progress payments at the end of each month after having read the correspondence up through the day it was submitted or approved.

**Highlight important text.**  When reviewing the source documents, highlight important text to facilitate locating important information later. Highlighting should be nonreproducible, as you may wish to copy the document to use as an exhibit or you may decide to emphasize entirely different parts of the text in your presentations. A yellow marker is recommended as it doesn't copy on most reproduction equipment (test it first). If making multiple copies for distribution, use a marker that leaves a noticeable shadow on copies but leaves the text readable.

If you must mark up source documents (e.g., to correct a wrong date), use an erasable colored pencil and a distinctive writing style. Date and initial the note. Then it will be recognized on both the original and copies as having been made later and can be erased if needed to be copied for use as an exhibit. Be especially careful when using working copies of source documents as exhibits to erase all notes, as they could be used to cross-examine you in arbitration or litigation.

**Tab key documents.**  Tab important documents in the chronological document file for easy access. A small note tab on the right margin, extending $1/4$ inch out from the edge, works well. If there are multiple issues, stagger the tabs down the right margin with a brief description on the top tab for each issue, so you can quickly review all documents relating to that issue. If there are more than five or six issues, group the less important ones in the top space with a brief description on each tab.

**Optionally, set up document files by issue.**  If convenient, create separate files of the important documents for each issue. Copy the documents in the chronological files that have been tabbed and place them in a separate file for each issue. The most important of these documents can be used as exhibits in an appendix to the narrative text.

**Summarize the facts as summary narrative text notes or detailed as-built schedule.**  While reviewing the chronological document file, summarize and record the relevant facts in the summary notes within the narrative text. Record daily construction operations in the detailed as-built schedule. Post relevant information between the chronological summary notes and the detailed as-built schedule.

**Benefits of chronological review.**  The most important result of the chronological review is that you learn how the project was built—as it was experienced by the project team. Reviewing all the documents concurrently gives you a better understanding of the project than the project team members—who probably had only part of the information and were too busy building the project to grasp the overall picture. If you also have the owner's records, you will have a superb grasp of the project, as you will see the project from the owner's viewpoint as well.

The second important result of the chronological review is creation of the chronological summary notes. When accurately and succinctly summarized, with key words coded to the appropriate issue(s) and dated and referenced to the source documents, these notes will be invaluable for your subsequent analysis of each issue. The act of summarizing and paraphrasing is an excellent aid to understanding and retaining facts for later recall.

The third important result of the chronological review is creation of a detailed as-built schedule. This schedule is the most powerful tool available for determining what really happened and why.

**Enter comments or questions.**  List questions or comments when creating the summary notes. Note them in capital letters (so they stand out), after the chronological note with a search code (Q: or C:) for easy retrieval. Optionally copy and post the comments and questions to the Administration Action subsection (A.5), of the narrative text, under the sub-subsection for the individual responsible for the question.

## 8.  Establish entitlement

The disputed issues must be analyzed to establish entitlement, with stated facts or opinions referenced to source documents and the analyses described in a logical manner. This requires that you:

1. Determine the facts. Describe and reference the initiating factual event, action, or condition to the supporting documents. Deduce or infer essential facts not clearly established from those available.

2. Identify the applicable contract provisions or law—the theories of recovery.

3. Establish liability. Confirm whether the proposed recovery theory appears viable, note the facts and pertinent contract clauses or law supporting your conclusions, comment on any uncertainties, and note the need for legal review by counsel.

4. Prove causation. Trace the linkage from the initiating event, action, or condition to the resulting costs incurred by the contractor.

The protocol for analyzing each issue for the detailed analysis is similar to but considerably more extensive than that used for the preliminary analysis described in Sec. B.12 of Chap. 7.

**Differences between conventional claims analysis and analysis using narrative text.**
Claim preparation as described in *HOW TO GET PAID* is a logical, integrated process. Analysis focuses on a single issue at a time, after gathering and organizing all pertinent information related to that issue. Review only relevant information, in the same order as events occurred, by reading the chronological

summary notes and examining the relevant portion of the detailed as-built schedule.

The conventional method of analyzing construction disputes doesn't use narrative text or detailed as-built schedules, nor is it based on a chronological review of the documents. It is not as effective. Claim preparers often start writing the analysis from an outline and develop the analysis from memory as they write, while considering many issues concurrently. They refer to the documents readily available but write from their recollection of material read earlier. They seldom support their statements in the text with references to source documents. Thus it is difficult to verify the facts stated or the basis of their analysis.

**Determine the facts.**    The first step in establishing entitlement is to determine the facts and to reference those facts to the supporting documents.

1. *Review the summary chronological notes for the issue.* Read the chronology for each issue, making notes as you read and optionally changing the most important text to **bold** to facilitate finding the information later.

2. *Review the summary review notes related to the issue.* Check the review notes in section F of the narrative text that are applicable to the issue.

3. *Examine the detailed as-built schedule.* Study the portion of the detailed as-built schedule relevant to the issue. This includes the activities affected, the weather, crew size, major events, etc.

4. *Eliminate irrelevant data and organize logically to reveal patterns or obscured facts.* The value of the chronological summary notes and the detailed as-built schedule is that they (1) condense the facts into concise statements, (2) suppress irrelevant information, and (3) organize and display the facts in a logical (chronological) sequence that reveals patterns and the underlying rationale for why things happened. They are essential to an effective claim analysis.

5. *Resolve problems with inadequate documentation.* When reviewing the documents, you will encounter problems, including insufficient documentation, conflicting or otherwise unreliable statements, confusing terminology, etc. Use the techniques described in Sec. D.2 to resolve these problems.

6. *Establish whether the asserted events and activities occurred.* When reviewing the files, look for confirmation or contradiction of asserted facts. Frequently the asserted events happened, but differently, or with significantly different results than alleged. When you are confirming facts, don't rely on unsupported statements by the project manager or owner's representative. Get confirming evidence, such as a superintendent's diary entry or a concrete pour report. Correspondence is often reliable only to establish when a party was aware of something or what their stated intent was at the time. A comment in the superintendent's daily report is far better evidence, but an annotated foreman's crew card is best. A corroborating comment in a subcontractor's diary or a test report would be conclusive.

7. *Determine what else happened and why.* A thorough review of the summary notes for each issue and other records will probably reveal many new facts

and some that weren't initially recognized as important. These need to be identified and investigated.

8. *Establish facts without direct documentation.* The source documents may omit critical facts. These must then be established from the available documents. You will have to determine unstated facts based on indirect evidence, industry practice, recognition of patterns, elimination of alternates, and logic. This is part of an expert's recognized expertise. Although it can be described and taught, mastery requires extensive knowledge of the industry and years of experience. This creates an almost subconscious awareness of patterns and indicators that accompany certain types of problems.

   You may have to use inference, deduction, and reasonable probability in determining the facts. Some things cannot be proved with the available data, but an expert can build a viable case based on a "significantly higher probability than the alternative." This requires a knowledge of logic and statistics, and rigorous thinking. For example, a contractor claimed costs for restricted site access. However, we found their estimator's notes from the prebid conference which recorded the statement that contractors had to approach the job "from...and then down...". Only one of the possible approaches (the restricted one) was downhill, as we pointed out and the arbitrator's award duly noted.

   In another example, the owner's attorney claimed that a handwritten, undated note on a supplier invoice proved there wasn't enough pipe to build the project. We had not noticed the note in our review, but after examination determined that the amount was exactly equal to the total job quantity less the amount installed the first month. We deduced that the note referred to the material still on hand *after* the first month's progress payments, as we were able to demonstrate.

**Develop a legal theory of recovery and determine liability.** Next, examine the contract to determine the applicable contract provisions, and consider any contract law that might supersede, modify, or clarify the contract provisions. This requires a broad familiarity with construction contract law and practice and should be supported by a legal review if there are any doubts.

If delay is alleged, determine if it is compensable or excusable. This requires a thorough knowledge of the contract and pertinent contract law. Determine if proper notice was given. If not, decide whether the owner was prejudiced by failure to give notice and whether the fact finder is likely to uphold the notice requirement. If a claim is related to a previously accepted change order (e.g., impact from the change order), verify that it hasn't been precluded by the release language on the change order.

**Prove causation—the linkage from initiating fact to damages.** Although causation as a legal term is primarily used in tort law, this book uses it to describe the process of linking the initiating event, action, or condition to the ultimate costs incurred by the contractor. Causation can be a complex chain of events. It often requires identification of additional facts using the techniques described above. It also requires knowledge of construction processes and a logical analysis to create the linkage.

Determine the cause for the event or activity. If a delay is alleged, the fact that the project was completed late doesn't confirm that a specific owner-caused delay caused it. Other reasons (change order work incorporated into the contract without a time extension, unexpected bad weather, contractor error, etc.) may be responsible for late completion.

Examine what happened just before the delay (or other event/action) occurred and until the delay ended. Did the alleged cause continue past the delay or end before the delay ended? Examine the as-planned schedule and schedule updates through the period of delay—in addition to the daily diaries, timecards if available, correspondence, etc. Compare the as-built schedule with the as-planned and the cost budget with the actual costs to determine the differences and probable causes. Determine if the contractor was partly responsible for the cause and if there was concurrent delay.

**Tools and techniques.**  There are a number of tools and techniques that help with determining the facts and analysis. Several are commonly used as exhibits but are also useful for the claim preparer to study and analyze an issue. They include:

- *Spatial layouts.* Mark up site plans and foundation and floor plans to show work areas and problems at various times during the project. For example, color the foundation plans to show the sequence the footings were poured, how the work progressed, what work space was available at any one time, any conflicts between trades, etc. Use a different color for each month or week, or transition through shades of one color or the color spectrum to communicate the process of construction.

- *Photographs.* Sort the photographs chronologically (pasting them onto $8^1/_2 \times 11$ inch sheets). This shows the work accomplished, working conditions, and problems.

- *Videos.* Videos of actual activities can be studied for progress, working conditions, etc. They can also be analyzed with time and motion studies to determine productivity and impact.

- *Miscellaneous other techniques.* Flowcharts or computer simulation of complex procedures are sometimes helpful for determining what actually happened or estimating productivity and impact.

## 9.  Complete the analysis of each issue

After determining entitlement for each issue, complete your analysis of the issue. Analyze the effect of that issue on the schedule and estimate the costs. Alternately, analyze scheduling impacts and costs concurrently. Finally, write up your findings for each issue. Prepare a brief summary of the facts, analysis, and conclusions. This will help gel your thoughts and aid in reporting to the claim support team or the contractor's decision maker on how to proceed.

When preparing the summary, be careful of wording to avoid potential problems in discovery, as noted in Sec. B.13.

### 10.  Analyze the schedule

After completing the chronological document review, creating the summary notes and analyzing the issues, you will be ready to complete the as-built schedule and determine the cause of delays and schedule impact. Schedule analysis is described in detail in Chap. 10 and is briefly summarized below. In some cases, however, you will have already created a detailed as-built schedule while creating the chronological summary notes.

**Review and analyze the as-planned schedule and the schedule updates.**  If not done before, review the original, as-planned schedule, modify it to correct obvious errors, and examine the schedule updates.

**Finalize the as-built schedule.**  The last schedule update is normally used as the as-built schedule, adding start and completion dates for those activities completed after it was prepared. If a detailed as-built schedule was created during the chronological document review, summarize it as a condensed as-built schedule so that it has the same level of detail as the as-planned schedule.

**Create the "would have been, but for…" schedule.**  Compare the as-planned schedule with the as-built by creating a comparison bar chart. Then create a "would have been, but for" schedule, showing what would have happened but for the issues for which the owner is responsible. Noncompensable but excusable delays will be included in the would-have-been schedule.

**Compare the as-built schedule to the "would have been, but for…" schedule.**  To determine the amount of delay for which a time extension should be granted, compare the as-built schedule to the would-have-been, but for schedule. The difference in time is the compensable delay, which is used to compute delay costs, as described in Chap. 11. The difference in working conditions (due to weather, crowding, sequencing, etc.) for a given activity determines the impact to that activity. That information will be used as described in Chaps. 11 and 12 to compute impact and inefficiency.

### 11.  Identify and analyze potential additional claims and owner counterclaims or defenses

During document review and analysis, you will identify additional issues that need to be investigated. These may include:

- *Claim issues not previously identified.* Project teams often overlook significant claim issues, some of which may have greater impact than those that have been identified. Project personnel are so close to the work and so concerned with getting the job done that they can miss the overall picture or major elements of it. Further, many are not expert in critical path schedule analysis, impact analysis, contract law, etc.

- *Potential counterclaims and owner defenses.* Although you will spend less time on potential counterclaims and owner defenses than on your claim against the owner, check for possible concurrent delay, contractor error, defective work, exculpatory contract clauses, notice requirements, etc. This information is sensitive and should be treated accordingly.

## 12.  Compute damages

The next step is to compute and substantiate damages, sometimes characterized as an equitable adjustment. Some experts[15] recommend computing damages concurrently with the analysis of entitlement. This is appropriate in many cases, as a cost analysis can be invaluable for determining the link between an event and subsequent costs. Usually, however, it is better to do only preliminary cost computations during analysis and wait until entitlement is firmly established before doing detailed cost computations.

Regardless of when costs are analyzed (concurrently or after establishing entitlement), cost computations should follow proof of entitlement in the claims document. Keep entitlement arguments and cost analysis separate. It is difficult enough to convince the owner's representative that they are responsible for extra costs without showing them the cost computations on the same page.

The steps in computing damages, detailed in Chaps. 11 and 12, are normally as follows:

1. *Review the bid and budget.* Determine how the project was estimated and if the estimate was adjusted when it was entered into the cost accounting system as budget. Note any estimating errors, determine what the budget should have been, and adjust it, if necessary, to reflect what the project should have cost when bid. Then be prepared to credit the owner for any bid errors if submitting a total cost claim.

2. *Review the change orders.* If the budget hasn't been adjusted for approved change orders, do so now.

3. *Identify work that was performed in a different manner than planned.* Examine the costs of those items that were budgeted to be performed with the contractor's own forces but then subcontracted to others or accomplished in different weather or other working conditions than anticipated. Also determine whether materials were changed or other actions taken that would create a variance from the budget without being a claim. Adjust the budget as needed.

4. *Review the final cost accounting report.* Determine the cost overrun and which line items (cost codes) were high or low and by how much. Focus efforts on the biggest cost overruns, instead of wasting time on minor issues.

5. *Review the monthly cost reports and weekly labor reports.* Examine how the cost overruns varied over time and in relationship to the alleged cause, in order to determine if the actual overruns are consistent with your recovery theories.

6. *Relate the cost overruns to entitlement.* Determine if the evidence supports the assertion that the owner-caused problems were responsible for the identified cost overruns.

7. *Make in-depth cost analyses and estimates.* Determine the direct costs for each issue, if not done earlier.

8. *Estimate delay and impact costs.* Determine the compensable delay from the schedule analysis and compute the delay costs, including escalation, extended jobsite overhead, and home office overhead. Determine the difference in working conditions between the as-built and the would-have-been schedules, and compute other impact costs from delay, acceleration, and disruption.

9. *Compute markups, interest, contract balances due and other damages.* Determine jobsite and home office overhead, profit, insurance and taxes, bond premiums, etc. Add below-the-line costs including retainage, interest, claim preparation costs, etc., for a total claim amount.

10. *Create a cost summary sheet.* Prepare a one-page summary that clearly describes, lists, and totals all of these costs. Each cost should be referenced to a detailed cost estimate sheet.

11. *Do a reality check.* Take a second look at your numbers, using an entirely different and simpler approach, to verify that they are reasonable. For example, after computing labor inefficiency, compare the total hours or dollars claimed with the actual overrun to determine if the numbers are compatible. You can claim better than estimated productivity, but it can't be too much better. You must have compelling reasons for the better than expected productivity and must be ready for a difficult sales job.

## 13.  Meet periodically with the claim support team to review status and action needed

Meet periodically with the claim support team to review the status of the claim. Compare actual progress to planned, and actual costs to budgeted, examine how well initial strategies and preliminary conclusions have been maintained, decide whether the claim preparation work plan needs to be changed, and revise the work plan if required. Present any required decisions to the claim support team, help make decisions on how to proceed, and document the results.

## 14.  Summarize your draft findings and conclusions

The next-to-last step in the analysis is to write a draft of your findings and prepare draft exhibits. This will be presented to the claim support team as noted below. After approval and authorization to proceed, it will be used to prepare the final claim document and exhibits, as described in Chap. 13.

1. *Summarize and check your damage computations.* Prepare a draft narrative and table listing the major cost items. Check the computed numbers with a simpler, alternate method to determine if the results are in the ball park. For example, a total cost estimate will verify whether the number from the detailed cost analysis is too large to be realistic. Summarize all costs on a single page, using descriptive headers, bold text, and other features to make it easily understandable.

2. *Write up your findings and conclusions.* Organize and write up your findings as a comprehensive report, with embedded figures, attached exhibits, and references to other supporting documentation. Identify any areas needing further investigation, resolution of legal questions, or actions by others. This will be entered in section B of the narrative text. Although not a polished narrative, it should be readable and reasonably well organized.

3. *Prepare draft exhibits.* Prepare draft exhibits as identified in the claim document, and indicate any additional exhibits that may be needed.

### 15. Verify your findings

Verify your analysis by having project team members and key subcontractors review it. It will be difficult to defend a position unless they agree with it. All team members should agree with the basic conclusions (or at least have no serious objections), especially if they will be testifying in arbitration or litigation. Obtain a signed agreement from the subcontractors that the claim fully represents their damages, or exclude subcontractor claims from the claim settlement.

If possible, review the claim with at least one expert not involved in its preparation. This should reveal where the draft needs to be strengthened and clarified.

### 16. Obtain authorization to proceed with the next phase

The final step is to present your work product to the claims support team, answer any questions, and help decide how to proceed.

1. *Present your findings to the claim support team and/or decision maker.* Give a clear and concise statement of your analysis and findings. This will probably require presenting graphics, charts and tables, and summary text. Provide a single page of text summarizing the facts and findings on entitlement, and a second sheet with a summary of damages.

2. *Jointly reach a decision on how to proceed.* Help the claim support team decide what to do next. This will be based on your analysis, legal counsel's opinion, and the team's assessment of the business and financial situation. Prepare a work plan to complete the claim and obtain approval before proceeding.

The next step will be to assemble the claim document, edit the text, and prepare final exhibits, as described in Chap. 13.

## D.  Dealing with Common Problems

The most common problems to overcome in claim preparation are the quality and extent of documentation and the lack of notice. Other problems include conflicts between the subcontractors and general contractor, contractor error, financial pressures, release of claim, expiration of lien rights, and inability (or unwillingness) of the owner to pay. How to solve these problems will be addressed in this section.

### 1.  Poor documentation

Construction disputes often involve mountains of conflicting, fragmentary evidence. Memories may be vague, incorrect, or self-serving. Correspondence may be of limited value, indicating little beyond proof of notice, intent, and the date the parties were aware of an event. Schedules are often incomplete, insufficiently detailed, or simply incorrect.

Daily reports and diaries (normally the most important records) often fail to include important information or may not consistently record it. Accounting records seldom identify the extra cost of impacts and may fail to separate impacted work from nonimpacted work. Often the time sheets upon which accounting records are based are too numerous to review or have too little information. Cost records may fail to note the quantity of work accomplished during each reporting period, so that separate productivity rates cannot be established for impacted versus unimpacted work. Photographs, if taken, often lack dates and rarely include a clear explanation of what is depicted. In short, it is often difficult to determine what actually happened, let alone who was responsible, the impacts, and their costs.

**Insufficient documentation.**  To cope with the problem of insufficient information, obtain information from the owner. File an informal request or make a Freedom of Information Act request from public agency owners. If in arbitration or litigation, obtain the information through discovery. Another source of information is subcontractor records, although subcontractors usually keep fewer records than general contractors.

One method of dealing with insufficient documentation is to focus on stronger issues—usually on entitlement instead of on damages. Make a modified total cost claim in lieu of proving detailed damages.

**Confusing terminology and lack of identification of individuals.**  Project personnel from the various organizations on a project tend to use different terms for the same issue. This varies with the author of the document, the project phase (issues are frequently recharacterized or renamed), or the subissue being addressed. Or they will use the same term for different issues. To deal with this problem, set up a cross-referencing system, identify which issue is being described, cross-correlate to ensure accuracy and consistency, and select a standard terminology for your narrative text and other documents.

Multiple coding systems create another problem, for each has its own numbering sequence for the same basic issue—request for information, notice of change, change order request, change order proposal, field directive/work order, and change order. Subcontractors will have their own coding sequence (or no system at all), and the owner and designer may use different numbering codes and sequences. In addition, the description of the change item may vary from one organization to another, and within the same organization as the perception of the problem evolves. Correspondence and other documents may refer to a change by its current description, to a number from one of the change numbering sequences, or even to "...our previous conversation." In addition, there may be several RFIs referring to the same or a related issue, and multiple NOCs and COPs referring to different parts of the issue or jointly to different issues.

To bring order out of chaos and to aid the reviewer in understanding the dispute, sort out and cross-index the various coding sequences and descriptions. List the coding sequences and descriptions for each issue, and use the most common (or most appropriate) description in your work. Do this as soon as possible, and refer to it while reviewing the chronological source document file.

Complicating the situation is the fact that one code (say for a notice of change) may refer to several different issues. If there are several notices of change on the same issue, it becomes difficult to trace one issue through the maze. The project team sometimes makes mistakes in numbering and may refer to an issue with the wrong description and no code, thus further complicating the task. Subcontractors will have their own numbering sequence, different from the general contractor's and without a cross reference. A well-organized cross-reference index is needed to clarify the coding. For example, item 4 of RFI-123 may be the same issue as CN-89, NOC-159, COP-24, item 12 of CO-12, and the contractor's cost code 15562. It may also be part of RFI-145.

As noted, the index can also include the "cast of characters" can be another problem. Job records may refer to an individual's first name, last name, or job title when two individuals share the same name or the same title. Develop a list of the parties involved [designer(s), owner departments, subcontractors, etc.] and the individual assigned to each position, plus the time frame they were on the job if multiple people were assigned to the same position. Samples of handwriting are helpful in determining the author of important documents.

**Poorly organized documentation.**    Job records are often out of order, documents are undated, and the author of a document is often unidentified. Ask the project office or someone in the home office to organize the files chronologically.

**Excessive information.**    Recommendations for dealing with too much information, especially on large projects, include:

- Limit the review to key documents, and return to the other records later if time allows.
- Have staff organize the files and flag critical documents.

- Use narrative text to collect and organize your analysis.

- Only examine the most important issues—the items with the most cost impact.

- Import data from the client's files to various data processing programs and use the ELIPSE schedule and other techniques described in Chap. 10 to display and analyze the information.

## 2.  Evaluating the reliability of source documents

Source documents are sometimes conflicting or unreliable. Keep this in mind and know how to identify incorrect or unreliable information.

**Use test reports, delivery records, and other neutral documents.**  If the facts stated in the source documents are unclear or conflicting, refer to another document to confirm the correct facts. These documents include test reports, delivery records, and other documents prepared by the owner, contractor, or others.

Concrete cylinder test results and concrete placement inspection reports include the date and description of the work placed. They are the most reliable record of concrete pour dates and are easy to locate, although they are occasionally wrong, as the preparer may not know the project work areas or may enter erroneous data.

Soil test reports indicate when compaction was complete and forming for footings can begin. They may also indicate problems with compaction, which may be the reason for delays.

Safety meeting reports can be helpful. They may indicate what the crews were doing, hazards from other crews (which indicates what work was being done), and startup operations (at the end of the job) that wouldn't otherwise be recorded.

Delivery records and any other document not normally used to posture for claims are generally factual.

**Use records maintained by field personnel.**  The records of field supervisors are usually more reliable than those by project managers and project engineers (who may be posturing for pending claims). However, field personnel may miscode crew time in order to conceal cost overruns, because they don't know the proper cost code or because they have insufficient time to verify that all data are correct. Instructing them on the importance of accurate recordkeeping will help minimize these problems.

**Be careful when using correspondence.**  Correspondence, especially notice of change letters, often has little value beyond establishing when a certain fact was known or a position was taken. When using facts quoted in letters, reference the letter as the source document to protect your credibility if the letter is later proved invalid.

**Use the other party's records.**  They will find it hard to dispute what their own people recorded.

**Cross checking for validity.**   If concerned about the reliability of an individual's records, or if conflicts exist, check other sources. Check the validity of that person's documents with known facts that can be verified to establish a pattern of reliability or unreliability. Look at (1) their knowledge of the facts, including what they should have known, (2) any reason for bias or self-interest, and (3) prior examples of exaggeration or sloppy thinking.

Cross-check statements of fact to other documents. Look for patterns and contrary statements in unimpeachable source documents. For example, NOAA weather records showing the weather on a given day can be used to confirm the date of a daily report. In another example, two daily report had the same date and a report was missing from the previous week. Examination of the reports revealed that one was completed with a different pen than used on the other reports that week, but matching those prepared the week with the missing report.

**Interview project participants.**   Ask them to explain their statements or documents by others.

### 3.   Failure to give contract-required notice

Many claims are filed without having met all the contract-specified notice requirements. Contracts often contain unreasonable requirements, not only for the initial notice but also for the submission of documented costs. In some cases, it may be nearly impossible to meet the notice requirements.

Consult an attorney when not in compliance with contract notice requirements. For federal contracts, the courts generally hold that constructive notice (where the government knew about the problem) is adequate, especially if there is no prejudice. That may not be true for some state, municipal, and private contracts.

### 4.   Disputes between the general contractor and subcontractors

Many disputes between the general contractor and subcontractors are due to mismanagement by either or both parties, or due to inadequate performance by a second subcontractor. When a contractor has a claim against a subcontractor, the normal action is to backcharge the subcontractor and withhold a portion of the subcontractor's payment. In response, subcontractors can file a construction lien or bond claim; but collecting from a general contractor is far more difficult than collecting from an owner and often results in the rupture of important business relationships. Consequently many subcontractors are reluctant to pursue claims against the generalcontractor.

The best means of dealing with a dispute between a general contractor and a subcontractor is to focus on a partnering approach with an emphasis on equity and business issues. For example, the settlement of a subcontractor claim against a general contractor might include a significant credit for using the subcontractor on future bids.

## 5. Evidence of contractor errors

Usually, neither party is entirely free from blame. Contractors sometimes have problems with concurrent delay, insufficient bid amounts, subcontractor disputes, change in jobsite management, defective work, etc. Owners will usually use these issues as their defense against paying due compensation. If there are significant problems that are the responsibility of the contractor, a contractor should be very cautious in making claims. Again, a partnering approach is highly recommended.

Thorough analysis, such as a detailed as-built schedule, may be required to show that the contractor's problems did not cause the disputed delay and damages. The contractor should acknowledge any relevant problems and demonstrate that the claim does not include those delays and costs.

## 6. Financial pressures

A contractor submitting a claim may be in financial difficulty—due either to the dispute or to problems on other contracts. This may severely limit the contractor's ability to fund an adequate claims preparation effort or to wait for negotiation of a fair resolution. Even if funds are available, however, many contractors are reluctant to expend money.

Suggestions on how to deal with this problem include:

- Train and use an in-house expert and project team members, as explained in Sec. B.4 of Chap. 4 and Chap. 6.
- Break the claim preparation work into phases, to better define costs and allow the possibility of using an interim work product to settle the dispute.
- Use company staff to help consultant experts, to save costs and train your staff.
- Focus on the large, easy-to-prepare claim issues and methods.
- Focus your effort on entitlement and use a modified total cost approach for costs.

## 7. Release of claim

When a contractor signs a change order for the direct cost of extra work, it usually includes a release for any other damages, including impact and delay. Unfortunately, contractors are often unaware of the extent of impact when signing a change order or of their contract rights.

The owner may assert that a signed change order is a release of claim for delay and impact. The counter would be that the change order doesn't cover issues that the parties didn't intend it to cover. In the future, carefully study the language of a change order before signing, and ask that it specifically exclude delay and impact—unless you have included these items in the change. If the owner will not exclude these items, consult with an attorney. If these items are not specifically included and you do sign the change order, sign it under protest and state in the transmittal letter that it doesn't include delay and impact.

In some cases, the owner refuses to release other, nonrelated funds until the contractor acquiesces to the owner's position on the disputed issue. If the amount involved in the dispute is large and needed to pay bills the contractor may be forced to accept an inadequate settlement in order to meet payroll. This is economic coercion and may be used as an argument against the owner defense of release of claim. If forced to sign a change order under such conditions, write the protest on the change order form if possible and note that the negotiated amount included only direct costs. If the owner will not accept an annotated change order, state the protest in the cover letter transmitting the change order. And, check with an attorney before signing the change orders.

## 8.  Expiration of lien rights

The ability to file a lien against the owner's property is a powerful lever to obtain a settlement and ensure payment. When preparing a claim, check the date when project work was complete, as it determines the cutoff date for filing a lien under applicable state law.

## 9.  Inability or refusal of owner to pay

The owner may be unable to pay or refuse to pay without being directed to do so by a fact finder in binding arbitration or litigation, even if the award may be more than would be paid in a negotiated settlement.

In this situation, mediation is one alternative. Retain a strong mediator on whose recommendation the owner's representatives can rely to protect them from political fall-out. Another alternative is to convince the owner's attorney of the certainty of losing and the resulting damage to their reputation.

In one recent case, an impoverished school board agreed to pay part of the cost over a period of 5 years and assigned their own claim against the architect (for contributing to the problem) to the contractor.

## E.  Countering Owner Defenses and Defending against Counterclaims

Owners will often respond to claims with counterclaims and various defenses that must be refuted. These should be addressed during the investigation and analysis, even if not yet raised by the owner. The analysis of the issues should be included in the narrative text but should not be included in the claim document unless the issue has already been brought up in earlier discussions.

## 1.  Concurrent delay defense

Alleged contractor delay concurrent with owner delay is one of the most common owner defenses to a scheduling claim. One reason it occurs is a contractor's natural tendency to back off on what now appears to be noncritical work when faced with an owner delay on the critical path. This "near-critical" path may then be an alleged concurrent delay.

The counter to this defense is to explain it as a mitigation effort to minimize damages. Your argument will be greatly aided by written evidence supporting a conscious decision to slow down. Notice to the owner of this decision is the best evidence. Notice should always be given when slowing the pace of construction to reduce costs while waiting for an overriding delay to end.

## 2. Contractor error defense

Contractors inevitably make mistakes in the course of building a project, and owners object to paying the extra costs of these mistakes. This is especially true for subcontractor errors, which are often well documented in the general contractor's records and sometimes exaggerated. However, in many cases, the errors would not have occurred except for the conditions resulting from the delay. For example, excessive hours spent resolving changes often results in supervisor error from fatigue or lack of time to properly plan.

Some errors are expected, and owners should expect to pay for these as well as for work that goes extremely well. Other errors can be defended as being a result of changes and overwork. Others may be shown as resulting from poorly documented owner-caused changes and delays. The rest should be acknowledged and credited when making a modified total cost claim.

## 3. Inaccurate or impractical as-planned schedule defense

When defending against a scheduling claim, owners often allege that the as-planned schedule was not attainable. This is almost certain to happen with early-completion schedules, when the contractor has scheduled project completion before the contract-specified completion date. Owners will cite late completion of critical path activities as evidence that the work couldn't be completed on time, even when those delays are a result of owner delays and changes.

Owner approval, or nonrejection, of the submitted schedule is a good rejoinder to this defense. If the schedule was impractical, why did the owner approve or accept it? Another response would be to note those activities that were not delayed by the owner but completed as scheduled.

Countering inaccurate and impractical arguments is more difficult when the as-planned schedule is obviously incorrect, lacks detail, or wasn't used to manage the project. In that case, use portions of the as-built schedule to flesh out or correct the as-planned schedule and reconstruct "reasonable" as-planned activities for the critical work items.

## 4. Inaccurate or insufficiently detailed as-built schedule defense

If the last schedule update is used as the as-built schedule, the owner may allege that it is inaccurate and insufficiently detailed to support the scheduling claim. This argument may have some validity, as updates seldom correct the logic but simply add actual start and finish dates, which are sometimes incorrect.

The solution is to correct logic errors and incorrectly entered actual dates. If necessary, create a detailed as-built schedule as described in Chap. 10.

### 5. Improper allocation of joint delay defense

When both parties share responsibility for delay or have a partial concurrency of delay, owners will allege that the allocation of delay to the owner is improper.

The response is a more detailed analysis of the reasons for the delays. The best tool for this is the detailed as-built schedule and chronological summary notes referring to the issues causing those delays.

### 6. Incorrect would-have-been schedule logic or activity duration defense

If the logic or the activity durations of the would-have-been schedule are different from the as-planned schedule, owners will often allege that they are incorrect and that the would-have-been schedule is therefore invalid. This occurs when the would-have-been schedule corrects errors in the as-planned schedule or is based on a different strategy than had been planned.

Changes to the as-planned schedule should be incorporated into a "modified as-planned" schedule, with supporting analysis and documentation. This separates, and simplifies, the arguments in favor of your changes. The next step is to clearly document the reasoning for the changes, using as-built logic and duration whenever possible. In addition, you may need to show what the project team "would have done" based on their correspondence or internal documentation, had conditions been as projected at that point on the would-have-been schedule.

### 7. Poorly supported damage computations defense

Contractor records may be inadequate to support many of the cost computations, and owners will assert that the costs are "not proved." This happens frequently for impact and inefficiency costs. A more detailed cost analysis, supported by nonaccounting records as noted in Chap. 11 and the use of rational analysis for computing inefficiency as explained in Chap. 12, will aid in overcoming these objections.

### 8. Late completion and liquidated damages counterclaim

Troubled projects are often completed late, and owners sometimes use the threat of liquidated damages or actually assess liquidated damages to force contractors into settling claims.

A thorough schedule analysis will determine whether the owner's allegations of contractor responsibility for the late completion are true. Even if the contractor is responsible for late completion, concurrent owner delay usually excuses the contractor from damages. Another defense is to show that the beneficial occupancy date, or the date when the owner could have taken beneficial occupancy, was earlier than alleged by the owner. This establishes the substantial completion date and limits assessment of liquidated damages, although specific damages (for continuing contract administration and inspection) may continue until final completion.

## 9. Consequential damages counterclaim

Consequential (actual) damages for the contractor's late completion can be assessed in the absence of liquidated damages. They can be enormous and can include:

- Extended construction interest and fees.
- Interest rate on the construction loan and permanent financing, if completion is pushed past some agreed-upon date.
- Extra licensing costs.
- Lost revenues and profits, including loss of market share.
- Costs of late opening (wasted advertising, etc.).
- Extended costs for inspection and salaries for the owner's representatives.
- Costs of operating personnel hired prior to occupancy.
- Additional architectural and engineering fees.
- Claims by follow-up or start-up contractors.
- Storage costs or delayed shipment charges for installed equipment, furniture, and furnishings.
- Extended insurance premiums.
- Additional utility costs.
- Escalation of costs for owner-furnished equipment.
- Diminished market value due to loss of a tenant.
- Attorney fees, costs of litigation, or awards to third parties.

Besides refuting responsibility for the delay and its costs, as noted above in Sec. E.8, your best defense is a legal defense contending that (1) assessment of actual damages isn't supported by the terms of the contract and the actions of the parties, and (2) the amount claimed is beyond the contemplation of the parties when entering into the agreement. In addition, you should refute the damages on an item-by-item basis. This starts with an examination of their documented losses. For example, revenues are often far less than projected on the pro forma forecasts of profits. In addition, many counterclaims are for gross revenues, not net profit. The owner's failure to mitigate damages is another defense.

The threat of a counterclaim for actual damages should always be analyzed as one of the downside risks when you submit a claim. Not only are counterclaims difficult to refute, the amount may vastly exceed the contractor's claims.

## 10.  Defective work counterclaim

Few projects are completed without some deficiencies. When faced with a claim, many owners will become nitpicky about accepting minor deficiencies that would normally be overlooked. In addition, operating problems, design defects, damage by others, and other problems for which the contractor is not responsible will often be included in a counterclaim. Undocumented repairs (or improvements) by the owner sometimes complicate the issue and make it difficult to determine who really was responsible unless the contractor is given ample opportunity to examine and document the conditions before repair.

A litany of alleged deficiencies may be posturing and an indirect allegation of mismanagement. Or it may be an emotional response that must be addressed before rational discussions can proceed. Start by listening, by paraphrasing owner statements to confirm that you understand, and by showing empathy. Then accept responsibility for those problems that are the fault of the contractor or subcontractors. Oftentimes, especially in home building, this will defuse the situation. But insist on full investigation by an independent expert before the alleged defective work is changed. Put the owner on notice of the obligation to allow such an investigation and to allow you an opportunity to repair the deficiencies.

Contractors are liable for the reasonable and necessary cost of replacing or correcting defective work and work that doesn't conform to the contract. However, if the work is nonconforming and the cost of strict compliance is too high, this would be legally described as economic waste. The damages would then be limited to a reasonable amount.

## 11.  Failure to complete some work counterclaim

The damages for failing to complete work on a lump-sum contract will be the cost to the owner to complete it. On unit price contracts, the unit price is offset against the cost of finishing the incomplete work.

## 12.  Other counterclaims

Carefully analyze each counterclaim and determine if there is provable validity. If so, estimate the reasonable minimum cost or identify other action that can be taken, and admit responsibility. Attempt to negotiate alternatives to making repair or payment.

# Analyzing Schedule Delay and Acceleration

Scheduling claims are frequent in construction and difficult to prove. This is especially true if the contractor hasn't prepared and maintained an accurate construction schedule or if the claim preparer doesn't fully understand critical path method (CPM) concepts and tools. Proving scheduling claims will be easier if the procedures and recommendations in this chapter are followed.

*Scheduling is an art as well as a science.* This chapter provides an overview of CPM concepts and construction scheduling practices. It explains how to analyze a scheduling claim and prepare scheduling exhibits. It describes some relatively new and improved techniques for scheduling and for analyzing scheduling claims. When implemented, these techniques will reduce scheduling problems and improve recovery of delay, acceleration, and impact damages.

This information is provided in the following sections of Chap. 10:

If you are familiar with industry practice and CPM techniques and are preparing a scheduling claim, proceed directly to Sec. E. Sections A through C are written for those unfamiliar with scheduling or who want more detailed information. Sections C, G, and H discuss advanced techniques for schedule analysis and presentation, and Section I addresses ancillary tools and techniques. If reading this chapter from start to finish, break your review into related sections (A-C, D, E-F, G-H, and I).

This chapter is based on my paper, "Proving Entitlement for Construction Scheduling Claims"[16] (published in *The Construction Lawyer* by the American Bar Association), twenty-plus years of scheduling experience, references to several authoritative treatises on scheduling, and input by friends and industry associates.

## A. General Introduction and Overview of Project Management Concepts

A quick review of a few basic project management concepts will help explain scheduling concepts and their application to claims analysis.

### 1. Project objectives—cost, time, and scope

Project objectives are usually defined in terms of *cost, time,* and *scope,* with scope being the quantity and quality of the finished project as specified in the contract documents. All three objectives are interrelated. Failure to achieve one objective (e.g., late completion) usually results in problems with the others (extra costs or reduced quality). All three depend on (1) the resources (people and equipment) available to accomplish the work and (2) the management and leadership skills of the project team.

Figure 10.1 illustrates this interrelationship and two additional points:

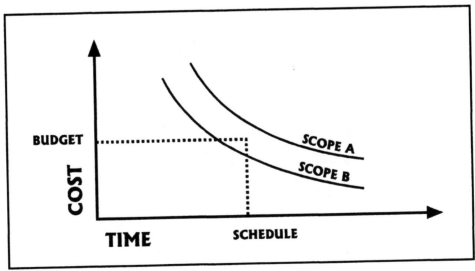

**Figure 10.1**  Interrelationship of cost, time, and scope.

- Costs increase as the project duration is accelerated. If the project duration is being reduced, the budget must be increased (follow the line for scope of work A up and to the left).

- If a project (scope of work A) cannot be accomplished for a given budget and schedule (the dotted lines), the quantity and/or quality of the project can be decreased (from scope of work A to scope of work B).

Customer satisfaction, sometimes listed as a fourth objective, is often overlooked but is a major factor in reducing disputes and achieving equitable compensation for extra work. Similarly, interpersonal relations between key players of the joint project team (owner, designer, contractor, subcontractors, and suppliers) can facilitate or hamper progress and the resolution of disputes. Attention to these issues will help avoid and resolve disputes.

## 2.   Least cost expediting and cost-time tradeoffs

*To reduce the project duration, accelerate only critical path activities.* Critical path activities are those that must be completed on time to avoid delaying the project. Accelerating noncritical activities would be a waste of money unless they become critical owing to acceleration of critical activities.

*When accelerating critical activities, first accelerate the activities that are the least costly to accelerate* (not the least expensive activities) before accelerating the more-costly-to-accelerate activities.

**Least cost expediting procedure.**   The cost to accelerate an activity one day is called the cost slope, which is generally constant for some period of time. The least cost expediting procedure is to estimate the cost slope of each critical path activity and accelerate the least expensive until it can no longer be accelerated at that cost. Then accelerate the next least expensive until it cannot be accelerated and continue until the project duration is reduced the desired amount. Although a precise calculation of each critical activity's cost slope is impractical, a rough estimate to guide acceleration efforts is sufficient.

**Cost-time tradeoff analysis.**   Figure 10.2 is an extension of Fig. 10.1 that shows how reduced indirect costs (overhead) offset increased direct costs as a project is accelerated. The figure also shows how indirect cost is added to direct cost to create the TOTAL COST curve. Moving from $t_{normal}$ on the horizontal TIME scale to $t_{optimum}$, reduces the total cost to the MINIMUM TOTAL COST value. This occurs because the increased direct cost of accelerating each day is less than the savings of reduced indirect cost. At the optimum time, the extra cost to accelerate one day equals the savings from reduced indirect cost (i.e., the slope of the DIRECT COST curve is equal and in the opposite direction to the slope of the INDIRECT COST line).

If the owner's lost income (for example, earlier rent from an office building) is added to the owner's indirect costs during construction, the INDIRECT COST line becomes quite steep and additional acceleration will be cost-effective (for the owner). Contractors can use this cost-time tradeoff concept to propose a value engineering change to accelerate and minimize the owner's total costs. If accomplished under a value engineering change proposal clause, the owner pays the contractor's acceleration costs and the owner and contractor would share the balance of the owner's savings.

Figure 10.2 also demonstrates that delay past the normal time (e.g., to $t_{\text{delay}}$) increases direct costs (through impact and inefficiency), indirect costs (extended overhead), and therefore total costs.

**Improve efficiency to reduce project duration.**   Contractors should consider using work improvement techniques to improve efficiency, which will increase productivity and therefore save time—while actually reducing costs. This is an alternative to conventional acceleration practice, which increases costs. For more information, contact the Lean Construction Institute at www./constructioninstitute.com.

3.  **Cash flow forecasts, percent complete curves, and earned value**

Project schedules can be used to forecast cash flow (of costs or billings). Cash flow forecast curves can be converted to percent complete curves and used to track progress (i.e., earned value) or to analyze and help prove scheduling claims.

**Forecasting cash flow with cumulative curves.**   To forecast cash flow:

1.  Cost load a schedule by allocating costs (or contract billings/revenue earned) to the activities. Enter the costs into the scheduling software; the rest is accomplished by the software.

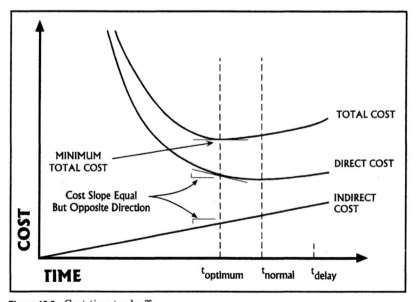

**Figure 10.2**   Cost-time tradeoff.

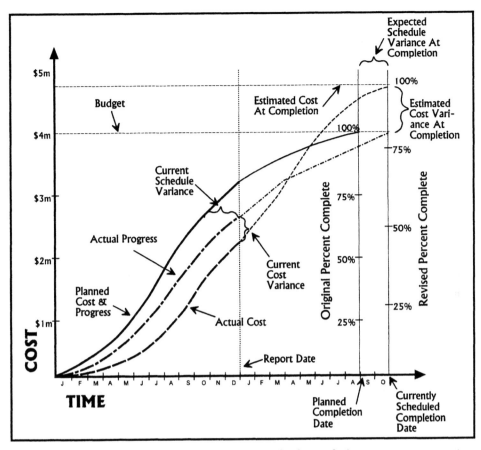

**Figure 10.3**  Cash flow/percent complete curves for earned value analysis.

2. Spread the estimated cost over time for the duration of each activity.

3. Add the total cost for each period of time (monthly) to obtain the total cost for each period.

4. Plot the forecast cost for each time period as a cumulative curve like the solid line labeled Planned Cost & Progress in Fig. 10.3.

Cash flow forecasting by contractors is normally done with billings (e.g., the schedule of values), as it is difficult to accurately assign estimated costs to activities. Contractors' costs are tracked by cost code in the cost accounting system, and duplicating that effort in the scheduling system would be too much work.

**Converting cash flow curves to percent complete curves.**  Convert cumulative cash flow curves to percent complete curves by simply setting the value at the top of the curve to 100 percent and adding a percent complete scale. This is the Original Percent Complete scale on the right side of Fig. 10.3. Other examples are Figs. 10.5 and 10.34.

Cumulative percent complete (cash flow) curves are generally shaped like an S and are often called S curves. They start off slowly, speed up through the middle of the project, and taper off at the end.

**Comparing planned progress with earned value (actual progress).**   To compute and plot actual progress (earned value):

1. Record the percent complete of each activity each time period (e.g., monthly) and enter into the scheduling software; the rest is accomplished by the software.

2. Compute the earned value of each activity as the product of the percent complete times the planned cost (or billings) allocated to the activity.

3. Compute the earned value to date of the project (actual progress) by summing the earned values of all activities.

4. Compute the project percent complete, which is the project earned value to date, divided by the project budget, times 100.

5. Plot the Actual Progress (Earned Value) curve (the dot-dash line in Fig. 10.3) on the same chart as the Planned Cost & Progress curve (solid line in Fig. 10.3).

If the Actual Progress curve is below the Planned Cost & Progress curve, the project is behind schedule (as in Fig. 10.3). If above, the project is ahead of schedule. The time ahead or behind (the Current Schedule Variance) is the horizontal distance between the curves at the report date. The chart shows the effect of changes and impact—if costs are accurately allocated to the correct activities.

Remember that the Planned Cost & Progress curve is normally based on all activities starting on their early start dates. Actual progress is usually slower but should be ahead of a curve based on all activities starting on their last start dates. In fact, most software can plot an envelope of curves consisting of both the early start schedule and the late start schedule within which actual progress needs to stay. Figure 10.7, for example, shows both curves, and Fig. 10.34 also shows the average.

**Earned value and forecast to complete.**   Earned value analysis compares the Actual Progress with the Actual Cost (the dashed line in Fig. 10.3). The Current Cost Variance indicates the difference between planned productivity and actual as of the report date. This information is needed, along with the availability of resources and a corrective action plan, to forecast actual cost and actual progress through to completion (the lighter lines in Fig. 10.3 to the right of the Report Date line). The example in Fig. 10.3 forecasts a late finish and a cost overrun, as indicated by the Expected Schedule Variance At Completion and the Estimated Cost Variance at Completion.

Although it isn't practical for contractors to allocate and collect costs by activity, labor hours can be used (or converted to labor cost) to generate all three curves. Allocate the estimated labor hours (or labor costs) to activities and the scheduling software can generate the Planned Cost & Progress curve and the

Actual Progress curve. The percent complete for the Actual Cost curve can be obtained from the cost accounting reports and manually plotted on the chart for each month: divide the actual labor hours to date by the total estimated hours and multiply by 100.

**Application of the concepts to construction.**  Although somewhat difficult to prepare, these charts help track progress and costs at an aggregate level. Even if not used, the concepts should be understood by schedulers and claim preparers. With the right facts, the three curves can be a powerful exhibit for substantiating labor impact and inefficiency.

## B.  Construction Scheduling Practices, Problems, and Solutions

Improved construction scheduling practices will reduce delays, disputes, and claims. A review of current practices and problems will reveal how to implement improved practices and will help in analyzing scheduling claims.

### 1.  Contract scheduling requirements

In construction, "time is of the essence" clause and a completion date are normally part of the contract. Most contracts also require the contractor to submit and maintain a progress schedule. This ensures timely completion and serves as the basis for resolving time-related disputes. However, contractors should go beyond minimal contract requirements for better project control and to provide better information for claim preparation.

*Contractors should always prepare a detailed schedule and update it monthly—* even if not contract required. If the schedule is submitted for approval, and not rejected, the owner will have a difficult time trying to refute it later. When updating schedules, record actual start and finish dates in addition to the percent complete or days remaining of ongoing activities. This information is essential for preparing a scheduling claim.

**Owner-controlled schedules.**  Some contracts require the contractor to follow an owner-provided schedule. Owners have no business preparing contractor's schedules, as it interferes with the contractor's fundamental right to manage the project. Be cautious if bidding such projects, and prepare and maintain your own schedule. Other contracts require that specific milestones be included in the schedule or specify mandatory milestone completion dates for interface with work by other contractors to ensure on-time completion.

A few project owners (e.g., the U.S. Veterans Administration) require the contractor to submit schedule data to the owner or an owner-retained consultant for processing. Some contractors rely on the government-generated schedule for running their projects, a poor practice. If owner scheduling is mandated, the contractor should maintain a separate schedule to ensure accuracy and prompt turnaround and to allow "what-if" analysis of alternatives. Scheduling is a fundamental management responsibility and should not be delegated.

**Cost-loaded schedules.** Some owners require contractors to prepare cost-loaded schedules (actually owner costs, i.e., contractor billings). Cost loading is used for computing progress payments and reduces monthly disputes over progress payments. However, a cost-loaded schedule takes time to generate and can be front-end loaded similar to a conventional schedule of values. Cost-loading schedules can lead to distorted progress reporting if contractors overreport progress to increase early billings.

Some owners require all activities to be cost-loaded, an unreasonable requirement since some activities have no costs, or refuse to pay until an activity is 100 percent complete. Other owners won't allow a completed activity with minor discrepancies to exceed 99 percent complete. This distorts the computed dates of succeeding activities and can cause negative float even if the project is on schedule, unless the schedule logic is deleted or the scheduling software is allowed to override the logic.

**Resource-loaded schedules.** Some owners require resource-loaded schedules to (1) ensure sufficient resources are available to complete all activities, (2) protect the owner against unwarranted claims, and (3) ensure timely completion. It takes considerable effort to generate a resource-loaded schedule, but the result may identify potential problems with crowding or inadequate resources.

Resource loading (or including "crew chases" as activity relationships) is recommended for the most critical resources, even if not required by the contract, but total labor resource loading (combining different trades) has little value. As a minimum, general contractors should require each subcontractor to provide their average crew size to confirm if the subcontractors' activity durations are realistic and to help identify understaffing.

## 2. Legal rights and obligations regarding scheduling

See Sec. E.3 of Chap. 3, Sec. F below, and Wickwire, Driscoll, and Hurlbut's book, *Construction Scheduling: Preparation, Liability, and Claims*[17] for a discussion of scheduling contract clauses and law.

**Owner failure to enforce scheduling requirements.** If owners fail to enforce the scheduling requirements, contractors should still comply with the reasonable requirements, as schedules can save time and money and help a contractor recover delay damages.

**Owner failure to resolve time extensions contemporaneously.** Many owners (and contractors) are reluctant to resolve time extensions at the time they occur, and some owners forbid the contractor from incorporating changes in the schedule before they are approved. The result is an inaccurate schedule that can't be used for managing the project. If this happens, maintain a separate schedule with the correct information.

## 3. A trend toward more scheduling disputes

Historically, the majority of scheduling claims have been filed by general contractors against the owners of publicly bid projects. Although mediation, partnering, and other Alternative Dispute Resolution (ADR) procedures reduce construction litigation, the trend continues toward more scheduling disputes with an increasing number between the general contractor and the subcontractors, sub-subcontractors, or material suppliers. This trend is also evident on private projects involving negotiated contracts. All parties are more aware of the cost of delay and impact, are more sophisticated in their scheduling techniques, and have tighter budgets and schedules. In addition, the relationships between general contractors and subcontractors are more contentious.

To reverse this trend, implement better scheduling practices and project partnering. General contractors should include subcontractors and key vendors in partnering and scheduling efforts. Subcontractors should insist on being included in partnering with the general contractor, should partner with their sub-subcontractors, and *must* have input into the schedule.

## 4.  The origins of most scheduling disputes

There are many reasons for delays and cost disputes. Owners want to reduce construction time, but the designers don't know how much time to allow. Designers are pressured into inadequate design fees, minimize the level of design, and in some cases pull details out of their Autocad library that are "close." Delays result from request for information (RFI) exchanges and cost discussions. General contractors seldom prepare a detailed schedule when bidding a project. Subcontractors commit to a fixed price without knowing the time that will be allotted for their work or the working conditions (such as trade conflicts, congestion, or winter weather) that can materially affect cost. Schedules are often prepared without input or commitment from subcontractors. If consulted, some subcontractors agree to deadlines without understanding the schedule or knowing the total crew requirements for all of their contracts. Some subcontractors take on more work than they can staff.

During construction, errors are found in the plans, unexpected site conditions are encountered, or changes are made by the owner that delay or impact the work. Failure to promptly order materials can cause delay. Tasks proceed slowly, and the work gets pushed into winter weather. Schedules are disrupted and commitments on other projects supersede. Rain and cold cause working conditions to deteriorate. In an attempt to finish on time, crew sizes are increased past an efficient level, working conditions deteriorate, efficiency plunges, costs spiral out of control, and the result is a scheduling claim.

At this point, you may discover that there is no original schedule or that it is incomplete, inaccurate, and insufficiently detailed. It may be a bar chart instead of a critical path schedule or may lack approval. Updates may be sporadic or may only indicate the activities' percent complete as of the report date. Actual start and completion dates may not be provided or are inaccurate, logic changes are not indicated, and no record was made of intermittent progress, insufficient staffing, or the reason for delays.

Although many contractors are aware of these problems, few have been able to avoid them. This chapter guides you in how to proceed.

## 5.  Types of scheduling claims and costs

The three basic types of scheduling claims are delay, acceleration, and disruption.

1. *Delay.* Delay claims can be compensable (due to owner/designer error) or excusable but noncompensable (e.g., for unusual weather). Excusable delays excuse the contractor from liquidated damages, while compensable delays also entitle the contractor to reimbursement for extended overhead, escalation, and impact costs.

   Delays can be either partial or total and may or may not affect critical path activities. Project delays cause escalation and extended overhead costs and normally cause inefficiencies in contract performance, resulting in impact costs. Extra work may also be required (e.g., wet weather access roads). If a subcontractor is delayed, they may have extended overhead costs even if the entire project is not delayed. In addition, delay to a noncritical activity can result in impact costs if that activity is pushed into different working conditions.

2. *Acceleration.* Acceleration claims are compensable if owner-caused. Claims can be for direct acceleration (the owner directs a contractor to accelerate) or constructive acceleration (e.g., the owner's failure to promptly grant a justified weather time extension indirectly forces the contractor to accelerate to avoid liquidated damages). Acceleration causes inefficiencies (impact) to original contract work and often requires extra work tasks to reduce the project duration.

3. *Disruption.* Owner-caused disruption is compensable. Disruption can result from owner interference (e.g., directed sequence of work), excessive change orders, design errors, or other actions that change either (1) the working conditions (e.g., crowding or inadequate access) or (2) how the work is performed (e.g., out-of-sequence construction). Disruption causes inefficiency and impacts costs, usually delays some activities, and may delay the project.

## 6.  Computer scheduling

Although scheduling relies heavily on computers, the software and basic scheduling concepts are often misunderstood. Many contractors buy a project management (scheduling) program for their computer but they fail to train their personnel in the software and CPM concepts, or they don't provide management input and control. The result is an inaccurate schedule, insufficiently detailed and unusable for managing the project. The schedule only fulfills contract scheduling requirements, and the superintendent controls the job with short-interval bar charts unrelated to the official schedule.

The courts and arbitrators tend to give a great deal of credence to computer analysis. Therefore, computer analysis is vital to ensure the credibility of your analysis or to disprove an invalid computerized counterclaim. Do not, however, rely on a computer expert to schedule or analyze scheduling claims; construction experience is essential.

Incorrect use of scheduling software features is a frequent problem. Many users aren't well versed in the software. They use the wrong features and incorrect logic to get the desired activity start and finish dates (e.g., using mandatory start dates instead of the correct logic). Then, when the schedule is changed, those features and logic errors give the wrong critical path with erroneous dates and float. Instead of correcting the schedule, the inexperienced user again "adjusts" it by distorting the logic to get the desired dates. By now, the schedule is invalid and will not support a legitimate claim analysis. If this happens, discard the original logic and features and enter the correct logic to create a modified as-planned schedule. Then correct the as-built logic on the last update and proceed with the analysis.

### 7.  The art of scheduling

*Schedules are simply estimates of what can and will be done.* Schedules vary greatly in accuracy, depending upon the skill and project knowledge of the scheduler, the complexity and determinability of the project, and the level of the scheduling effort. More importantly, they are subject to the commitments made by the individuals or groups who accomplish each task as well as to the persistence and effort of those managing the project. In addition, the schedule evolves as the project unfolds, and changes should be recorded at each update.

A good schedule can be accomplished with the available resources, within the allotted time, and at a reasonable cost. There may be several alternative schedules that would be as good. Optimistic schedules are best, as they "push" the project team, if not so tight that the team can't catch up by working a little overtime. A project that continually falls behind is disruptive and more expensive to build than one that is generally on time.

### C.  Critical Path Scheduling Concepts, Techniques, and Tools

Critical path method (CPM) scheduling was developed in the late 1950s and has been available on relatively easy to use computer software for more than a decade. Yet few schedulers fully understand the concepts and techniques or how to correctly use the software.

CPM is an elegant technique—sophisticated, yet simple. The important concepts and basic techniques can be taught in a one-day seminar. Most are presented in this section and must be understood to correctly schedule a project or prepare a scheduling claim. This section also describes several different scheduling techniques and their relative advantages, as claim preparers may encounter these techniques. As noted below, the recommended scheduling technique is precedence computing and any of several types of timescale network diagraming.

### 1.  A review of bar chart scheduling

The bar chart (also called a Gantt chart) was developed in the early 1900s by Henry Gantt. It was the preferred method of scheduling until the development of CPM and is still used today by many project personnel. It is a viable scheduling technique for many applications. In addition, bar charts are often used to display CPM-scheduled projects since the format is easy to understand. Therefore, claim preparers must understand bar chart scheduling. In addition, the following review of bar chart scheduling explains many of the techniques used for CPM scheduling.

**Description.**   A bar chart consists of:

- A calendar scale across the top of the drawing.
- A list of tasks (activities) in a column on the left, in general sequence from start to finish of the project.
- A timescaled "bar" on each line representing the activity, from its start date to its finish.

The bar chart in Fig. 10.4 illustrates the activities' start and finish dates, durations, and the general order of work. It was generated from a schedule that had been CPM-scheduled and therefore differentiates between critical and noncritical activities and shows the late finish dates of the noncritical activities as small inverted triangles.

**Strengths and weaknesses.**   Because they are timescaled, bar charts are easy to prepare and understand. However, they often lack sufficient detail (unlike Fig. 10.5). Owing to their layout, with only one activity per line, it is difficult to have more than 100 activities per large drawing, and a bar chart of more than 50 activities becomes difficult to read. More importantly, bar chart schedules normally fail to show the relationships between tasks, which determines the impact of a delay, and do not show the critical path, which tells whether delay to an activity will also delay project completion. This lack of detail results in critical activities being overlooked. The lack of relationships results in schedules with overlapping activities which are, in fact, sequential. In both cases, the result is delay.

Another drawback of bar charts is the tendency of many users to use the desired project start and finish dates as the beginning and ending dates for the first and last activities, and adjust the other activities to fit within the available time. The result is insufficient time to accomplish the work unless additional resources and management effort are committed.

**Bar chart applications for scheduling.**   Bar charts can be a schedule tool (for noncomputer scheduling) or a graphical display technique for computerized/CPM-scheduled schedules. Bar chart scheduling applications include the following:

- *Preliminary schedules.* Bar charts are good for prebid scheduling by estimators or for initial project planning by project managers and superintendents, because they can be quickly roughed out at a summary level on paper before creating a more detailed schedule on computer. Or a pre-

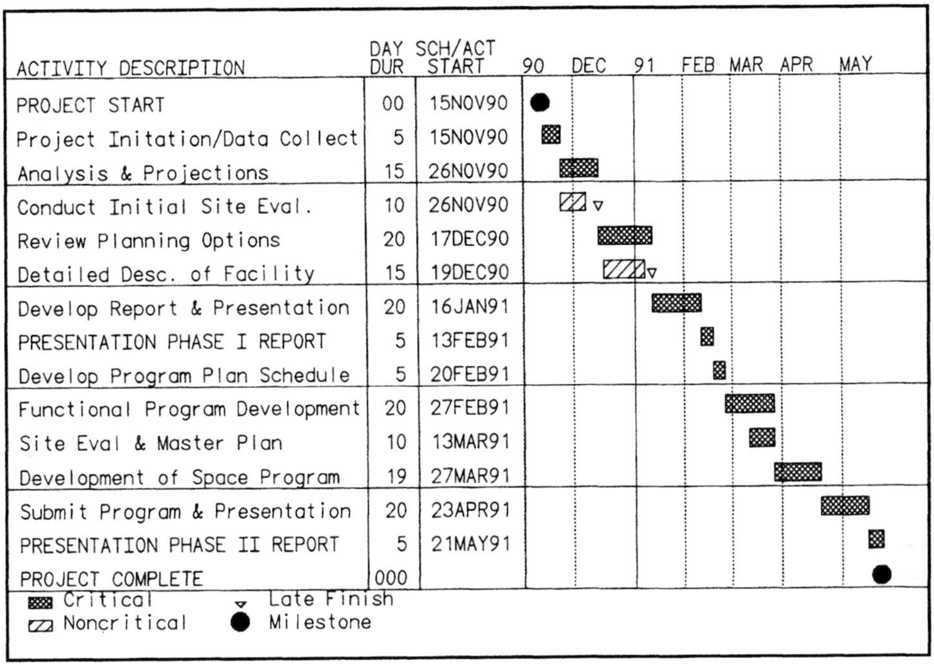

| ACTIVITY DESCRIPTION | DAY DUR | SCH/ACT START | 90 | DEC | 91 | FEB | MAR | APR | MAY |
|---|---|---|---|---|---|---|---|---|---|
| PROJECT START | 00 | 15NOV90 | ● | | | | | | |
| Project Initation/Data Collect | 5 | 15NOV90 | ▨ | | | | | | |
| Analysis & Projections | 15 | 26NOV90 | | ▨ | | | | | |
| Conduct Initial Site Eval. | 10 | 26NOV90 | ▨ ▽ | | | | | | |
| Review Planning Options | 20 | 17DEC90 | | ▨ | | | | | |
| Detailed Desc. of Facility | 15 | 19DEC90 | | ▨▽ | | | | | |
| Develop Report & Presentation | 20 | 16JAN91 | | | | ▨ | | | |
| PRESENTATION PHASE I REPORT | 5 | 13FEB91 | | | | | ▨ | | |
| Develop Program Plan Schedule | 5 | 20FEB91 | | | | | ▨ | | |
| Functional Program Development | 20 | 27FEB91 | | | | | | ▨ | |
| Site Eval & Master Plan | 10 | 13MAR91 | | | | | | ▨ | |
| Development of Space Program | 19 | 27MAR91 | | | | | | ▨ | |
| Submit Program & Presentation | 20 | 23APR91 | | | | | | | ▨ |
| PRESENTATION PHASE II REPORT | 5 | 21MAY91 | | | | | | | ▨ |
| PROJECT COMPLETE | 000 | | | | | | | | ● |

▨ Critical     ▽ Late Finish
▨ Noncritical    ● Milestone

**Figure 10.4**  Simple bar chart.

liminary schedule can be drawn by hand as a bar chart, and adjusted, before computerizing.

- *If scheduling software not available or not mastered.* If the scheduler isn't familiar with scheduling software, hand-drawn bar charts can be used.

- *Resource-constrained schedules.* Bar charts work well for projects that are largely resource-limit-dependent instead of job-logic-dependent. A critical path often doesn't exist and the job must be staffed with sufficient workers to complete the multiple concurrent activities on time.

Figure 10.5 is an example of a resource-constrained bar chart for the main steam piping installation for a large coal-fired power plant. This schedule was generated by hand in 1977 and would be computerized if done today.

- *Short-interval schedules.* Superintendents use bar charts for their weekly short interval (look ahead) schedules, as bar charts are easy to prepare and work well for this application. CPM scheduling of short-interval schedules is not required, although computerization facilitates updating. Regardless of how they are generated, short-interval schedule tasks should tie into the master schedule activities.

- *Schedules to completion.* Bar charts are useful near the end of a project when scheduling the many small remaining tasks. The relationships between tasks are no longer distinct and are often too numerous and complex to show.

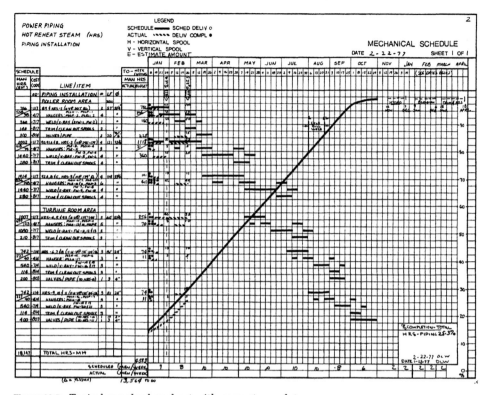

**Figure 10.5**  Typical complex bar chart with percent complete curve.

The project is usually constrained by the number of available workers for each craft and the work space available rather than by clearly definable job logic.

**Bar chart applications for displaying CPM-scheduled activities.**  Bar charts are easier to understand than most network diagrams and can be used in lieu of network diagrams to display computerized CPM schedules.

- *Comparison schedules.* Bar charts are the only practical means for comparing two schedules, to show their differences. It is impractical to create a comparison network diagram, as the lines overlap.

- *Summary schedules.* Bar charts are especially useful as summary schedules for presentations and as exhibits for demonstrative evidence, because they are simple and easy to understand.

**Grouping (sorting) bar chart activities for maximum clarity.**  List bar chart activities in sequential order by subnet, to make the bar chart more understandable. This facilitates following each chain of related activities down the drawing. If the activities are computerized and CPM-scheduled, sort by float and early start before printing. This displays the critical path first, followed by each chain of sequential activities with the same float.

**Displaying float and relationship line.** Computerized bar charts can display activity float if the activities are CPM-scheduled. Activities are plotted at their early start and finish dates, and the scheduling software can indicate activity float by a dashed line from the end of the bar to the late finish date or by an inverted triangle at the late finish date (as in Fig, 10.4).

Scheduling software can show activity relationships, as in Fig. 10.6. It is usually better to plot only the controlling relationships, to avoid overlapping relationships which make the schedule hard to read.

**Bar chart updates.** There are several ways to manually indicate progress on a bar chart: fill in the bar, note the percent complete below the bar at the update date, or draw a status line. The status line is recommended. When drawing a status line, start at the update date on the calendar at the top. Draw a vertical line down to the first activity and then jog left or right depending whether the activity is behind or ahead of schedule. Pass through ongoing activity bars at a point where the scaled duration of the bar to the right of the status line equals the days remaining of the activity. Continue to the bottom of the drawing, returning the status line to the update date for activities in the future or completed and at the bottom of the drawing.

The status line shows not only the current status but also the estimated days remaining of ongoing activities. In addition, the status lines for prior

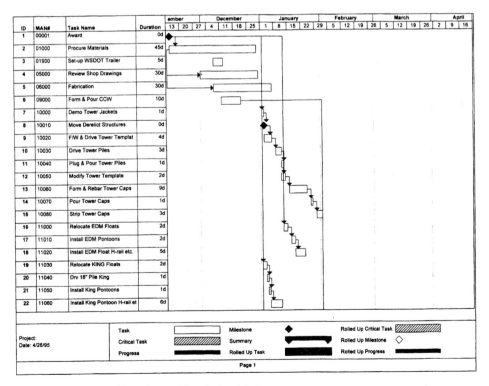

**Figure 10.6**  Connected bar chart with relationship lines.

updates clearly indicate the status of each update, and the relative distance between status lines indicates trends. If the status lines are closer together than the update frequency, the project was falling behind; if farther apart, progress was better than expected.

When manually updating a bar chart, record actual start and finish dates, just as you would on a CPM schedule, and annotate the schedule with comments on progress, problems, etc.

**Enhancements.** Bar charts can include percent complete curves or resource histograms, as in Figs. 10.7 and 10.33.

## 2. Summary of critical path scheduling concepts

Critical path scheduling is the process of planning or diagraming (determining the activities comprising a project and their sequential relationships) and computing (determining the start and finish dates and float) to determine the "critical path."

**Definitions.** A *project* consists of a series of related tasks (called activities) that lead to some objective or end point within a defined period of time. The

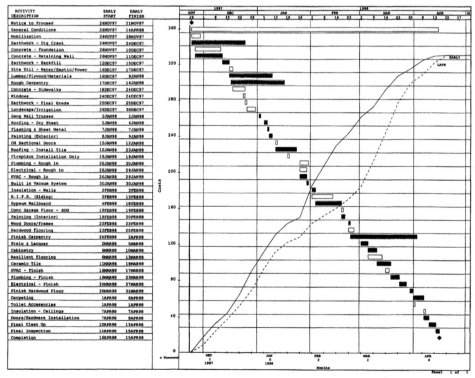

**Figure 10.7** Bar chart with additional graphics.

*objectives* for a construction project are defined by the contract documents. An *activity* is a discrete task with a definable beginning and end that must be accomplished, along with other tasks, in order to complete a project. Each activity has two events—its beginning and end. Significant events that are tracked for management purposes are called *milestones. Hammocks* are summary activities, created by "rolling up" a number of related activities to identify project phases or major work areas.

The start of an activity is constrained by the finish of its *prior activities* (predecessors). In some cases, an activity can overlap the finish of its prior or is controlled by the start of the prior. The normal breakpoint between activities is when work is handed off to another craft or when the crew moves to another work area. Subsequent activities following an activity are called *successor activities.*

The activities and their relationships are shown on a *network diagram,* which displays the activities and their relationships and can have one of several graphical formats. A *subnet* (subnetwork) is a portion of the network that is grouped together, usually representing the same work area or performed by the same organization. Organizing a network by subnets makes it more readable. A *fragnet* is a smaller portion of the network including only those activities needed to display specific information (such as substantiating a time extension request). Fragnets are also used to generate repetitive networks.

*Computing the critical path consists of two steps* accomplished by the scheduling software—the forward pass and the backward pass. The *forward pass* starts with the first activity on day one and computes the early start and early finish of each activity. The *early start* is the earliest an activity can start, provided all its priors are started and completed on schedule. The *early finish* is the earliest an activity can finish, provided it starts on time and takes no longer than planned.

The *backward pass* computes the late start and late finish dates of the activities. The *late start* is the latest an activity can start without delaying the project, and the *late finish* is the latest an activity can finish without delaying the project.

*Float* (also called total float or slack) is the number of days an activity can be delayed without delaying the project. It equals the late start minus the early start of an activity (or late finish minus early finish). *Critical path activities have zero float.* The number of days an activity can be delayed without delaying another activity is its *free float.* Only activities at the end of a network "chain" have free float.

The *critical path* is the longest continuous chain of activities through a project network. There may be more than one critical path at any point in time. The critical path determines the minimum duration of the project. A project cannot be completed faster than its critical path activities can be completed. To avoid delay, every critical path activity must start on time and take no longer than planned unless other critical activities are accelerated to make up for the delay.

The original schedule is referred to as the baseline, or version or update number 1 (or 0), and subsequent updates (normally done monthly) are numbered sequentially. *Updates* include reporting of progress (activity percent completes or days remaining, actual start and finish dates) and may include revisions. *Revisions* are changes in network logic, added or deleted activities, or changed activity durations. A revision is usually, but not always, combined with an update.

**CPM versus PERT.** Critical path scheduling was developed as two independent efforts that led to two different techniques—CPM (Critical Path Method) and PERT (Program Evaluation and Review Technique). Today, only CPM scheduling is used. However, some of the PERT concepts are important for understanding CPM.

1. *PERT scheduling.* PERT was developed on the Polaris missile program for the Navy Bureau of Ordinance by Lockheed and Booz-Allen-Hamilton. The method addressed the needs of a research and development effort for a new weapons system, without well-defined tasks, and focused on major events (milestones). PERT was developed for an extremely large program and used a statistical approach with three estimates of elapsed time between milestones.

2. *Estimating CPM activity durations while considering probability.* CPM users should understand one important concept from PERT, which is probable elapsed time. PERT requires three estimates of the elapsed time between milestones: $a$ = optimistic, $b$ = most likely, and $c$ = pessimistic. The software computes the probable elapsed time $e_t$ using the formula

$$e_t = \frac{a + 4b + c}{6}$$

This recognizes that the probability curve for the duration of an activity is not a symmetrical bell curve. The likelihood of finishing before the most likely time is small and the left side of the curve is short and steep. However, there is no limit to how badly things can go, and the right side of the curve may extend far into the future, as in Fig. 10.8. The probable elapsed time is therefore longer than the single most likely time. An experienced CPM scheduler who understands the PERT concept can pick the most probable activity duration directly by adjusting for the skew of the curve.

3. *CPM scheduling.* CPM was developed for a large capital extension program at the DuPont Company by Remington Rand, on the Univac I computer. It focused on activities, each of which had a single estimated activity duration. The use of milestones for major events was added later.

**Precedence scheduling versus i-j scheduling.** CPM scheduling originated with the i-j method of computing the schedule and displaying the diagram. i-j scheduling will eventually be abandoned but is still used on some projects, and scheduling claim analysts should understand both methods.

1. *i-j scheduling.* i-j scheduling assigns sequential numbers to the beginning and the end of each activity. The beginning number of an activity is called its i-node and the ending number is its j-node. If the j-node number of one activity

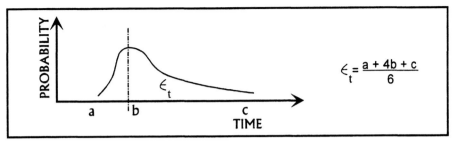

**Figure 10.8**  Activity duration probability curve and the PERT formula.

is the same as the i-node number of a second activity, the first activity must complete before the second can start. The computer scheduling software uses the i-j node numbers to determine relationships between the activities when computing the schedule.

i-j diagrams use nontimescaled arrows to represent activities that are linked together to form a network. Each activity arrow has nodes (circles) at the beginning and ending of the arrow that contain the i-j numbers and are shared with the preceding and succeeding activities. When relationships between activities cannot be displayed by linking the arrowhead of the prior to the beginning of the successor, the relationship is shown by dummy arrows which are plotted as dashed lines. Figure 10.9 is an i-j diagram (also called activity-on-arrow).

Unfortunately, the computing method is limited. Activities have only a finish-to-start relationship (e.g., an activity's prior has to finish before the activity can start) and no logs. In addition, many of the relationships between activities must be indicated with a "dummy arrow," which increases the number of activities to schedule and review. Even worse, it is extremely difficult to revise an i-j network without renumbering the activity nodes.

2. *Precedence scheduling.* Precedence CPM scheduling was developed by the H. B. Zachery Company and IBM in 1964. Precedence diagraming uses boxes instead of arrows to represent activities with lines between the boxes showing relationships. The boxes may contain additional information: the activity's start and finish dates, float, and duration. Figure 10.10 is a precedence diagram (also called activity-on-node).

Precedence computing identifies activities by a single activity number and the relationship between activities by listing each activity's priors or successors. Precedence computing allows more scheduling options (start-to-start relationships, finish-to-finish relationships, and overlap or delay between activities). Precedence computing also allows scheduling constraints (earliest possible start dates, latest allowable finish dates, and mandatory start and finish dates). Precedence computing is used by all PC-based scheduling programs, although some programs can also do i-j scheduling.

**Mixing arrow diagraming with precedence computing.**  Precedence computing is the only suitable way to compute a schedule and should always be used.

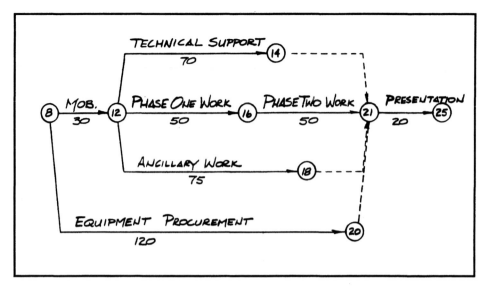

**Figure 10.9** Arrow diagram with i-j nodes (also called i-j diagram).

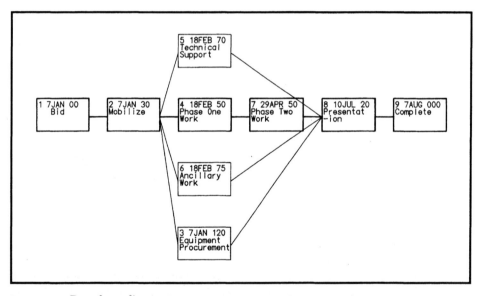

**Figure 10.10** Precedence diagram.

However, any method of diagraming can be used for creating the initial network diagram (before inputting information into the computer) and for displaying the network after computing the scheduled dates. Simply identify each activity on the network diagram with an activity number and use the activity numbers to identify priors and successors. The timescale arrow diagram is recommended.

3. **Steps in CPM scheduling**

The five steps in preparing a CPM schedule are shown in Fig. 10.11.

1. **Planning—drawing the network diagram.** Planning is the process of identifying the activities required to accomplish a project and determining the sequential relationship between those activities. The primary tool for planning and the final product of the planning process is the network diagram. The diagram displays the activities, their durations, and their relationships.

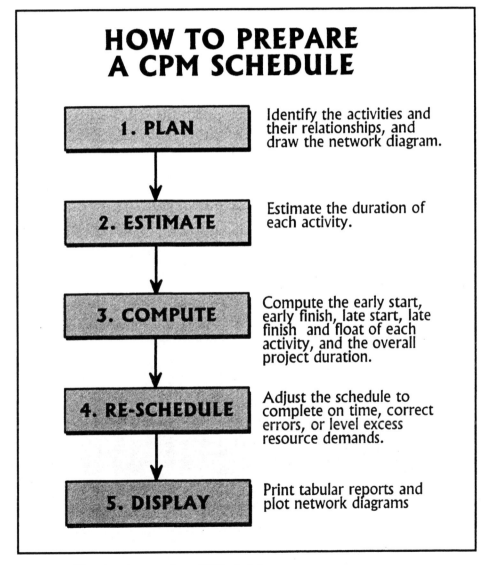

**HOW TO PREPARE A CPM SCHEDULE**

| **1. PLAN** | Identify the activities and their relationships, and draw the network diagram. |
| **2. ESTIMATE** | Estimate the duration of each activity. |
| **3. COMPUTE** | Compute the early start, early finish, late start, late finish and float of each activity, and the overall project duration. |
| **4. RE-SCHEDULE** | Adjust the schedule to complete on time, correct errors, or level excess resource demands. |
| **5. DISPLAY** | Print tabular reports and plot network diagrams |

**Figure 10.11**   Five steps in preparing a CPM schedule.

Initial diagraming of large projects should be done with pencil and paper before entering the schedule information into a scheduling program for computing dates. Simple projects can be entered directly into the computer without drawing a diagram. Whether to hand-draw a network diagram for a medium-sized project or input the schedule directly into a computer depends on the size and complexity of the project, the scheduler's experience, and the ease of using the software. My recommendation is to draw the diagram. After computing, the resulting information (activity start and finish dates and float) can be plotted on a network diagram or printed as tabular reports.

Diagrams can be nontimescale i-j diagrams like Fig. 10.9, nontimescale precedence diagrams like Fig. 10.10, bar charts with connecting lines like Figs. 10.6 or 10.22, timescale logic diagrams like Fig. 10.12, or timescale arrow diagrams like Fig. 10.18 or 10.20. Another technique, not often used, is linear scheduling, which is based on line-of-balance analysis of manufacturing operations.

Timescale logic diagrams are like connected bar charts in that each activity is represented by a bar with relationship lines between activities. However, the activity bars are placed throughout the diagram instead of one activity per line. This allows many more activities per drawing than a bar chart and is more practical for showing large networks. If there are many relationship lines, they tend to overlay, making it difficult to determine priors and successors without referring to a tabular report.

Timescale arrow diagrams are similar to timescale logic diagrams except that arrows instead of bars are used to represent activities. In addition, the arrow extends from the end of the prior activity to the beginning of the next activity instead of using relationship lines to connect the bars. The advantage over timescale logic diagrams is that the diagram is far more readable, since the relationships are easily differentiated.

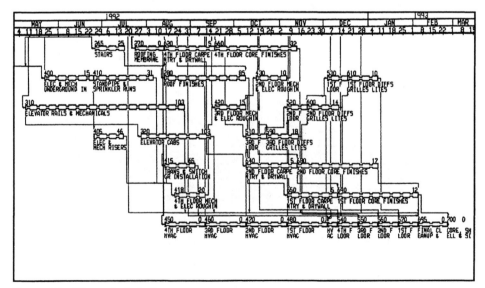

**Figure 10.12**  Timescale logic diagram.

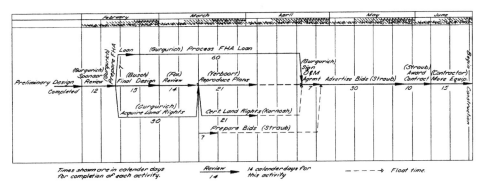

**Figure 10.13**  Hand-drawn arrow diagram, timescaled and without nodes.

Figure 10.13 is a hand-drawn timescale arrow diagram.

**2.  Estimating activity durations.**  After creating the network diagram, determine the duration of each activity. Recall the PERT probable elapsed time concept and look at similar activities on prior projects. If necessary, prepare a quantity takeoff of the work required for each activity, assume a crew, estimate their daily production rate, and compute the duration (quantity of work divided by daily production rate). Allow time for setup, learning curve productivity improvements, interruptions, bad weather, and miscellaneous ancillary tasks such as testing and cleanup.

When estimating activity durations, remember that the "work" duration can be different from the "allowance" duration. For example, asphalt paving in late fall may require 3 "work" days but may be scheduled with an "allowance" duration of 10 working days to ensure that at least 3 days of good weather will be encountered. That same activity scheduled in midsummer may have only a 4 working day duration (with the extra day allowed for setup, unexpected problems, etc.).

**3.  Computing start and finish dates and float.**  Next compute the early start (ES), early finish (EF), late start (LS), and late finish (LF) dates of each activity. Also compute the float and project duration.

Computing could be done manually, as a simple but tedious process of addition and subtraction, or graphically with a timescale arrow diagram as described in Sec. D below. Normally, however, computing is done on a computer by inputting the data from the network diagram and running the scheduling feature of the software program.

Figure 10.14 shows the standard finish-to-start relationship between activities with the prior finishing before the activity can start and the successor starting after the activity finishes. The figure also displays the following:

■ Activities as arrows.

■ Activity descriptions above the arrows. (The descriptions are represented by the letters A through G in Figs. 10.15a and 10.15b.)

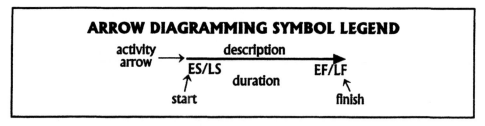

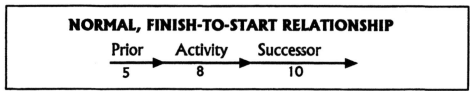

Figure 10.14   Arrow diagraming symbol legend, and finish-to-start relationship.

- Relationships between activities by linking arrowheads of prior activities to the beginning of succeeding activities (or by a dashed relationship line between priors and successors).

- Activity durations in working days below the arrow.

- Early start/late start and early finish/late finish dates (ES/LS and EF/LF) below the arrows. For this illustration, the dates are working day numbers, not calendar dates, and represent the beginning of the day. Thus an activity that starts on (the beginning of) day 1 and lasts 5 days ends on (the beginning of) day 6.

Figures 10.15a and 10.15b illustrate the computations for calculating the critical path on a sample network.
    The forward pass (Fig. 10.15a):

1. Assigns working day 1 to the beginning of the first activity (A) as its ES day (day 1).

2. Adds the duration (5 days) to get the EF date of activity A (the beginning of day 6).

3. Assigns the EF of activity A as the ES of all succeeding activities (B, C, and D).

4. Adds the durations of each succeeding activity (B, C, and D) to the ES to get the EF.

5. Assigns the EF of activity D (the beginning of day 14) to be the ES of activity F.

6. Assigns the latest EF of the activities preceding activity E to be its ES (e.g., day 14 from activity D). Even though activity C finishes on day 12, activity E cannot start until day 14 when activity D is complete.

7. Adds the durations of activities E and F to their ES dates to get their EF dates (e.g., days 19 and 16).

8. Assigns the latest EF of the activities preceding activity G to be its ES date (e.g., day 19 from activity E). The EF of activities B and F are (the beginning of) days 11 and 16.

9. Adds the duration of activity G (4 days) to get its EF date (day 23).

Activity G finishes on the beginning of day 23, and therefore the project duration is 22 days.

   The backward pass (Fig. 10.15*b*):

1. Assigns the EF (day 23) of the last activity (G) to be its LF date.

2. Subtracts the duration (4 days) of activity G to get its LS date (day 19).

3. Assigns the LS of activity G (day 19) to be the LF of all its priors (activities B, E, and F).

4. Subtracts the durations of each prior activity (B, E, and F) from its LF to get its LS.

5. Assigns the LS of activity E (day 14) to be the LF of activity C.

6. Assigns the earliest LS of the activities succeeding activity D to be its LF (e.g., day 14 from activity E).

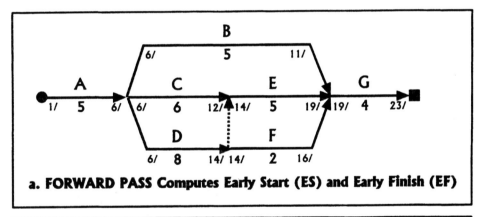

**a. FORWARD PASS Computes Early Start (ES) and Early Finish (EF)**

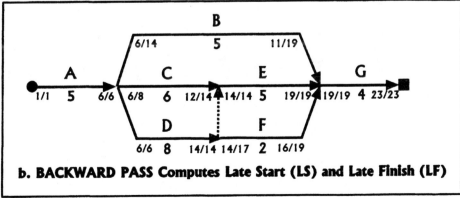

**b. BACKWARD PASS Computes Late Start (LS) and Late Finish (LF)**

**Figure 10.15**  Forward and backward pass for critical path scheduling.

7. Subtracts the durations of activities C and D (6 and 8 days) to get their LS dates (days 8 and 6).

8. Assigns the earliest LS of the activities succeeding activity A to be its LF (e.g., day 6 from activity C).

9. Subtracts the duration of activity A (5 days) to get its early start (day 1).

The float is the difference between the ES and LS (or the EF and LF) for each activity. The float of activity A is 1 minus 1 (or 6 minus 6) which is zero. Likewise the float of activities D, E, and G is zero. They are on the critical path. The float of activity C is 2 days (8 minus 6), the float of activity F is 3 days (17 minus 14), and the float of activity B is 8 days (6 minus 14).

After running the scheduling feature of the software, print tabular reports, plot a network diagram, and view the results.

**4.   Rescheduling.**   If necessary, reschedule to achieve a desired completion date, to correct errors, or to level resource demands. If accelerating to achieve a desired completion date, use the least-cost expediting procedures described in Sec. A.2. If resource-leveling, use the techniques described in Sec. C.5. Rescheduling may require committing more resources to reduce activity durations, changing activity relationships by overlapping priors to save time, adding activities, etc. It often takes several iterations to achieve an optimal schedule.

**5.   Displaying the results.**   The final step is to display the results. This can include tabular reports, bar charts, or network diagrams. If resources have been leveled, a resource histogram may be plotted. If the schedule has been cost-loaded, a cash flow/percent complete curve may also be plotted.

## 4.   Lag/lead, alternative relationships, and constrained dates

Precedence computing offers alternative relationships (to finish-to-start), lag/leads, and constrained dates that can be extremely helpful in creating a realistic schedule. They can also give misleading results if used incorrectly. These features can be illustrated with arrow diagram fragnets in Fig. 10.16-1 through 10.16-7.

**1.   Finish-to-start (FS) with lag (delay).**   Some activities have a lag or delay between the finish of their prior and the start of the activity. This may be for an unnamed activity that isn't tracked separately (such as cleaning forms between stripping and erecting). Alternately, the lag can be changed to a separate activity, but this increases the number of activities by 50 percent, which requires more data input, longer reports, more review time, etc.

**2.   Finish-to-start (FS) with lead (overlap).**   Many activities can start before their priors are complete. The amount of overlap depends upon the activities, the available working area to separate the crews, and the additional supervisory effort required to avoid interference. Alternately, the prior activity can be broken into two activities, but this requires more activities and doesn't represent the true relationship or the fact that the prior is actually a single activity.

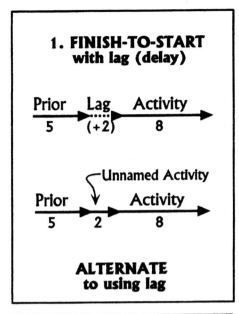

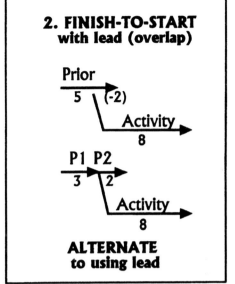

16a

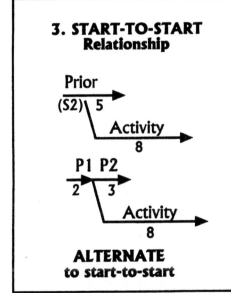

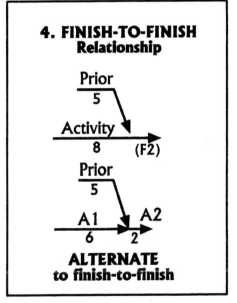

**Figure 10.16**  Lag/lead, alternative relationships, and constrained dates with arrow diagraming. (*Continued*).

**3.  Start-to-start (SS) relationships (with lag/lead).**  The start of some activities is tied more to the start of the prior than the finish of the prior. An example is pipeline work where trenching can start a week after clearing starts (i.e., a lag of 1 week), pipelaying can start 1 day after trenching starts, and backfill can start 2 days after pipelaying starts. The lag provides adequate working space for each crew and a buffer of completed work in case one crew encounters delays (provided the crew production rates are balanced).

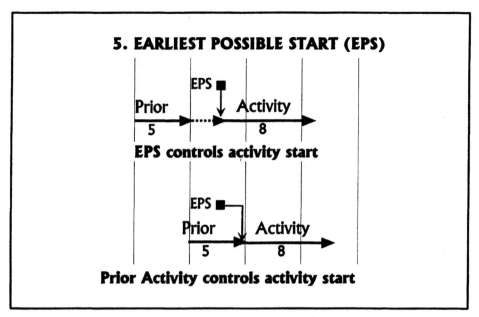

**Figure 10.16** (*Continued*).

The same result can be achieved with either a finish-to-start relationship and lead or a start-to-start relationship with lag. However, different computer scheduling routines compute somewhat differently under some circumstances, after the prior activity starts. Start-to-start relationships should be used only if applicable.

**4.  Finish-to-finish (FF) relationships (with lag/lead).**   In a few circumstances, the relationship between activities is governed by the finish of the two activities (with the start not necessarily related). For example, pipeline backfill may have a finish-to-finish relationship to lay pipe with 1 week delay, meaning backfill can't be completed until 1 week after laying pipe is complete.

**5.  Earliest possible start (EPS) date.**   An earliest possible start date is a constraint on the early start of an activity. It can prevent the activity from starting until long after all of the activity's priors are complete. If the priors push the start of the activity past the earliest possible start date, the constraint has no effect on the schedule. The early start of the activity is then controlled by the early finish of its latest prior.

Earliest possible start dates are used to control the start of activities with no priors, to start weather-sensitive work in the spring, to prevent work from starting before an environmental window, or for other reasons to prevent an activity from starting before a specified date. An earlier possible start can result in a discontinuous critical path. Earliest possible start dates are sometimes used incorrectly to force activities to start on a desired date, without using job logic. This can result in an inaccurate schedule after updating and make it difficult to prepare an acceptable scheduling claim.

6. **Latest allowable finish (LAF) date.**  A latest allowable finish date constrains the late finish of an activity. It is used by the scheduling program on the backward pass when computing late finish dates. It can cause the required late finish (i.e., the LAF) of an activity to be earlier than its computed early finish. In other words, the activity must be finished before it *can* be finished—it has negative float.

Latest allowable finish dates are used to control the contract completion and contract-specified or contractor-selected interim milestones that must be complete by a certain date. If a project has negative float, it may be caused by a latest allowable finish date on either the contract completion or an interim milestone. For example, if a bridge contract specifies a limited period of time for work in the river (i.e., an environmental window), the start of an in-river

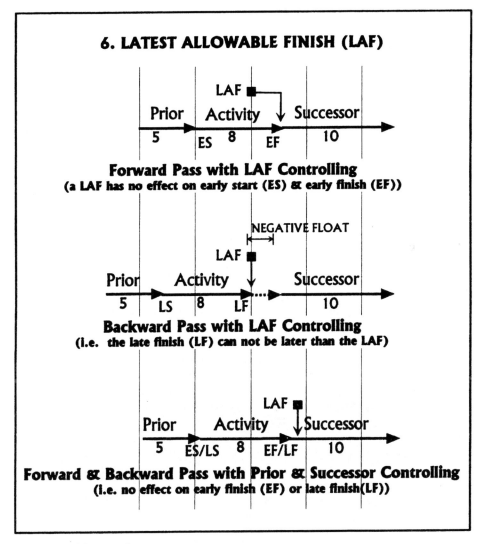

Figure 10.16  (*Continued*).

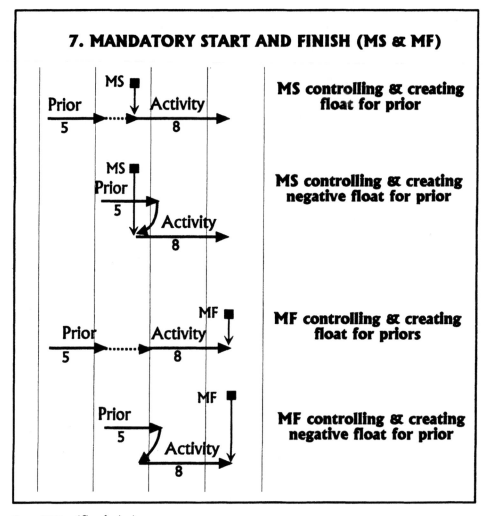

**Figure 10.16.** (*Conclusion*)

activity can be given an earliest possible start date and the end a latest allowable finish date to guarantee the work is scheduled within the window.

**7.  Mandatory start (MS) and mandatory finish (MF) dates.**   Mandatory start and mandatory finish dates force an activity to start or finish on specified dates, regardless of the start and finish of any priors. The computer scheduling routine ignores priors and schedules the activity to start on its mandatory start date and finish on its mandatory finish date if one is given. For example, work within an environmental window can be given a mandatory start date and a mandatory finish date to have the work continue for the entire period of the window. This may cause the activity's priors to have negative float.

**5. Resource loading, forecasting, and leveling**

Activity durations are dependent upon the amount of work required and the resources (labor and equipment) available to do the work. Resource availability is a major factor determining activity durations and network logic (e.g., whether activities can be concurrent or sequential). Resource forecast and use curves are a powerful tool for project planning, control, and claims analysis.

**Use crew chases to show resource constraints as activity relationships.** One means of resource leveling is to include "crew chases" as part of the network logic. Create specific activities for the most critical crews and equipment and link those activities with finish-to-start relationships to create crew chases. This ensures that the crews have work at all times and aren't scheduled for two locations at the same time. Alternately, for example, check "boom time" on high-rise building construction to verify the crane is able to service all crews without excessive overtime.

**Resource forecasting and analysis with histograms.** Often, however, resource use is more dispersed, as when a subcontractor works several crews of fluctuating size and workers are shifted from one crew to another. It may not be efficient to add more people (especially for a short period of time), additional experienced workers may not be available, or supervisors may be stretched too thin to control more workers. In addition, wide fluctuations in resource use should be minimized, which may result in apparent schedule discontinuities but can be defended on the basis of efficiency.

When the use of critical resources is more dispersed than can be controlled by crew chases, resource forecasting is needed, for at least the critical resources. It is accomplished by:

- Estimating the resource use (e.g., labor hours or crew size) for each activity.
- Spreading the estimated resource use over time for the duration of each activity.
- Adding the total resource use for each period of time (e.g., weekly) for all activities ongoing during each period, to obtain the total resource use for each period of time.
- Plotting the forecast resource use for each time period as a histogram (periodic curve) as in Fig. 10.37.

As discussed in Sec. I.5 and illustrated by Fig. 10.38, resource histograms of actual resource use can show delay, impact, and what work was being performed. Histograms comparing actual to planned resource use, as in Fig. 10.39, are even more powerful for illustrating delay and impact.

**Resource leveling with scheduled start (SS) and scheduled finish (SF) dates.** Scheduled start and finish dates can have one of two meanings. The most common meaning is the computed early start and finish dates as determined by the scheduling routine. Another meaning, which is addressed here, refers to the dates assigned to activities by a resource-leveling routine that reschedules an activity to start sometime between its early start date and late start date—in order to level resource use.

Figure 10.17 illustrates how an activity's scheduled start might be delayed past its early start date. Resource-leveling software (or field supervisors) can also stretch out or compress an activity's duration or split it into two activities in order to minimize overall resource demands or wide fluctuations in resource use.

## D.  Timescale Arrow Diagrams

In the early 1970s, Pinnell/Busch developed a noncomputer method of critical path scheduling called timescale arrow diagraming (TAD). TAD is a process of creating a network diagram on paper that combines the timescale of the bar chart with the relationships of a network diagram. An arrow diagram (Fig. 10.18) is used because the arrows can be timescaled, but the i-j nodes are deleted since precedence computing is used if computerizing the schedule.

### 1.  Advantages of timescale arrow diagraming and timescale arrow diagrams

Timescale arrow diagraming (TAD) is superior to using bar charts or other hand-drawn network diagraming techniques (connected bar charts, non-timescale arrow diagrams, precedence diagrams, or timescale logic diagrams) or (except for smaller projects) generating the schedule directly on a computer.

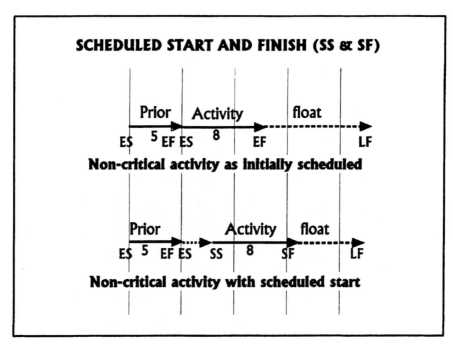

Figure 10.17   Resource leveling showing scheduled start and finish dates.

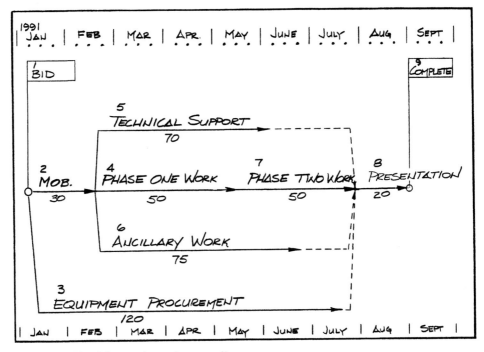

**Figure 10.18** Hand-drawn timescale arrow diagram.

**Advantages of preparing schedule with a TAD diagram.**    TAD is a flexible, powerful, interactive tool to plan and control projects. The scheduler can "see" an entire project on a single drawing, while simultaneously adding or changing tasks and considering resource constraints, work conflicts, limited space, access, weather impacts, overcrowding, etc. The scheduler also sees the critical path and the relative criticality of the activity and chain of activities being added. If an activity won't fit within the time available, the scheduler can decide to change the logic or the duration and assigned resources to maintain the completion date.

Initially TAD is somewhat more difficult and time-consuming to prepare than a conventional nontimescale network diagram, but the advantages are great. The TAD diagram needs no revisions to level resource demand, resolve work conflicts, reduce overcrowding, or avoid scheduling weather-sensitive work in the winter. That is all done during the initial preparation. The project manager or superintendent preparing the diagram "builds the project on paper."

Simple projects can be scheduled directly on the computer. But it is faster and easier to schedule large projects on paper with the TAD technique than to create a complex schedule directly on the computer. Once developed, the network can be computerized for updating and tracking progress. Or forgo computerization and control the project with a noncomputer TAD network by using status lines and manually drawing in changes. Eventually, scheduling software will be superior to manually preparing a TAD diagram on paper even for large projects. However, this will require easier software and larger monitors than are currently available.

**Advantages of using a TAD diagram to analyze a schedule or to track and control progress.** TAD diagrams are more usable than nontimescale network diagrams. Instead of reading two separate diagrams (i.e., a bar chart and a nontimescale arrow or precedence diagram or a tabular list of dates), you have a single document that displays everything. The critical path is identified at a glance, and the float of a chain of noncritical activities, whether a few days or several weeks, is easy to see.

TAD diagrams are also more usable than connected bar charts or timescale logic diagrams. The TAD drawing displays (1) far more information, (2) in much less space, (3) with greater clarity, and (4) with a sense of organization and flow that communicates more information. A TAD diagram for a given project is much smaller, easier to read, and far easier to understand. TAD diagrams also display float, the critical path, and subnetworks better than other network techniques. The primary advantage of a TAD diagram is that all relationships are clearly displayed on the diagram. The other methods often require use of a tabular report to determine relationships, as the relationship lines on the diagram tend to overlap.

**Advantages for claims analysis and presentations.** Replotting an existing schedule as a TAD diagram speeds reviewing and understanding a schedule or a fragnet of a delay. Most importantly, fact finders (arbitrators, judges, and juries) "see" the critical path and understand the logic.

## 2. Preparation of hand-drawn timescale arrow diagrams

The start and finish dates of each activity are graphically computed as the TAD is prepared, and the float displays as horizontal relationship lines at the end of each chain of noncritical activities. Figure 10.18 is an example timescale arrow diagram.

**Preparing the initial diagram.** A TAD schedule is normally prepared on fade-out grid paper (Clearprint) to aid in the layout and timescaling of the diagram. Grid paper with a 10 by 10 grid (per inch, with heavier lines at 1-inch increments) and a scale of 1 inch per week works best for most projects. The horizontal grid aids drawing straight horizontal lines 1 inch apart and the vertical grid aids timscaling for a 5-day week between the 1-inch heavier lines.

On large long-duration projects 1 inch per month and an 8×8 grid works better. Two grid lines approximate 1 week and activities are planned in monthly, weekly, or 5 workday increments.

For details on preparing hand-drawn TAD diagrams, see Sec. F of Chap. 1 or refer to my articles in *Civil Engineering Magazine* (August 1980) and *Pacific Builder & Engineer Magazine* (February 1981).

**Updating.** Update a hand-drawn TAD by drawing status lines at the update date from the calendar at the top of the drawing down through the activities. Jog to the left or right depending upon whether and how much an activity is behind or ahead of schedule, just as with a bar chart. Box and hatch out activities progressing out of sequence and ahead of schedule. Box and leave open out-of-sequence activities behind schedule that are far to the left. Figure 10.19 is a hand-drawn TAD fragnet showing a status line update.

3. **Computerized timescale arrow diagrams**

In 1980, my firm introduced project management software called PMS80 that generates a TAD diagram. The software is similar in scope and features to the Primavera software, but generates a TAD diagram instead of a timescale logic diagram. It contains special features for claims analysis, including the ELIPSE schedule described in Sec. H below. The software operates on MS-DOS only, but a condensed version that generates a TAD diagram, after importing computed schedule data from other software, may be available on MS Windows in 1999. PMS80 solves the dilemma of using other types of timescale graphs (connected bar charts or timescale logic diagrams), instead of TAD diagrams, for computerized schedules. It also allows far more detail on the size of drawing than hand-drawn networks, as evidenced by Fig. 10.20.

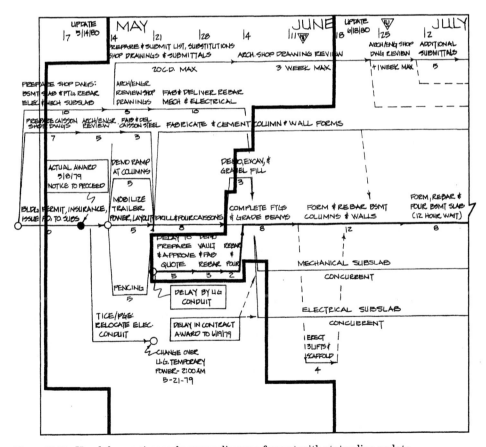

**Figure 10.19**  Hand-drawn timescale arrow diagram fragnet with status line update.

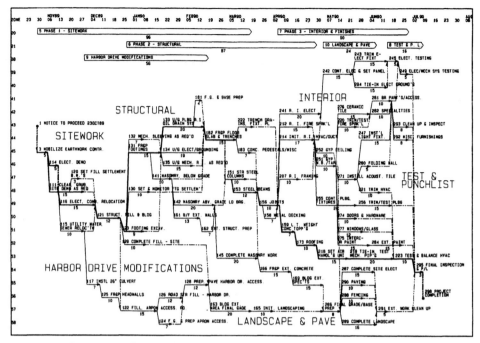

**Figure 10.20**   Computerized timescale arrow diagram.

Primavera's and SureTrak's timescale logic diagram and Microsoft Project's connected bar charts are alternatives to the TAD diagram. However, they require much larger drawings to show the same amount of information and are more difficult to read.

**4. Fragnets with timescale arrow diagrams**

TAD diagrams are especially useful as fragnets since they clearly show relationships and display far more information than other types of diagrams. Reviewers easily "see" the impact of changes and the change in float on one page. Refer to Fig. 10.31 for an example, and read Sec. I.1.

**E.   Schedule Analysis Procedures**

If delays or acceleration occurred on a project, the claim must include a schedule analysis. Although a preliminary schedule review is needed before preparing the claim analyses (as described in Chap. 9), the detailed schedule analysis should be done after understanding the project and examining all the issues in dispute. The results of the schedule analysis are then used to complete the issue analyses.

The four basic steps to analyzing scheduling claims are as listed in Fig. 10.21 and are described in detail below. Three schedules are needed: (1) the as-planned, (2) the as-built, and (3) the schedule that would have been accomplished but for changes by the owner. The would-have-been schedule includes the excusable delay (for noncompensable time extension) but not the compensable delay. Comparison of the as-built schedule with the would-have-been schedule determines the compensable delay (for escalation, extended overhead, equipment standby, and compensable time extension). The difference in working conditions between the as-built and would-have-been schedules determines the impact and is used to compute efficiency.

To perform the analysis, follow the basic steps in Fig. 10.21.

## 1. Proving delay or acceleration

If work is delayed for excusable reasons, the amount of delay must be proved to obtain a time extension. If work is delayed or accelerated for compensable reasons, the amount of delay or the extent of acceleration must be proved before damages can be recovered. Historically, this was accomplished by establishing that the project was completed late or by directly calculating the length of the owner-caused delay. However now CPM scheduling must be used to prove delay, acceleration, and impact. It is the only reliable means to determine if an action or event delayed the project.

Substantiating the "intent" of the project team is a major problem in schedule analysis. Intent is commonly represented by the as-planned schedule, but this schedule is often inaccurate and insufficiently detailed. Few schedules reflect the superintendent's detailed job planning. Many only minimally comply with the scheduling specifications.

Documenting what actually happened is another problem. Although appearing to be a straightforward inquiry, it can be difficult to determine what happened and why.

Determining what "would have happened, but for" the actions of the owner is the greatest challenge. A single delay or series of delays can change the circumstances and working conditions so extensively that it is nearly impossible to determine what the parties would have done and what really would have happened if the delay had not occurred. The methodology described below resolves these problems and generates the most likely scenario.

## 2. Identify the scheduling issues in dispute

Examine the contract documents, the schedule, and the scheduling issues in dispute.

1. *Review the contract scheduling provisions.* Review the contract scheduling clauses: the contract duration, specified milestones or construction sequences, schedule reporting requirements, liquidated damages, etc.

2. *Identify the major functional phases and milestones.* Identify the major project milestones. These are events that either control the start of significant phases of construction or indicate the accomplishment of key activities. For building construction, the major milestones include:
- Notice to proceed
- Start excavation

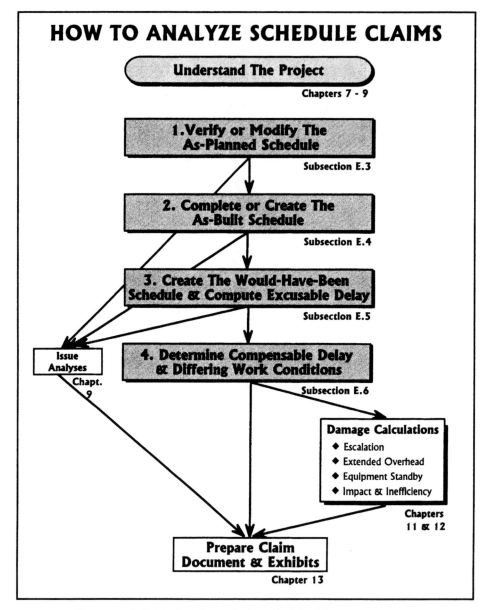

**Figure 10.21**  Recommended procedure for analyzing schedule claims.

- Issuance of building permit
- Delivery of key materials, if procurement is an issue
- Start foundation concrete
- Concrete pour dates for critical elements of the structure
- Start structural framing
- Start and finish roofing
- Building enclosed (roofing, siding and windows or temporary enclosure)

- Inspection of structural work, plumbing, and electrical rough-in
- Start insulation
- Start sheetrock
- Start carpeting, floor covering, and painting
- Finish system commissioning
- Permanent power and heat available
- Final inspection
- Certificate of occupancy
- Substantial completion
- Move in

Although as-planned schedule activities usually have finish-to-start relationships without overlap, actual construction includes considerable overlap and start-to-start relationships. Therefore, tracking start dates is more important than tracking finish dates, since the critical path often proceeds to the next trade long before the first trade finishes all of its work.

3. *List and document the scheduling issues in dispute.* Prepare a list of the known scheduling issues to be investigated, analyzed, and confirmed or disproved. Establish a file for each with relevant source documents and record all facts regarding each issue.

## 3. Confirm the as-planned schedule

Next, confirm the adequacy of the as-planned schedule. If necessary, computerize the original schedule and correct any errors. Verification of the as-planned schedule normally includes the following steps:

1. *Convert bar chart schedules into CPM schedules.* If the as-planned schedule is a bar chart, convert it into a CPM schedule. Draw relationship lines between the activities on the bar chart, most of which will be finish-to-start relationships as in Fig. 10.22. The relationship between bar chart activities is often obvious, as the end of one activity coincides with the beginning of another. In other cases, the relationship is unclear and analysis is needed to discover why an overlap or gap exists.

The relationships in Fig. 10.22 were added, based on the reviewer's knowledge of the work. The hand-drawn arrows show relationships, starting with the notice to proceed. Float is represented by horizontal dashed lines, followed by a relationship arrow to another activity. For example, activity 2 (mobilization) has over 1 week's float and must be complete by 2 weeks after the start of activity 6 (general conditions) to avoid delay by activity 6. However, the relationship between the end of activity 3 (demo kitchen) and the beginning of activity 5 (asbestos removal) has no float and the activity appears to be on the critical path. In fact, the critical chain of activities can be traced from activity 1 to activity 27. Incidentally, the schedule was already CPM scheduled and noncritical activities display differently from critical activities.

2. *Computerize the as-planned schedule.* If not computerized, computerize the as-planned schedule for processing and comparison with the as-built

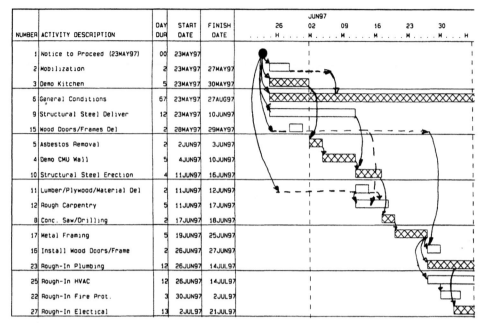

| NUMBER | ACTIVITY DESCRIPTION | DAY DUR | START DATE | FINISH DATE |
|---|---|---|---|---|
| 1 | Notice to Proceed (23MAY97) | 00 | 23MAY97 | |
| 2 | Mobilization | 2 | 23MAY97 | 27MAY97 |
| 3 | Demo Kitchen | 5 | 23MAY97 | 30MAY97 |
| 6 | General Conditions | 67 | 23MAY97 | 27AUG97 |
| 9 | Structural Steel Deliver | 12 | 23MAY97 | 10JUN97 |
| 15 | Wood Doors/Frames Del | 2 | 28MAY97 | 29MAY97 |
| 5 | Asbestos Removal | 2 | 2JUN97 | 3JUN97 |
| 4 | Demo CMU Wall | 5 | 4JUN97 | 10JUN97 |
| 10 | Structural Steel Erection | 4 | 11JUN97 | 16JUN97 |
| 11 | Lumber/Plywood/Material Del | 2 | 11JUN97 | 12JUN97 |
| 12 | Rough Carpentry | 5 | 11JUN97 | 17JUN97 |
| 8 | Conc. Saw/Drilling | 2 | 17JUN97 | 18JUN97 |
| 17 | Metal Framing | 5 | 19JUN97 | 25JUN97 |
| 16 | Install Wood Doors/Frame | 2 | 26JUN97 | 27JUN97 |
| 23 | Rough-In Plumbing | 12 | 26JUN97 | 14JUL97 |
| 25 | Rough-In HVAC | 12 | 26JUN97 | 14JUL97 |
| 22 | Rough-In Fire Prot. | 3 | 30JUN97 | 2JUL97 |
| 27 | Rough-In Electical | 13 | 2JUL97 | 21JUL97 |

**Figure 10.22** Marked-up bar chart with hand-drawn relationship lines.

schedule. Computerization allows comparing two schedules using a comparison bar chart, creating special charts such as the ELIPSE schedule, and plotting network diagrams that are easier to understand than hand-drawn networks or bar charts.

If using a scheduling software program different from that used to generate the as-planned schedule, create an ASCII file (by sending reports to a disk file) for importing to the preferred program. Most scheduling software programs import several different data formats, and the major programs comply with the Corps of Engineers' NAS Data Exchange format (ER 1-1-11, March 15, 1990) for the transfer of scheduling data between different programs. Others interface with Microsoft Project, a popular scheduling program for smaller projects. If the original schedule is unavailable on disk, reentering the data isn't especially time-consuming or difficult for a skilled secretary.

Plot a network diagram and print tabular reports. Verify that that the start and finish dates are the same as on the original schedule and that the network logic is correct.

*3. If necessary, modify the as-planned schedule (or create one if none exists.* Hopefully, the as-planned schedule is complete and accurate. If it contains obvious material errors that would have been corrected as construction proceeded (without additional delays), make the corrections. If the as-planned schedule lacks detail, modify it to better compare the as-planned schedule with the as-built. Add missing activities, correct logic errors, break some activities into multiple activities and add milestones and project phase hammocks, combine activities, etc.

When modifying an as-planned schedule, converting a bar chart into a CPM schedule, or computerizing a noncomputerized schedule, record all assumptions and the reasons for changes (the better scheduling software programs allow attaching notes to activities and printing a report with the notes). Include these data with the schedule presented to the claim reviewer. It is *essential* that all modifications be justifiable and fully documented. Otherwise, your entire schedule analysis may be rejected in arbitration or litigation.[18]

The modified as-planned schedule must be consistent with the estimate and with the actual work (except when prevented by delays or changes). For example, do not propose a would-have-been schedule requiring special equipment unless there is some evidence the project team planned to use that equipment when creating the as-planned schedule.

If no as-planned schedule exists, reconstruct what it probably would have been from the project records. Determine the required start and finish dates and interim milestones from the contract documents. Interview the project team members to determine their expectations (document the interviews) and examine other project records. Create the most probable schedule and note any likely variations.

4. *Convert the as-planned schedule to a timescaled network.* Timescaled network diagrams are *far* easier to analyze than tabular reports, nontimescale diagrams, or conventional bar charts. If the original schedule doesn't include a timescale network (or connected bar chart), create one of the following to facilitate analysis.

- Connected bar chart as produced by Microsoft Project.
- Timescaled logic diagram as produced by Primavera or SureTrak.
- Timescale arrow diagram produced by PMS80.
- Hand-drawn timescale arrow diagram as described in Sec. D.
- Hand-drawn relationship lines on a bar chart.

At a minimum, if scheduling software isn't available, hand draw a timescale arrow diagram of the critical path and activities in dispute. Use grid paper with a 10 by 10 grid and a scale of 1 inch per week for a typical project. Draw the critical path activities, either head to tail on the same line or on adjacent lines, identify the critical path, and add relationship lines using the techniques described in Sec. D.

5. *Review the as-planned schedule.* Review the as-planned schedule to determine the initial plan of construction. Identify major phases and critical milestones. The phases can be either natural elements of the work (e.g., excavation, foundation, structural, enclosure, rough-in, and finish work for a building) or known problem periods.

Trace the critical path(s) from notice to proceed to final completion. Then trace near-critical paths and disputed work activities.

6. *Be prepared to verify resource demand and availability.* Resource demands for the as-planned schedule may be high, especially for an early-start schedule that hasn't been resource-leveled. The claim reviewer may assert that the as-planned schedule wasn't practical because of excessive resource demands. If that is likely, resource-load the activities (for the most critical crafts only) and

check the resource demands by craft before submitting a scheduling claim. Then resource-level the project, while maintaining the project completion date.

Contractors need not be concerned if the initial as-planned schedule has excessive resource demands and wide fluctuations, as long as it can be resource-leveled to reasonable limits. Field supervisors make these adjustments with short-interval schedules and weekly and daily job planning.

## 4. Create or complete an as-built schedule

Next create an as-built schedule showing what actually happened. The as-built schedule is normally based on the last schedule update maintained during construction. However, in some cases it may be necessary or advisable to create a detailed as-built schedule from the basic job records.

**Create an as-built schedule from the schedule updates and other information.** An as-built schedule can be created from the final schedule update, if:

- The project team prepared a reasonably detailed and accurate as-planned schedule.
- The schedule was maintained during construction, updated monthly, revised for changes, and actual start and finish dates were entered.
- You add start and finish dates for the activities uncompleted on the last update and make any minor corrections.

To create an as-built schedule from the schedule updates, follow these steps:

1. *Review the schedule updates.* Briefly examine each schedule update in sequential order to understand how the project was built, when delays occurred, and the intent of the project team during construction. Check for changes in the completion date and major milestones. Then review the final update in detail, since it is normally used to create the as-built schedule.

Verify several key dates of the last schedule update before converting it to the as-built schedule. Compare the recorded actual dates with dates from other records. If differences exist on more than one significant activity, examine the update in more detail and correct inaccuracies. Optionally, use a comparison bar chart to review the schedule updates. Compare each update with the original schedule or prior update (or with the most recently approved schedule). When analyzing the differences between two updates, use the software's tabular schedule comparison report (similar to Fig. 10.23) to identify the differences. If a timescaled diagram wasn't generated during construction, optionally plot progress on a timescaled diagram.

2. *Review the short-interval schedules.* Review the short-interval schedules in sequence, correlating the short-interval schedule tasks to the master schedule activities and adding activities to the last update if needed.

3. *Review any schedules to completion.* Near project completion, some project teams no longer update the master schedule and prepare a bar chart showing the remaining tasks to substantial completion. These tasks are seldom correlated to the activities on the main network and are shown in far

```
C1130                                                 PMS80                                    Update: 13 Page: 1
Compare As-Built (4Mar) to Would Have Been But For   SCHEDULE COMPARISON - REPORT NO 11        Update date: 4MAR91
PROJECT FILE:C1130 RECORD:    13 SORT:Activity        PINNELL/BUSCH, INC. - PORTLAND, OREGON USA Print date: 03MAY91
-----------------------------------------------------------------------------------------------------------------
                           (------------------28AUG90------------------) (------------------4MAR91------------------) (---CHANGE---)
    NUM ACTIVITY DESCRIPTION   DUR EARLY ST EARLY FN T/F REMARKS          DUR D/R EARLY ST EARLY FN REMARKS           DUR  ES  EF
-----------------------------------------------------------------------------------------------------------------
       3 PERMIT DELAY                                                      13      A 3SEP90 A19SEP90 1d .04
    1000 FOOTING RT. WALL        3  4SEP90  6SEP90 14 5>3d same as act.     3      A18SEP90 A20SEP90 0d                 0   10  10
    1002 RETAINING WALL - CONC   8  7SEP90 18SEP90 14 10> W 5 act. -1d      9      A21SEP90 A 30CT90 3d .42             1   10  11
    1004 CURE WALL               5 19SEP90 25SEP90 14 7plan>5 act=11w/2"+  11      A 30CT90 A170CT90 8d2.81            6   10  16
    1006 WATERPROOF              2 26SEP90 27SEP90 14 same as plan          5      A180CT90 A240CT90 3d1.36            3   16  19
    1008 EXCAVATE FOOTINGS       2  7SEP90 10SEP90 19 3>2 same as act.      2      A 80CT90 A 90CT90 1d .01             0   21  21
    1010 SLAB FOOTING FORM & POUR 4 11SEP90 14SEP90 19 8d act w/1.73" rain  8      A 80CT90 A170CT90 5d1.73            4   19  23
    1012 SLAB - 1ST FLOOR        2 17SEP90 18SEP90 19 3d act w/1.36" rain   3      A180CT90 A220CT90 3d1.36            1   23  24
    1014                         2  7SEP90 10SEP90 23 plan=2d act=10d                                                 -2  122 120
    1016 FOUNDATION DRAINS       2 19SEP90 20SEP90 19 =plan act=10d w/2"+  10      A150CT90 A260CT90 7d2.57            8   18  26
    1018 PLUMBING UNDERGROUND    3 11SEP90 13SEP90 20 =plan act=6d w/1"+    6      A100CT90 A170CT90 4d1.72            3   21  24
    1023 CMU - 1ST FLOOR                                                    1      A 1DEC90 A 2DEC90 0d .52
    1024 BACKFILL                3 28SEP90  20CT90 14 =plan act=8d w/1.6"   8      A240CT90 A 2NOV90 6d1.59            5   18  23
    1026 CMU - 1ST FLOOR         2 30CT90   40CT90 14 3d1.43                2      A 5DEC90 A 60EC90 1d .02             0   43  43
```

**Figure 10.23**  Schedule comparison report of two updates.

greater detail. The completion schedule may be updated weekly (often not), and several updates may exist.

To convert the last update of the master schedule into an as-built schedule, obtain the start and finish dates of the uncompleted activities from the schedule to completion. Unfortunately the tasks on the completion schedule are dropped when completed without recording a completion date. This makes it difficult to create an as-built master schedule from the completion schedule unless it is updated frequently.

4. *Review any fragnets or what-if analyses.* Review fragnets and other special schedule analyses for additional information as to what actually happened and why.

5. *Review schedule-related correspondence and other records.* Letters, meeting minutes, or other project records may refer to the schedule. Review them and incorporate significant information into the as-built schedule. Check whether late submission or review of shop drawings and late material delivery delayed construction.

6. *Correct minor errors and add information.* Correct minor errors, supply missing information, change relationships, and add milestones and project phase hammocks, etc. Make the as-built schedule compatible with the modified as-planned schedule (or vice versa), to allow comparison and analysis using a comparison bar chart.

7. *Document your work.* Record sources for all as-built dates, added information, and other schedule adjustments.

**Alternately, create a detailed as-built schedule from basic project records.**  You will need to create a detailed as-built schedule from the basic project records:

1. If a detailed and accurate schedule wasn't prepared and maintained during construction.

2. If more information is needed to prove responsibility for delays or the extent of impact. The as-built schedule created during construction seldom explains disruption, impact, or the reason for delay. Instead it shows activities taking longer than planned or not starting when their prior activities are completed.

The process of creating a detailed as-built schedule is described in Sec. G. The detailed as-built schedule provides a road map of the entire project in a form that facilitates analysis and understanding. It differs substantially from an as-built schedule created from the last update because it contains more detailed information revealing essential facts that are otherwise unobtainable. This more detailed information is used to create the would-have-been-but-for schedule and to analyze scheduling issues.

**Optionally record actual resource use.** When creating the detailed as-built schedule, optionally record actual crew sizes or labor hours. Display the information on the detailed as-built diagram or on the as-planned/as-built comparison schedule. The information can be obtained from the superintendent's daily reports, or the timecards if necessary. Create a table at the bottom of the as-built schedule in tabular (or graphic) format, with each crew or trade on a line and the daily crew size (or equivalent crew size if the base data are in hours) for each day.

**Record significant as-built scheduling information in the narrative text.** The completed as-built schedule is essential for analyzing many of the issues that appear in the narrative text, as discussed in Chap. 9. It is also used to compute delay and acceleration damages, as explained in Sec. E.6 below and in Chaps. 11 and 12.

## 5. Create a "would-have-been, but for..." schedule

Next create a schedule of what would have happened absent the compensable delays but including the excusable/noncompensable delays. This requires the following steps:

1. *Decide how to create the would-have-been schedule.* Creating the would-have-been schedule requires identifying responsibility for certain events or actions, linking the events or actions to an ultimate impact (through an often convoluted chain of cause-and-effect relationships), and determining what would have happened absent the initiating events or actions.

Although the would-have-been schedule may be created immediately following creation of the as-built schedule, more facts must often be determined, assembled, and analyzed. To do this:

- Plot an as-planned versus as-built comparison schedule and examine the differences.
- Examine the detailed as-built schedule, if one was prepared.
- Prepare and analyze an ELIPSE schedule with accounting, extra work, and productivity data as explained in Sec. H.
- Examine the schedule issue analyses in the narrative text, or review key documents in the chronological source document file for more information. If necessary, interview project team members.
- Optionally, perform "what-if" analyses or schedule simulations to determine what would have happened had certain events or actions not occurred.

■ Optionally, prepare a Time Impact Analysis of each change order, as explained in Sec. E.8.

2. *Plot an as-planned versus as-built comparison bar chart.* Compare the as-planned schedule with the as-built, using a bar chart comparison diagram as in Fig. 10.24. This requires both schedules have generally have the same activities, although the as-built may contain more. Two or more as-built activities may be associated with one as-planned activity, to be plotted together.

Creating a meaningful comparison diagram without omitting activities or distorting the facts requires understanding how the software works. It may also require adding "linking" activities or making other modifications to the schedule. Most scheduling software compare only activities in the two schedules one-to-one, using the activity number. Normally when comparing two schedules, the as-built schedule is the "current" schedule (it will have some activities not included in the as-planned schedule) and the as-planned schedule is the "referenced" schedule.

Another way to ensure compatibility between the two schedules is to "roll up" the extra activities in the as-built schedule into summary activities that can be compared with equivalent activities in the as-planned schedule.

To allow comparing two schedules (as-planned to as-built or as-built to would-have been), *do not* reuse activity numbers of deleted activities or change the number of the existing activities.

| NUMBER | ACTIVITY DESCRIPTION | DAY DUR | START DATE | FINISH DATE |
|---|---|---|---|---|
| 2 | Notice to Proceed | 00 | A 3JAN96 | |
| 3 | General Conditions | 71 | A 3JAN96 | 10APR96 |
| 4 | Mobilization | 8 | A 3JAN96 | A13JAN96 |
| 5 | Demolition | 9 | A 3JAN96 | A15JAN96 |
| 6 | Earthwork | 18 | A 7JAN96 | 7FEB96 |
| 10 | Lumber Deliver | 3 | A15JAN96 | A17JAN96 |
| 7 | Concrete/Building | 5 | A20JAN96 | 6FEB96 |
| 20 | Plumbing Rough-in | 8 | 2FEB96 | 13FEB96 |
| 13 | Manufactured Wood Trusses | 1 | 6FEB96 | 6FEB96 |
| 12 | Rough Carpentry | 13 | A20JAN96 | 20FEB96 |
| 19 | Roofing | 1 | 9FEB96 | 9FEB96 |
| 22 | Vinyl Windows | 1 | 12FEB96 | 12FEB96 |
| 20 | Flashing & Sheetmetal | 1 | 13FEB96 | 13FEB96 |
| 30 | Electrical Rough-in | 6 | A29JAN96 | 19FEB96 |
| 8 | Concrete/Sidewalks | 3 | 14FEB96 | 16FEB96 |
| 11 | Vinyl Siding | 2 | 14FEB96 | 15FEB96 |
| 17 | Insulation/Walls | 1 | 20FEB96 | 20FEB96 |
| 9 | Structural Steel Handrails | 5 | 21FEB96 | 27FEB96 |
| 23 | Gypsum Wallboard | 10 | 22FEB96 | 6MAR96 |
| 26 | Painting | 10 | 7MAR96 | 20MAR96 |
| 24 | Floor Covering/Resilient | 3 | 14MAR96 | 18MAR96 |
| 21 | Doors/Frames/Hardware | 1 | 18MAR96 | 18MAR96 |
| 14 | Finish Carpentry | 5 | 21MAR96 | 27MAR96 |
| 15 | Millwork Materials | 1 | 21MAR96 | 21MAR96 |
| 16 | Cabinetry | 5 | 21MAR96 | 27MAR96 |
| 18 | Insulation/Ceiling | 1 | 21MAR96 | 21MAR96 |
| 27 | Toilet Accessories | 1 | 28MAR96 | 28MAR96 |
| 29 | Plumbing Finish | 3 | 28MAR96 | 1APR96 |
| 31 | Electrical Finish | 2 | 2APR96 | 3APR96 |
| 25 | Floor Covering/Carpeting | 5 | 4APR96 | 10APR96 |
| 32 | Completion | 000 | | 10APR96 |

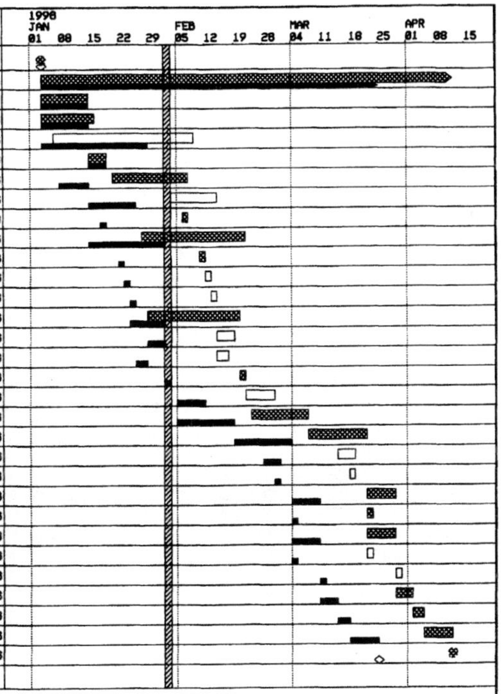

Critical  ▭ Hammock
Noncritical  ● Milestone
Comparison  ◇ Comparison

**Figure 10.24** Comparison bar chart.

To avoid confusion, plot only the controlling relationships between the as-built activities. Overlay planned and actual percent complete curves and resource histograms on the bar chart when analyzing the schedule. This is especially helpful when the events causing the impacts are identified on the diagram.

Many delays and the reasons for delay become obvious when examining the comparison schedule. Problems that are the owner's responsibility may be apparent and can be analyzed for their schedule impact. Concurrent contractor delay may also be evident, precluding both liquidated damages and compensation for extended overhead. If the causes and responsibilities are not obvious, continue with step 3.

3. *Examine the detailed as-built schedule, if one was created.* A detailed as-built schedule will provide the facts necessary to determine the responsibility for delays and impacts, and the extent of the impact.

4. *Optionally, create an ELIPSE schedule.* An ELIPSE schedule is a comparison schedule, as described in Sec. H below, incorporating additional information from other sources. It is used to develop the would-have-been schedule and can be used for analysis or as an exhibit.

5. *Optionally, review the narrative text issue analyses and the chronological notes and files.* If some scheduling issues have not yet been resolved, review the issue analyses in the narrative text and the chronological summary notes for additional information.

6. *Optionally, perform what-if analyses, computer simulation, or iterative analyses.* Conduct various what-if analyses or computer simulations to determine what would have happened. Focus on the essential elements. For example, on a large South American harbor project, I spent three weeks analyzing one activity (pile driving), breaking it into dozens of subsidiary tasks, and only three days on the remainder of the schedule. However, the pile-driving activity was repeated 400 times and controlled the duration of the entire contract.

7. *Create the "would-have-been, but for..." schedule from the above records.* Some experts start with the as-built schedule and subtract the claimed delays to generate the would-have-been schedule. This allegedly shows what would have happened "but for" delays by the owner and is commonly called a "But For" evaluation. One problem with this approach is that the subsequent additional delay caused by an initial delay is not removed. For example, a 20-day initial delay may push weather-sensitive work into winter conditions, resulting in another 10 days' delay. Simple subtraction will not identify those 10 days of impact delay unless the CPM calendar was modified to reduce the available number of work days during the extended duration.

The best approach is to build the would-have-been schedule one step at a time from the beginning. Determine what *would have* happened at the time a specific event occurred, based on the intent and knowledge of the project team and the working conditions at the time, not what *could have* happened. Correspondence can help establish this, but the as-planned schedule is more important than any other document in establishing intent. The next-best proof of what would have happened is a forceful assertion by the project manager or superintendent if it is supported (or at least not contradicted) by the records and common sense.

Adjust the duration of activities on the would-have-been schedule that would have been accomplished under different conditions. Consider weather, crowding, trade conflicts, changed work sequences, access, layout, etc. A comparison of similar activities accomplished on the project under unimpacted conditions, or a detailed analysis of similar activities, may be required. The results may be significant. On one project, such an analysis determined that a 12-day delay to a 6-month project increased the cost by nearly 50 percent. The analysis showed that those 12 days represented 42 percent of the available time to complete weather-sensitive earthwork before winter rains began. Using a separate calendar for weather-sensitive operations (e.g., earthwork), which automatically adjusts the overall duration, is an alternative to extending activity durations.

The would-have-been schedule should have the same level of detail and generally contain the same activities as the as-built schedule, to generate a comparison schedule.

The durations of unimpacted activities may be less than planned. A consistent pattern of better-than-expected progress of unimpacted activities is a strong argument for using more optimistic durations for impacted activities than the planned durations. On the other hand, slower progress than planned on unimpacted activities indicates an overly optimistic schedule. Although true for critical path activities, it may not be true for noncritical activities, as noncritical activities will usually be performed some time between their early start and late start dates. The superintendent may "borrow" resources from the nonimpacted, noncritical activities to overcome delays on the critical path.

8. *Account for excusable delays.* Differentiate between excusable and compensable delays. Record the number of days of excusable delay and include them in the would-have-been schedule. Include the direct excusable delays as well as the impact of excusable delays. Document the reason for the delay, the number of days, and why, and create a data table of excusable delay for use in the claim. For example, abnormal rainfall in early October may result in 5 days of excusable delay, which pushes weather-sensitive work into heavy rains in November, resulting in 4 additional days of delay. The total 9 days' delay is excusable and is included in the would-have-been schedule.

Owner-caused delays may be concurrent with contractor delays. If so, include both delays in the would-have-been schedule, as concurrency changes them from compensable to noncompensable but excusable.

9. *Record your assumptions, references, analysis, and conclusions. Always record your assumptions when creating a would-have-been schedule.* Figure 10.25 demonstrates the type of comments to record and how to display them.

## 6. Determine the compensable delay, time extension due, and differing working conditions

Next, determine the compensable delay and the difference in working conditions. Generate an as-built versus would-have-been schedule comparison. The difference in project completion dates is the compensable delay. The compensable delay, plus the excusable delay recorded when generating the would-have-been schedule, is the time extension due.

```
C1130                                                        PMS80                           Update: 15 Page: 2
Would have been, but for                      SCHEDULE COMPARISON - REPORT NO 11            Update date: 28AUG90
PROJECT FILE:C1130 RECORD:   15 SORT:Activity Numb  PINNELL/BUSCH, INC. - PORTLAND, OR      Print date: 24SEP91
--------------------------------------------------------------------------------------------------------------
NUM              WORK <AS-PLANNED SCHEDULE> < -------WOULD-HAVE-BEEN SCHEDULE-------> < -----CHANGES----->
BER ACTIVITY DESCRIPTION    AREA DUR EARLY ST EARLY  FN DUR EARLY ST EARLY FN REMARKS (Rainfall)   DUR  ES  EF
--------------------------------------------------------------------------------------------------------------

   1018 PLUMBING UNDERGROUND      A   3 14SEP90 18SEP90   3  7SEP90  11SEP90 1d .03 .        0   -5  -5
   1024 BACKFILL                  A   3 8OCT90  10OCT90   4 25SEP90  28SEP90 2d .09         1   -9  -8
        This activity was increased from 3 to 4 days, in spite of being scheduled
        in September when no rain fell. This was to adjust for the larger excavated
        area caused by the three sanitary sewer line taps (a change to the contract).
        This was probably overly conservative. The actual work took 8 days from 24Oct
        to 2Nov, but 6 of those days experienced rain with a total of 1.59" in
        addition to the substantial rains before work started.

   1026 CMU - 1ST FLOOR           A   3 11OCT90 15OCT90   3 1OCT90          30OCT90 3d1.43  0   -8  -8
```

**Figure 10.25**   Detailed notes on assumptions and analysis for would-have-been schedule.

The difference in working conditions for a given task on the two schedules determines compensable impact (from different weather, crowding, sequencing, etc.), for example, a foundation-forming activity would have occurred on four dry, cool fall days as shown on the would-have-been schedule but actually occurred on 6 rainy days in November as shown on the as-built schedule. This is the basic information needed to prove inefficiency as described in Chap. 12.

### 7. Conventional methods of analyzing scheduling claims

The conventional methods of schedule analysis are not recommended except in specific circumstances. They are often misused but may be acceptable if the budget precludes the better methods described above and if they are properly adjusted for contractor error. They may also be used to help explain a more detailed analysis. They include:[19]

1. *Global impact or "total delay" approach.* The global impact method of showing delay, acceleration, and impact is simple but often misused. It resembles the "total cost" approach of computing damages. It plots the as-planned and as-built schedules (and sometimes the would-have-been schedule) as highly summarized bar charts, and alleges the differences to be the owner's fault. Often no attempt is made to show contractor error or concurrent delay, or the logic of cause and effect on the overall project. A list of the alleged disruptions and delays is usually presented to substantiate the claim. The sheer number of problems and their "obvious" relationship to the overall delay is intended to convince the reviewer that additional time and money are due.

2. *"Net impact" approach.* The net impact approach resolves problems of concurrence by displaying the net effect of claimed delays. It displays the as-planned and as-built schedules as single bars or summary bar charts and shows the period of each alleged delay separately or embedded in the as-built schedule. Often the entire period of a change order (from issuance to completion of the work) is shown as a delay to the overall project. The total sum of the delays (which often exceeds the total actual delay) is either not considered or is alleged to be proof of constructive acceleration to achieve the as-built completion

date. This method does not consider that many of the delays are overlapping or to noncritical activities.

3. *"One-sided but for" approach.* The "one-sided but for" approach inserts the acknowledged delays of the party preparing the document into the as-planned schedule and alleges that all other delays are the fault of the other party. It often understates impacts from the admitted delays and makes no adjustment for circumstances and the schedule status at the time of the event. It often fails to provide adequate linkage between the cause and its alleged effect.

4. *"Adjusted as-built" approach.* The "adjusted as-built" approach uses a critical path network from the as-built schedule and inserts alleged delays as distinct activities restraining the project (while reducing as-built activity durations). It is similar to the "net impact" approach but purports to use critical path analysis.

## 8. Time Impact Analysis approach

Time Impact Analysis is an iterative process, using the then-current schedule (which must be valid for the approach to work) and adjusting it each time an impacting event occurs.[20] It is superior to the four methods described in part 7 above, and similar to the recommended approach in parts 1 through 6 above. Although the Time Impact Analysis procedure resembles the recommended method, this book uses the term Time Impact Analysis to describe a schedule analysis of a single change, accompanying a time extension request.

The steps in preparing a Time Impact Analysis are as follows:

1. Examine the scope of the change and delay, especially the planned and actual quantity and quality of work, performance, planned and actual procedures and sequence, etc.

2. Review all relevant documents including the contract, schedule, letters, and cost records.

3. Identify all parties affected and obtain their input.

4. Determine the activities affected by the change.

5. Examine the schedule and determine the change in scheduled or actual start and finish dates.

6. Identify the relevant facts of the change and its impacts.

7. Analyze the schedule using fragnets to show the planned and actual sequence of construction and administrative activities (e.g., notice).

8. Prepare a written narrative supporting the fragnet.

When performing a Time Impact Analysis, use a standardized format for a Time Impact Analysis report, similar to Fig. 10.26.

The form is completed as follows:

- Description of the work affected before the change, the plan for its accomplishment (both a narrative and a fragnet), and its status before the change (i.e., under way, imminent, etc.).

**XYZ Construction Company, Inc. – Time Impact Analysis (TIA)**

Job No/Title ______ Date ______ COR No ______ Page ____ of ____

COR Title ______ Ref Dwg/Spec ______ Scheduler ______

**DESCRIPTION OF PRE-CHANGE CONDITIONS**

Description of Affected Work (status, progress, etc.) ______

Fragnet of Affected Work:

**DESCRIPTION OF CHANGE**

Description (w/origin, development, resolution) ______

**DESCRIPTION OF CHANGED CONDITIONS AND WORK**

Description of How Change Affected Work ______

Fragnet of Changed Work:

Conclusions ______

References Attached ❑

**Figure 10.26** Time Impact Analysis form.

- Description of the change and any added work, perhaps including a sketch.
- Analysis of how the change is to be accomplished and how it will affect original work (or did affect original work).
- Description of the changed work and a fragnet of the revised CPM schedule.
- Conclusions on the impact to individual subcontractors, interim milestones, and project completion.
- Quantification and classification of delays (compensable, excusable, or nonexcusable).

- Description of impacts with information for computation of inefficiencies or extra work costs.
- References to source documents.
- Title, version, or update of the schedule used in the analysis.

### 9. Other schedule analysis techniques

There are numerous other techniques for analyzing scheduling delays and acceleration. Many are variations of the Time Impact Analysis approach. Another technique of interest is the collapsed as-built CPM analysis, which is based on the activity durations and relationships of the as-built schedule after removal of owner delays.[21]

## F. Schedule Analysis Issues

When analyzing schedules, you must consider the following issues:

### 1. Time-in-kind for time extensions

When computing a time extension, the contractor is due "time-in-kind," an equivalent time but not necessarily the same number of calendar days. For example, granting a 10-day time extension in winter is not equivalent to a 10-day delay to weather-sensitive work in summer. To prove the need for time-in-kind, plot environmental conditions (temperature, rainfall, etc.) on the as-built/would-have-been comparison schedule, and show the difference in working conditions.

### 2. Computing weather delays

Proving weather delay requires proof of three conditions:

1. The weather was abnormally severe.
2. Work was delayed by the weather.
3. The delayed work is on the critical path.

The most important condition is the effect of the weather on the work. One "gully washer" rainstorm in a month may impact mass excavation more than a light rain every day of the month with twice the total rainfall. The opposite is true for painting, which may be shut down the entire month because of steady light rainfall. Formula methods of computing delays should be applied with caution and adjusted as necessary.

### 3. Resource constrained schedules

Resource constraints (of workers or equipment) often dictate network logic and project duration without being shown on the network diagram as crew chases. Labor crews are seldom resource-leveled, and what appears to be float, for use by the subcontractors or owner, may be needed for resource leveling by the field supervisors.

Many specialty subcontractors have a limited number of skilled workers and supervisors. General availability of the workers in the trade may not be significant in the short term owing to the difficulty of integrating untested personnel into an established crew.

Delaying a crew or subcontractor past their scheduled start date may cause additional delay if that crew or subcontractor has a limited window available to perform the work. They may be committed to other work and be unavailable until some later time.

## 4. Concurrent delay

Concurrent delay occurs when both the owner and contractor delay the project or when either party delays the project during an excusable but noncompensable delay (e.g., abnormal weather). The delays need not occur simultaneously but can be on two parallel critical path chains.

**Contractor mitigation of delay (pacing) misinterpreted as concurrent delay.**  Contractors normally slow the pace of other work if critical path work is delayed, to avoid "grinding to a halt" and attendant inefficiencies. Workers are either shifted from noncritical to critical activities or less productive employees are laid off. Unfortunately, the owner is seldom informed, which leaves the contractor open to accusations of concurrent delay. Therefore, contractors should inform the owner's representative of their actions when attempting to mitigate damages.

**Segregation of concurrent delays.**  Some owners and courts lump concurrent delays together and deny compensation, even if owner delay exceeds contractor delay or is not truly concurrent with contractor delay. Creating a detailed as-built schedule and careful analysis avoids this problem.

## 5. Ownership of the float

Conventional wisdom, and legal practice, assigns the float to whoever uses it first. That is fair only under the specific conditions described in Sec. E.3 of Chap. 3.

**Contractor's right to schedule and control progress (e.g., means and methods).**  Unless otherwise specified, the schedule belongs to the contractor, who plans and executes the work in any reasonable manner, as long as the owner is:

- Informed of current status and future progress.

- Assured of timely completion.

- Enabled to perform owner duties and actions (inspection, construction staking, preliminary move-in, ancillary construction, etc.).

**The accuracy of schedules.**  *Schedules are models of what should happen, with estimates of the time required.* Schedules are less accurate than cost estimates. Cost estimates may be accurate within 3 to 5 percent for the overall project, but only 5 to 10 percent accurate for individual cost codes. Noncritical chains of activities often take longer than planned, and activity durations have little contingency. Therefore, some noncritical activities are likely to end up being critical, even if the owner does not use the float.

**Contractor's use of and need for float.**  Contractors need float for (1) noncomputerized resource leveling by the field supervisors, (2) resolution of the multiple small inaccuracies inherent in any schedule, (3) flexibility to switch crews to other work and maintain productivity when unexpected changes occur, (4) avoiding delay if noncritical work takes longer than planned or scheduling errors are made, and (5) optionally accelerating to finish early. The owner's use of float can preclude these very important options and cost the contractor more money, even if the reduced float doesn't lead to project delay.

Few construction schedules are resource-loaded and fewer are resource-leveled, owing to the effort required and the limited resource-leveling capability of scheduling software. If schedules were resource-loaded, the resource forecasts would show erratic fluctuations in resource demand. Yet wide fluctuations seldom occur in practice, because of intuitive leveling by superintendents when preparing short-interval schedules and by foremen in planning daily work assignments. Workers are shifted between crews, crews are shifted between tasks, and tasks are worked out of sequence (within limits).

The actual job logic, resource assignment, and number of tasks are an order of magnitude more numerous and complex than the most detailed schedule and the most sophisticated resource-leveling software. It isn't practical to prepare or maintain a master schedule with that level of detail. Therefore, owners should not commandeer the contractor's float without considering the contractor's needs for shifting resources and this greater level of complexity.

**Resource leveling to protect the float.**  Resource-leveling algorithms use the float to level resource demands, thus protecting the float from indiscriminate owner use. This is a strong argument for resource loading and leveling the as-planned schedule. If the owner has used the float of a schedule that wasn't resource-leveled and the project was subsequently delayed, resource-load and -level the schedule to prove the float never existed.

**Hiding float on the schedule.**  Contractors often "hide" some schedule float as a contingency or to prevent owner use, as activity durations are always uncertain and unexpected problems (unseasonable weather, accidents, etc.) might occur. If done in excess, the schedules become unrealistic and unusable as management tools and may be used against a contractor in a dispute. However, this practice will probably continue until a more realistic allocation of float is adopted, as recommended in Chap. 3.

## 6.  Delay claims for early-completion schedules

Contractors have the right to finish early and save overhead costs. Owners cannot unreasonably delay progress to prevent early completion without paying for the contractor's increased overhead costs. However, the owner must be notified of the planned early completion, and the submitted schedule must indicate the early completion. The owner is not obligated to comply with unreasonable demands (i.e., too little time for shop drawing review, construction staking, or delivery of owner-furnished materials). When claiming delay to an early-completion schedule project, it helps to present a prebid schedule and a general conditions bid estimate based on the shorter duration.

## 7.  Owner scheduling defenses and counterclaims

Section E of Chap. 9 describes how to deal with owner scheduling claim defenses and counterclaims.

## 8.  Reasons for acceleration

It is possible, and not infrequent, to have a delay claim and an acceleration claim on the same project. After delay by the owner, contractors are often forced into constructive acceleration to avoid either liquidated damages or pending inclement weather conditions.

## 9.  Recommended changes to improve scheduling practices and claims resolution

The following changes will improve construction scheduling practices and facilitate resolution of scheduling claims if problems occur.

**Implement improved scheduling practices.**   Eight recommended practices to improve scheduling are:

1. *Document simple but thorough techniques, procedures, and standards.* Establish and document techniques, standards, and procedures for scheduling, updating, and controlling progress. Give joint responsibility for scheduling to the project manager and the superintendent and provide them with the knowledge, tools, and support to do a good job.

2. *Train project personnel.* Train all supervisory personnel including foremen, superintendents, estimators, project engineers, and project managers in CPM concepts and techniques. This requires (1) a 1-day seminar by a skilled trainer, (2) breaking the class into teams, (3) 4 to 8 hours of subsequent team effort during the following 2 weeks to prepare a small schedule for an actual project, (4) a brief follow-up session by the trainer to review and critique each team's schedule, and (5) continuing verification by company management that the techniques and company-adopted procedures are being followed or that variances from the standard are acceptable or justified by circumstances.

3. *Train a part-time in-house scheduling expert.* Select and train an in-house expert to coach and advise the project managers and superintendents in scheduling projects. The expert also reviews all project schedules and helps prepare scheduling-related claims but generally should not prepare the project team's schedules. *The project team must prepare the schedule or have significant input*

*and commit to the project schedule.* For that reason, the term scheduler as used in this book refers to the individual who prepares the schedule, usually a project team member, and not to a company scheduler.

4. *Provide software support or train all schedulers.* Teach all schedulers to use the company's scheduling software or provide software support. Individuals unskilled in the software can be taught to create hand-drawn timescale arrow diagrams as described in Sec. D and to read computerized reports and charts. Office staff can computerize schedules for generating reports and updating for progress. It is seldom economical for all project team member schedulers to become expert in the scheduling software, but they should be supported by an expert.

5. *Prepare CPM schedules on all projects.* Regardless of the contract specifications, prepare and maintain a CPM schedule on all projects.

6. *Use fragnets for what-if analysis and time impact analyses.* Fragnets are used for what-if analysis and to show the impact of changes. Time impact analyses should include preparation of a fragnet of the affected activities to show whether the critical path was delayed.

7. *Submit time impact analyses for changes.* Promptly prepare and submit a time impact analysis, as described in Sec. E.8 above, for all changes. When delays are encountered, promptly submit a request for a time extension and press for early resolution.

8. *Use short-interval scheduling for weekly control.* Although the master schedule is the most important, most projects are built with short-interval bar chart schedules prepared by the project superintendent. This works well if the short-interval schedule ties each task back to an activity on the master schedule. Otherwise critical activities may be overlooked, the true project status is obscured, and preparation of an as-built schedule is difficult.

To minimize schedule slippage, display the past 2 weeks' progress and the next 3 to 4 weeks' planned work on the short-interval schedule. Plot a comparison schedule with the current plan and recent history compared to the previous week's (or baseline target schedule's) plan. Comparing each crew's planned progress with actual and distributing the information to everyone on the project generates pressure on subcontractors and supervisors to stay on schedule.

Computerization of short-interval schedules is recommended, especially if comparing the current week's plan with prior plans and actual progress. A simple scheduling program, a spreadsheet, or even word-processing software saves considerable time over generating the schedule manually. Do not prepare short-interval schedules on an erasable board unless a record is maintained and the subcontractors and field supervisors receive copies.

The superintendent's short-interval schedule is normally presented and discussed at weekly progress meetings. If possible, distribute the schedules before the meeting (by fax) so that the subcontractors can be prepared to discuss the plan and their ability to meet it. The general contractor should keep notes and distribute a summary of decisions, plus the finalized short-interval schedule if changes are made.

**Implement better scheduling standards.**   Company standards are needed to ensure that all schedules are of acceptable quality and detail, and to facilitate review by other company personnel.

1. *Identify all relevant job logic.* Carefully examine each activity and its relation to preceding and succeeding activities. This requires understanding the work sequence of each trade and the interface between trades. It also requires a reasonable level of detail.

2. *Use understandable activity descriptions.* Use verb-adjective-noun descriptions (e.g., "Pour East Footing Concrete") for clarity. Avoid milestone-type descriptions (e.g., "Complete East Footings") or overly general descriptions. The descriptions should be understandable by someone unfamiliar with the project. Use common abbreviations and standardized descriptions of work areas, responsibilities, and functions.

3. *Accurately estimate activity durations.* Activity durations are normally given in working days, although most scheduling software also accepts calendar day durations (for administrative actions, concrete curing, etc.). Unless a commitment is made to expedite with additional cost and management time, estimate durations independently of the need to finish early. Always confirm the estimated durations with the person responsible for the activity.

4. *Assign activity numbers (identifiers).* Avoid long or complex numbering codes relating to responsibility, work area, etc. Use a three- or four-digit number (avoid letters and do not exceed five digits) with the first one or two digits related to the subnetwork and the last one or two numbers sequential within the subnetwork (to aid in locating it on the diagram). Some limited coding within the activity number may be helpful, but use separate codes instead of complex activity numbering systems.

5. *Identify responsibility and work area codes.* Assign responsibility codes and work area codes to all activities and include the codes on the network diagrams and tabular reports to aid in tracking the work or preparing claims. Include the codes in the activity description or in special fields. Avoid overly long complex codes. Verify that each subcontractor has at least one activity and that the work in each work area is included in the network.

6. *Check for resource constraints.* To avoid delays, use "crew chases" as part of the network logic and resource-level the schedule if practical. Resource leveling is an extensive, time-consuming effort and few contractors are willing, or have the time, to do it.

7. *Ensure on-time delivery of materials and installed equipment.* Show delivery of key material items on the CPM schedule. *Always prepare and maintain a submittal log* that ties delivery of key materials to the early start date of the activity needing the material.

8. *Don't overlook significant work.* Ensure that each bid item on unit price contracts is included in an activity.

9. *Check other factors affecting working conditions, productivity, and activity durations.* Check weather, space limits, access, available work areas, safety constraints, spatial relationships, etc., for potential problems. For example, avoid paving in winter. If unavoidable, allow plenty of time to encounter, say, a

two- or three-day window of good weather. Also, avoid scheduling four pipefitters, two plumbers, an electrician, and a painter in a typical mechanical room at the same time. Check the specific sequence of foundation and utility work in restricted space, as it may require excavation, partial construction, backfill, and construction of the next element to avoid trench collapse.

10. *Achieve the needed level of detail.* Create enough activity detail to identify relationships with other activities and possible constraints, but not too much detail. There isn't enough time or information at the beginning of a project to prepare an extremely detailed schedule. Add more detail as the work proceeds, and allow for the fact that the superintendent's short interval schedule will contain five to ten times the number of activities of the master schedule. Provide enough detail to guide the superintendent, but don't try to micromanage.

The maximum level of detail should be a single crew, with activity durations averaging a week or longer, with a few exceptions. The minimum level of detail depends on how closely you control the project. For guidance, look at the average value of each activity. For example, do you want to separately identify tasks valued at $1/4 million, $100,000, or $10,000? Also, at what level do you need to track subcontractors? What is necessary to prove delay for a time extension? More detail is better but takes additional effort to prepare and to ensure against mistakes.

11. *Obtain input and commitment from those doing the work.* The superintendent should either prepare the schedule or provide major input and commit to accomplishing the schedule as planned. It is also essential that subcontractors provide input and commit to their portion of the schedule. Then all parties must continue to be involved in the updates and revisions.

12. *Sorting and selective printing tabular reports.* Selectively print tabular reports to limit the activities to only those of interest to the user (e.g., only that supervisor's activities and only those scheduled for the next 30 days).

Use one of the following sorts for easier review:

- Activity number to facilitate looking up a specific activity (e.g., as noted on the network diagram).
- Float/early start to facilitate tracing the critical path from beginning to end, and then tracing each set of sequential activities in a subnetwork chain of activities with the same float.
- Responsibility/another sort to examine the activities of a specific subcontractor or general contractor crew.
- Work area/another sort to examine the activities in a specific area of the project.

**Share most scheduling information with subcontractors, suppliers, and the owner.**
Sharing scheduling information is a double-edged sword. If not shared with the owner, contract rights are risked and contract obligations may not be met. Without subcontractor input, the schedule is unreliable, and the subcontractors cannot be held to it. If too much information is provided, you may be held to what was proposed. The owner may use the float that

you later need and subcontractors may claim extended overhead and impact if delayed past their scheduled dates. Sharing most scheduling information is recommended but requires a thorough and capable effort to avoid problems.

The recommended procedure for sharing schedule information is a partnering approach, as follows:

1. Prepare the draft schedule using your own plan and initial input by key subcontractors.

2. Send each subcontractor a copy of the schedule for review (with their portion highlighted), request a bar chart or CPM fragnet detailing their work, and call or meet with them to discuss it.

3. Discuss their work in detail, including (*a*) what must be complete before they can start and what work by others can't start until they are complete; (*b*) their activity work sequence; (*c*) the duration of each activity; and (*d*) the number of workers assigned to each activity, the quantity of work, and the productivity rate. Review these items in detail, check for validity with rough estimates based on your experience, and record the discussions for possible use in resolving future disputes.

4. Discuss how their work fits into the schedule, adjust the activities if needed, and negotiate a reasonable solution. If necessary, dictate their time frame to ensure on-time project completion.

5. Obtain a firm commitment from each subcontractor to meet the schedule, and note the consequences for failure to comply.

6. Require subcontractors to provide input for schedule updates, revisions, and forecasts to complete. If behind schedule, obtain their commitment to catch up and require a schedule recovery plan. Request timely input on schedule impacts of change order requests and claims.

7. To expedite and ensure schedule approval, jointly review the schedule with the owner's representative and your superintendent. Ask the owner's representative(s) to come prepared to review disputed issues in depth and to issue tentative approval at the meeting, if all significant concerns are addressed.

8. Jointly update the schedule monthly with all currently active subcontractors. Review progress, update actual start and finish dates and percent completes, revise completed and planned logic if required, develop corrective action plans, and identify improvements to save time and money. Use the meetings to build a sense of teamwork and to apply peer pressure on laggards.

Even with a partnering approach, some leverage is needed to ensure subcontractor cooperation. This can be a subcontract provision conditioning progress payments to approval of the subcontractor's schedule submittals.

**Update schedules monthly and revise when changes occur.** *Schedules must be updated monthly and revised promptly as changes occur* (i.e., due to delays, extra work, changes in sequencing, etc.). Update concurrently with progress payment requests. Record actual start and finish dates, days remaining or percent complete, logic revisions, etc. and the reasons. If a significant delay occurs, promptly give notice, prepare a time impact analysis, and revise the schedule. Otherwise the schedule becomes unusable for managing the project and analyzing changes.

**Be aware of shortcomings in your scheduling software.** Actual activity relationships are far more complex than assumed or as tracked by scheduling software. For example, a start-to-start relationship can limit the start of the successor activity to a number of days after the start of its prior. In reality, however, there may be a continuing relationship, with further progress of the prior needed for the successor activity to continue. Yet most software assumes no relationships between the activities once the successor activity has begun. To avoid this, use a finish-to-start relationship to tie the end of the prior (and therefore continued progress of the prior) to the end of the activity.

**Improve subcontractor scheduling practices.** Subcontractors need to schedule and resource-load all projects to ensure they have sufficient workers, supervisors, and equipment to meet their commitments. For smaller to medium-sized subcontractors, bar chart scheduling is sufficient, with scheduling one or more activities for each project, estimating the worker days per activity, and adding the total demand for each week. A simple scheduling program or a manual system works well. Larger subcontractors, however, need CPM schedules for all of their projects.

Subcontractors should provide input to the general contractor's project schedule, even if not requested. Insist on receiving a copy of the approved schedule and all updates, and provide input to all change order time impact analyses. If appropriate, subcontractors should maintain their own schedule and provide copies to the general contractor.

Resolve scheduling issues at the weekly meetings. Ask that the short interval schedule be related to the master schedule, or vice versa. If the general contractor abandons the master schedule, insist that it be revised, updated, and kept current as part of the general contractor's overall responsibility for project control and coordination. Keep a record of the weekly progress meetings. If an official record isn't maintained, prepare and submit your own version to the general contractor.

## G.  Detailed As-Built Schedules

Accurate, complete, and detailed schedule updates are often unavailable for use as an as-built schedule. Or more detailed as-built information may be necessary to refute alleged concurrent delay or contractor error. In either case, the technique described below is the best method of generating a detailed as-built schedule and then condensing the detailed as-built schedule to the same level of detail as a normal as-built schedule. Although not strictly "new," it is still relatively unknown.

## 1. Three alternatives if incomplete as-built data

When the schedule updates and other records are inadequate for a normal investigation to determine the facts with certainty and at a reasonable cost, three alternatives are possible:

- Pursue a "modified" total cost approach.
- Rely on an expert's judgment.
- Reconstruct the project with a detailed as-built schedule.

Modified total cost claims are used for a quick proposal, to avoid the cost of a more detailed analysis, or when the records are insufficient for analysis. They are easy to prepare but difficult to sell and are recommended only as a last resort or when the changes to the project constitute a cardinal change.

An alternative to a modified total cost approach is to supplement the limited as-built information maintained during construction with the project team's assertions or an expert's opinion. Preparation of such a claim is easy and reasonably economical, but claim reviewers question this approach and it often results in a credibility contest between opposing experts.

The best alternative is an integrated multisource document reconstructing the project on a day-by-day basis. This reconstruction synthesizes numerous sources, each by itself insufficient for determining the events that happened and the consequences, into a coherent single document that describes what happened and why. Reconstruction requires a medium able to accept hundreds of separate notes with references to individual source documents, brief narrative statements describing what happened, tabular information such as daily rainfall and crew sizes, graphic information such as resource histograms or percent complete curves, and timescaled as-planned and as-built schedules. All must be in a readily understood format.

## 2. How to create a detailed as-built schedule

The detailed as-built schedule can be hand-drawn or computer-generated.

**Hand-drawn detailed as-built schedule.**   Hand-drawn detailed as-built schedules are drawn on paper with a pencil and large eraser, similar to Fig. 10.27. Although the final product may be an E-size sheet (3 × 4 feet) or multiple sheets, it is more convenient to work with smaller sheets of paper (17 × 22 inches) and tape them together. Use Clearprint paper as recommended in Sec. D.2 above and a scale of 1 inch per week, which gives two small grid squares for each day of work if using 10 × 10 grid-per-inch paper. Weekend work can be shown as a large dot at the end of the week. If working on a regular 6-day schedule, a 6 × 6 grid can be used. A 10 × 10 grid is preferred, however, for a basic 5-day week with a little weekend work.

The best layout for a detailed as-built schedule is illustrated by Fig. 10.27. This is a portion of one sheet from multiple sheets and doesn't illustrate all the features. The features include:

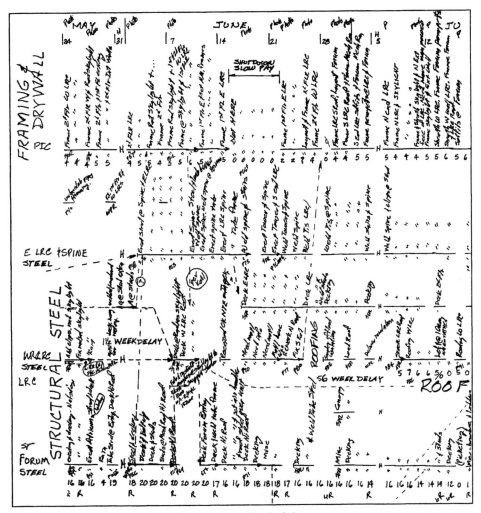

**Figure 10.27**  Part of a hand-drawn detailed as-built schedule.

- Calendar across the top.

- As-planned schedule (not shown), in condensed bar chart or timescale arrow diagram format, immediately below the calendar. Minimize the spacing between as-planned activities and utilize all unused lines in order to conserve space. The as-planned schedule shows the planned work at any point in time, helps determine what tasks are being accomplished, and ensures that your completed as-built schedule activities correlate with the activities on the as-planned schedule.

- General notes, mobilization, and administrative activities on the first as-built activity line (not shown).

- As-built activities in general order as the work progressed, with descriptions of the activities (in small capital letters in the example) and related work (in large capitals). The layout of the as-built schedule resembles a modified bar chart rather than a network diagram. Sequential activities are added on the same horizontal line if the preceding activity is complete or immediately adjacent if not.

- Daily "events" as a short line, that are joined with the preceding and succeeding daily event lines to form an activity. The first event for "*FRAMING & DRYWALL*" in the upper-left corner of the example is a continuation from the prior sheet. The description ("*Frame 2nd Flr W LRC*") is written vertically above the event line and the source document reference ("*SDR*" for superintendent's daily report) is written below the line. Repetitive descriptions are indicated as ditto marks (" " " ").

- The end of an activity as an arrowhead (on 9 June for the first activity in the example) with the following activity ("*Frame 1st Flr E & set H.M. frames*") starting immediately afterward. Periods of unknown progress are left blank. Note delays ("*SHUTDOWN SLOW PAY*" for 15–21 June) or known inactivity ("*NOT HERE*"). Indicate a known activity start as a short vertical line (first event of "*E LRC & SPINE STEEL*" on third activity line starting 2 June).

- Holidays are indicated by the letter "*H*" as for 31 May.

- Crew size (or hours by trade) matrix and cost accounting data toward the bottom of the sheet, or associated with individual activities.

- Rainfall and other weather records at the bottom, in either tabular or graphical form (not shown).

Guided by the as-planned schedule, plot each daily event from the source documents onto the drawing to form the as-built schedule. If convenient, code individual one-day events forming an activity (e.g., the letter "p" above the line to signify a concrete pour, "f" for form, and "r" for rebar as part of a "construct footing" activity). If a specific crew is assigned to an activity, print the crew size below the source document reference (as in the example). If not related to a specific activity or group of activities, print crew sizes or labor hours for each trade or cost code across the bottom of the drawing in a matrix format to help identify the start and finish of activities, periods of inactivity, and levels of effort.

Each activity on the detailed as-built timescale arrow diagram is built a day at a time until the start and finish dates are established, along with periods of inactivity. Each entry is referenced to a source document for verification. Uncertain progress or no progress is indicated by the absence of the short bar for that day. Periods of known inactivity are identified with a short vertical bar on the line at the beginning and end of each shutdown, a blank space during the shutdown, and the source document reference below the line. Sequential tasks, whether related by job logic or crew constraints, are linked by arrowhead to tail or by dashed arrows to show the flow of work. Use judgment to fill in the blanks when preparing the condensed as-built schedule. Use colored tape or highlighting to summarize and clarify the activities and to show that some progress is assumed but unconfirmed for days without a specific record of progress or lack of progress.

**Computer-drawn detailed as-built schedule.** A detailed as-built schedule can also be generated with Excel spreadsheet software or similar programs. This is faster than hand-drawn detailed as-built schedules (if using templates and macros), and the results are easier to read. In addition, graphics can be imported for resource and rainfall histograms or other elements affecting the schedule. See Fig. 10.28 for an example. If printed on a laser printer on 11×17-inch paper with a point size of four or five, the diagram displays four-plus months horizontally per page. Allocate calendar space at the top, a rainfall line, 10 concurrent as-planned activities, and 40 characters of vertical text for each event (day's activity), leaving space for nine concurrent as-built activities on the top sheet and more on additional sheets. By plotting multiple sheets vertically and horizontally and taping them together, E-size drawings can be created with 30-plus concurrent as-built activities and 18 months per sheet.

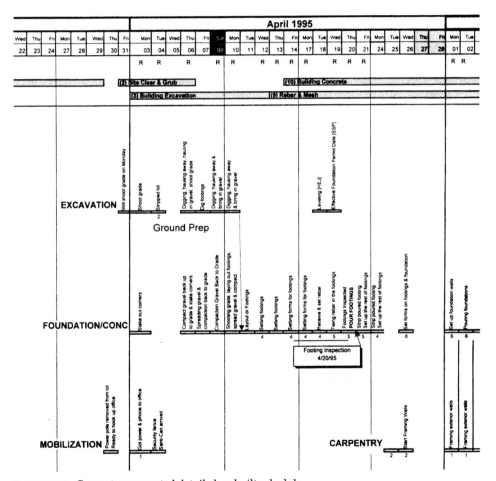

**Figure 10.28**  Computer-generated detailed as-built schedule.

**Templates and macros save time in creating computer-generated detailed as-built schedules.** For details on how to obtain copies and more detailed instructions, see Sec. F of Chap. 1.

### 3. General comments and guidelines

Creating a detailed as-built schedule can be difficult. Activities are often intermittent, with varying (and sometimes inadequate) crew sizes. The as-planned logic is often incorrect and grossly simplifies the actual complex relationships of tasks. Actual activities overlap and are frequently performed in a seemingly random order based on the project manager's preference, perceived efficiency, or some unknown reason.

The actual start and finish dates may be difficult to precisely determine. Should the dates include a subcontractor's mobilization, scaffold erection, material delivery and layout, or repair and rework? Or should these tasks be identified as separate tasks or as part of preceding or following activities? Judgment, experience, and consideration are required. One suggested criterion for determining the actual finish is when the next activity can start.

**Source documents.**  The detailed as-built schedule focuses on field activities rather than on contract issues, administration, and correspondence. The detailed as-built schedule can be prepared concurrently with the chronological summary notes but is usually prepared separately using field-generated documents. The superintendent's daily reports and job photographs are the primary source documents. Other documents include annotated timecards, cost accounting reports, foremen's reports, inspector's reports, requests for information, notices of change, daily diaries by project team members, certified payroll reports (if the superintendent's reports don't record crew sizes), weather records, and other records of field activities. Concrete pour records and test results, delivery tickets, and other more detailed records may provide specific information. Correspondence and minutes of meetings are also used, as they often refer to progress.

**Verifying data and resolving conflicts.**  Verify the reliability of source documents and resolve conflicting information in different source documents as described in Sec. D.2 of Chap. 9.

Inconsistencies and conflicts can be difficult to resolve. On one project, the back side of several daily reports copied during discovery were stapled to the wrong front sheet, creating a major inconsistency at a critical point. It took several weeks to identify and correct. Don't rely on project-prepared summaries without verifying accuracy, since they may contain numerous errors. Recorded facts on the detailed as-built schedule must be without significant errors, to avoid inconsistencies during analysis, loss of credibility, or possible exclusion as an exhibit (if in litigation).

**Using the document.**  The detailed as-built schedule is a very detailed document, containing hundreds of referenced events. It becomes a secondary source document and can be referenced by a written report or in testimony. If challenged, each element can be traced to an original source document or to the expert's judgment. Even if project records are grossly inadequate, a detailed as-built schedule is useful, since a large number of small judgments during an extensive analysis are more reliable and more credible than a few gross assumptions.

### 4. Conversion to a condensed as-built schedule

The next step is to condense the detailed as-built schedule into a more manageable form for computerization and comparison with the as-planned or would-have-been schedules. Depending upon the complexity and completeness of the detailed as-built diagram, a clear acetate overlay and colored pens can be used to define the tasks and flow of work. The result is the condensed as-built schedule, which can be input to a scheduling program and plotted either separately or in comparison to the as-planned or would-have-been schedules.

The condensed as-built schedule defines the actual progress and flow of work. Since it will be compared to the as-planned schedule, it should have similar activities and the same activity numbers. Normally it contains more activities than the as-planned schedule (unanticipated tasks, added work, and impact activities). It may show intermittent progress for some activities owing to inclement weather, contract delays, or insufficient personnel. Annotate the condensed as-built schedule to highlight delays, acceleration, impact, and their cause. Show the original contract completion date, the revised contract completion date with approved changes, the actual substantial completion date, and the final completion date. Also record the assumptions and judgments made when condensing the detailed as-built schedule tasks into the condensed as-built schedule activities.

Considerable construction experience and careful review of the as-planned schedule and the actual logic is required to condense a detailed as-built schedule. Establishing relationships between activities is especially difficult. The physical effort, however, is relatively easy since the computerized as-planned schedule is simply copied and edited for differences in duration, logic, and additional activities.

### 5. Advantages and disadvantages of detailed as-builts

A detailed as-built schedule is an accurate, complete, and detailed record of what happened on the project. The advantages over other tools include that it:

- *Is easier to understand.* The expert, the claim support team, and the owner's claim reviewer or a fact finder can quickly grasp the essential features of the project, the delays and impacts.
- *Refutes alleged concurrent delay.* The additional detail shows which activities are truly critical and whether the near-critical path could and would have been accelerated absent the delay.

- *Explains apparent contractor errors.* Reasons become apparent for previously unjustified delays to critical activities, longer-than-planned activity durations, intermittent progress, etc.

- *Identifies additional issues.* Issues previously overlooked by the project team that materially affected progress and costs will become apparent and can be claimed as extra work or impact.

- *Allows quick response to unanticipated questions.* You can quickly respond to questions in negotiation or cross examination, with detailed answers supported by source document references.

- *Establishes credibility.* The extreme level of detail (which is relatively easy to generate) and your quick response with specific dates and source document references for facts and events will lend credibility to your judgments. You will know more about the project than the project superintendent can remember.

- *Provides a foundation for conclusions and intimidates opposing witnesses.* The detailed as-built schedule can be introduced, its development explained, a few examples provided, and its conversion to the condensed as-built schedule described. You can quickly move on to your primary presentation without wasting time or weakening your presentation by defending the validity of the as-built schedule or refuting arguments from the owner's representative. Opposing witnesses are reluctant to attack your position or statements since you can readily call up source documents to support your position.

- *Generates productivity rates and damages.* Specific data on work quantities accomplished and descriptive data in the detailed as-built schedule can be quantified for productivity rate computations and a measured mile analysis for inefficiency.

- *Offers convenient display and access of other data.* The detailed as-built schedule can display both graphic and tabular data such as rainfall, temperature, crew size, equipment use or standby status, or critical event dates. It can display abstracts of key documents, the expert's comments, etc. And each data element is referenced to a source document.

- *Provides information for issue analyses and reports.* The detailed as-built schedule is used to prepare issue analyses or written reports while referencing each comment to either the original source document or a date/date range of the detailed as-built schedule.

- *Provides all the above simultaneously on one drawing.* Data from a wide range of documents are compiled on a single document, facilitating access and use of that information.

A detailed as-built schedule may be the only way to respond to an owner's allegations of concurrent delay when the contractor has caused some delay, and the delays and impacts are closely intertwined and difficult to segregate. In such cases the courts often deny all delay damages unless the contractor can reasonably allocate responsibility and takes responsibility for contractor-caused delays.

The primary problems encountered with the detailed as-built schedule are the cost of preparation and the difficulty in finding someone familiar with the technique. Reduce the costs by focusing on critical issues (or limiting the source documents to, say, the daily logs). For example on one claim, a detailed as-built schedule was prepared for just the first 60 days of construction, which was the crucial period in dispute. The last schedule update was used for the balance of the as-built schedule.

### 6. The cost of preparing a detailed as-built schedule

The cost of a detailed as-built schedule depends upon the quality of the available data and the size and complexity of the project. For example, the preliminary claim analysis to defend a transit agency from a $250,000 suit on a $900,000 project costs $10,000, and a hand-drawn detailed as-built schedule to confirm the initial conclusions costs about $15,000. Other analyses, exhibits, and testimony cost more, but our client won and was awarded $130,000 on a countersuit. In another example, a hand-drawn detailed as-built schedule for a $20 million advanced wastewater treatment plant costs nearly $50,000 owing to grossly insufficient conflicting records and a very complex project. It would have cost $25,000 to $30,000 had the records been adequate and even less had the computerized technique described above been available.

More recently, on a $500,000 claim for a $2 million multistory building, the computerized detailed as-built from the superintendent's daily report took only 45 hours for a scheduling analyst to prepare. It covered 10 months of construction and tracked the general contractor's administrative activities, framing, finish carpentry, miscellaneous, and general conditions activities, plus 18 subcontractor crews. It contained nearly 900 daily entries with a brief description of the work accomplished and the crew size, in addition to rainfall and other notes. It was invaluable for analyzing what happened and why.

The cost of claim preparation or defense should be proportional to the expected benefit. Balance the additional cost of a detailed as-built schedule and a more thorough would-have-been schedule and damage computations against the likelihood of a more favorable award. It will usually pay for itself many times over. If in doubt, compromise with a detailed as-built schedule for critical periods and a more traditional analysis for the balance of the project.

The most effective means to reduce the cost of preparing a detailed as-built schedule or any other claim analysis procedure is to maintain good construction records.

## H.    ELIPSE Schedules—A New Tool

Additional more voluminous data from the cost accounting system and the job logs is usually available and would aid schedule analysis. Pinnell/Busch coined the term ELIPSE schedule to describe this new tool, based on an acronym for the types of data including:

- **E**vents—RFIs, change order notices, start and finish dates of change order work, milestones, etc.

- **L**abor hours or crew size—by cost code, activity, work area, and crew or supervisor.

- **I**mpacts—including stop work orders, the start and finish dates of impacts, etc.

- **P**roductivity—units of work installed per day or per week, or per labor hour.

- **S**chedules—the as-planned and as-built schedules by activity and task (subactivity).

- **E**nvironmental conditions—including rainfall, temperature, wind, crowding, access limits, etc.

An ELIPSE schedule is simply a comparison bar chart with additional information not normally associated with a schedule. The combination of data makes it a powerful tool. It can be used to substantiate the effect of multiple changes and RFIs and clearly show intermittent progress or disrupted effort. By importing existing computer data or manually inputting from printed reports, the ELIPSE schedule allows economical processing and the display of large quantities of data in an easily understandable graphic form. It serves as both an analytical tool and an exhibit that graphically communicates disruption and impact.

## 1. Format

The ELIPSE schedule displays comparison bar chart activities with the subactivities (tasks), events, and milestones that comprise or impact the activity. It can also display labor hours or crew size, daily or weekly work quantities, and productivity. Intermittent activity progress or reduced effort due to disruption or delay are displayed as intermittent activity bars along with numeric crew size or hours. Impacts on an activity (RFIs, change order requests, architect supplemental instructions, and change order work) are displayed directly below the affected activity. The viewer can easily relate the impacts to delayed starts or disruptions. The ELIPSE schedule can also show weather, crowding, or access problems and note impacts and other conditions affecting the work.

The comparison bar chart portion of the ELIPSE schedule can be generated with most scheduling software. Some software (e.g., Primavera and PMS80) display intermittent progress and impacting events immediately after the affected activities. Some software (e.g., PMS80) can import RFI or change order logs, timecard hours, and production quantities. This information can also be added to the diagram manually or with separate graphics software.

An ELIPSE schedule need not display all of the possible types of information but only what is pertinent or readily available. In some cases, two separate ELIPSE diagrams can be more effective than one diagram containing too much information.

## 2. Example ELIPSE schedule 1—events and impacts

Multiple requests for information (RFIs) and extra work can be extremely disruptive, especially if initiated when work has started and the crews are interrupted. Proving causation and impact for each change is not cost-effective and is often impossible. However, an ELIPSE schedule comparing the as-planned versus as-built schedules plus all RFIs, architectural supplemental instructions, change order requests, and change orders associated with each activity clarifies the impacts and can be economically generated.

The following ELIPSE schedule example was generated when an owner's representative refused the contractor's request for impact payment, noting that most activities were completed later than scheduled. To respond, the contractor's last schedule update was imported from their scheduling program to our software. We also imported the RFI log (from a spreadsheet), coded each RFI with one or more activity numbers, and merged them with the activity file. The change order file was imported from a word processing program, start and finish dates for the change order work were manually entered from the cost accounting records, and a few key events and milestones were added. The resulting ELIPSE schedule was plotted on six E-size drawings with sharp contrasting colors clearly indicating that a majority of the activities were affected by numerous RFIs and changes.

An examination of activity 300, "PLUMBING Rough-In Mid-Level," in Fig. 10.29, confirms that the actual progress (the larger bar on top) took much longer than planned, as asserted by the owner. However, the reason is also apparent: (1) the numerous RFIs that began when the actual work started and continued throughout the duration of the activity, (2) the two change order tasks (81-0043 and 81-0054) which weren't done until just before the activity was completed, and (3) the last two ASIs which weren't issued until shortly before the work could be completed. Upon viewing this, the contractor's superintendent remembered the sequence of interference and could testify to their impact. He described what happened on a half dozen of the hundreds of impacted activities. We demonstrated that the impacted activities generally finished late and over budget while the unimpacted activities generally finished on time

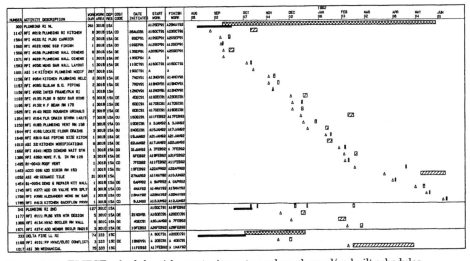

**Figure 10.29** ELIPSE schedule with events, impacts, and as-planned/as-built schedules.

and within budget. However, the image of the six large drawings with innumerable changes was the most impressive evidence presented. The modified total cost claim was accepted.

This approach works best and is most economical when the contractor maintains computerized information logs with actual event dates. The logs should include issuance and response dates for the RFIs and change order requests, change order approval dates, and the dates change order work was performed. The change order work dates can be obtained from the cost accounting system if separate cost codes are established for each change order. The information logs should be set up for easy export of data to the scheduling software or other program used to create the ELIPSE schedule.

### 3.  Example ELIPSE schedule 2—labor hours and intermittent progress

If labor is charged to specific cost codes and the weekly cost accounting data are available (on disk, printed reports, or timecards), import the information or manually input it into the program. Then plot the daily (or weekly) hours for each cost code, subtotal by work area or other category, and total for each day (or week). If weekly work quantities are recorded, compute and display weekly productivity per labor hour. If not, compute average monthly productivity per labor hour based on monthly progress payment requests. Importing computerized data is simple: print reports to a disk file (which puts it in ASCII format) instead of paper and import the ASCII file to the ELIPSE plotting software.

For this example (Fig. 10.30), the initial report was a comparison bar chart of the as-planned and as-built activities plus a numeric bar chart for each cost code associated with the activity formed by the daily or weekly hours from the time cards or weekly labor reports. Alternately, display the daily or weekly hours for each crew or work area from the superintendent's weekly labor report. Then input the start and finish dates of the work for each cost code to create a subactivity bar for each cost code. This can be either a solid bar from the first hour charged to the last, or it can be intermittent bars for only the days worked. Ignore days with less than a minimum number of hours worked. The bar chart with numbers indicating the level of effort for each day (or week) is the result.

The example ELIPSE schedule is based on daily hours worked. It was generated to show an arbitrator how work was disrupted and delayed by owner changes that caused reduced crew sizes and intermittent progress. The daily labor hours were manually input from the timecards (in a few hours) and plotted below the as-built schedule information. The original as-built schedule based on the schedule updates was inaccurate and didn't show limited crew size and intermittent progress.

If daily, weekly, or monthly production data (units of work accomplished) is available, combine it with the daily labor-hour chart. The ELIPSE schedule can then compute productivity rates for each period, which is a powerful tool for analyzing what really happened.

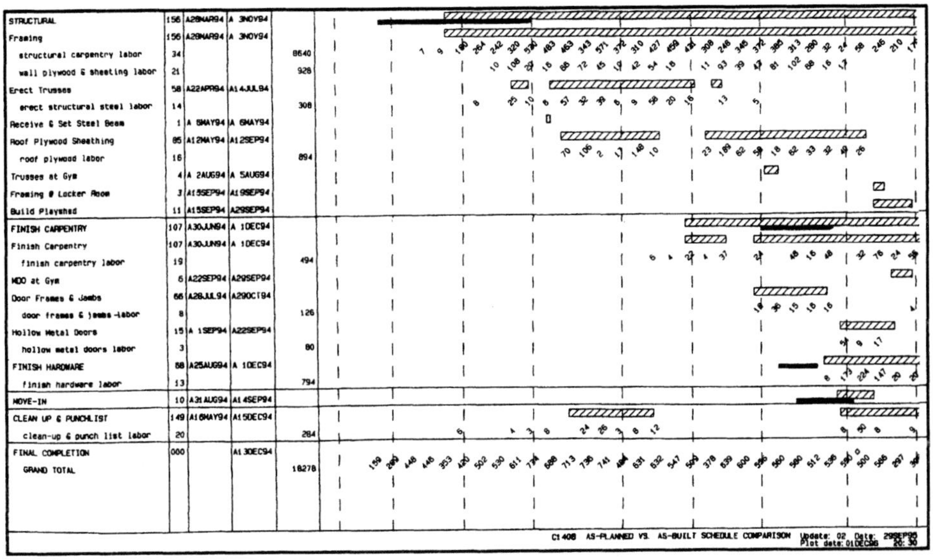

**Figure 10.30**   ELIPSE schedule with labor hours and as-planned/as-built schedules.

## I.   Other Tools for Scheduling Analysis and Exhibits

Numerous other tools are available for analysis and use as exhibits.

### 1.   Fragnets

Fragnet schedules include only the pertinent portions of a network diagram, for easier analysis and understanding. Extract them from the full network, enlarge, and display them as separate exhibits explaining specific events and issues (see Fig. 10.31). Some software does this automatically. Use fragnets for time impact analyses, for what-if analyses (when examining the scheduling effects of alternative actions), and for explaining complicated construction sequences to claim reviewers and finders of fact.

### 2.   Banded comparison bar charts

The banded comparison bar chart uses shaded envelopes around the as-planned and as-built activities to graphically communicate differences not otherwise apparent on a standard comparison bar chart. Overlap areas are hatched with the two different shadings. Banded comparison bar charts are simple in concept, easy to create with any scheduling program that creates comparison bar charts, and can be edited manually or with graphics software. The impact is stunning.

Figure 10.32 is part of an E-size drawing used in a negotiated settlement of a $1.6 million claim by a mechanical subcontractor against the general contractor for delays on a building complex. It displays the planned versus actual concrete work on the tower building in a complex, with each floor separated by

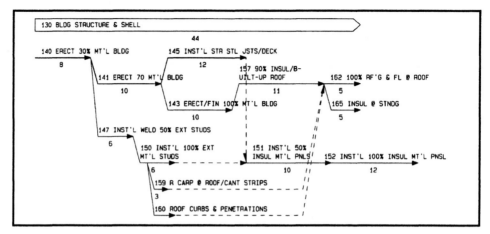

**Figure 10.31** Typical fragnet.

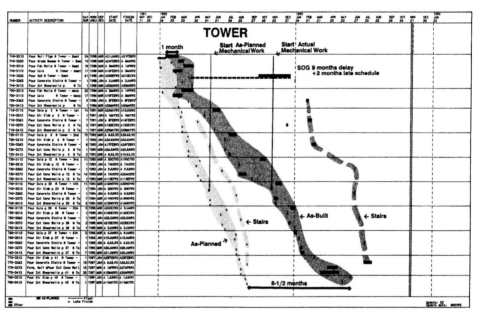

**Figure 10.32** Banded bar chart showing impact of late concrete work on mechanical work.

a horizontal line. It was created by printing a comparison report of the general contractor's as-planned and as-built schedules and manually drawing the envelopes, using different colors for the as-planned and the as-built schedules.

In Fig. 10.32, problems were apparent with the first activity, "Pour Wall Ftgs @ Tower—Basement," when the as-built activity (larger/upper bar) started on the day it was supposed to end and finished one month late. By the time the last activity finished, work was $8^1/_2$ months late. The two graphical arrows and notations at the top and bottom of the figure make this readily apparent.

Banding the activities provides additional information. The as-planned activity band is narrow and fairly steep, indicating that work was expected to go quickly, with average times from starting to finishing a floor being about 1 month. The as-built activity band is wide and flatter, indicating that each floor was tied up for 4 months.

Several other items are also apparent from Fig. 10.32: (1) The stairs were planned for completion 1 or 2 months after the other work on each floor (far too late). Yet they were actually completed 6 to 8 months late, while the subcontractors' crews stood in line waiting for the manlift. (2) The actual start of mechanical work in the basement, which was controlled by cure time and stripping/reshoring, is shown as $5\frac{1}{2}$ months late. Then the slab-on-grade wasn't poured until 2 months later, which made work more difficult. The end result was to reduce the time for mechanical work from 18 months to 12.

## 3. Summary bar charts or timescale arrow diagrams

Summary schedules quickly explain the overall plan and progress to the claim reviewer or fact finder. They can be either summary bar charts (preferably with connecting lines) or modified timescale arrow diagrams and can show either one schedule or a comparison of two schedules (as-planned to as-built or as-built to would-have-been). They should have 10 to 30 summary activities, milestones, and phase hammocks, plus graphical and text annotations to indicate delay and impact.

Summary as-built schedules should have the same layout, color, and scale as the detailed schedule they summarize. Superimposed resource histograms and percent completion curves show the level of effort and accomplishment as in Fig. 10.33.

## 4. S curves (cumulative cost, percent complete, and earned value)

Cumulative curves are frequently used to show planned or actual progress (by percent complete, cost expended, or earned value) and to compare them. An abrupt change in the slope of the curve (e.g., a flatter slope, indicating slower progress) coinciding with an event (i.e., alleged cause of the delay) followed by a steepening when the issue is resolved supports the contention that the event caused the change.

Percent complete curves for CPM-scheduled projects normally show the average as-planned progress. They can, as in Fig. 10.34, show both the early start and late start as-planned progress as an envelope within which actual progress must stay to complete on time. The average is also shown. Actual progress can be below the average as-planned line and still not be behind schedule. However, if actual progress gets too close to the late start line, delay is probable. In fact, a comparison of the early start/late start envelope before and after owner use of the float may demonstrate a marked difference in the breadth of the envelope and the flexibility available to a contractor. This will help justify limiting use of the float by the owner. A project with multiple critical paths and too little float

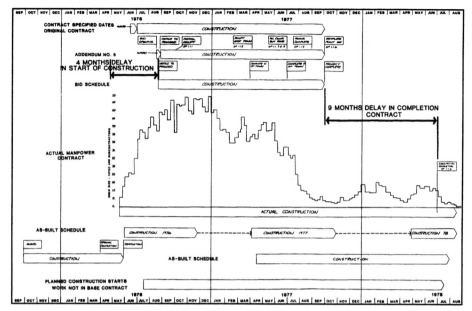

**Figure 10.33**  Summary as-built schedule with resource histogram and additional text.

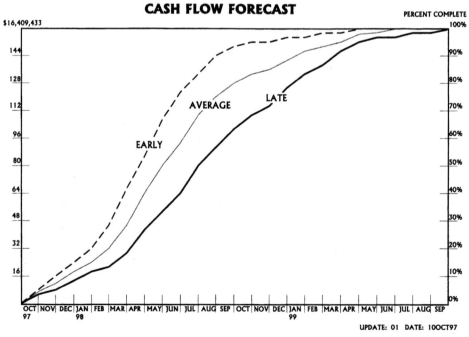

**Figure 10.34**  Cumulative progress curves.

is invariably more expensive to build, requires more management attention, and is more likely to be delayed than one with a single critical path and lots of noncritical activities.

Cumulative curves are often overlaid on summary schedules to graphically illustrate the pace of construction.

**Cumulative rainfall curves.**    Rainfall is normally displayed with a histogram, which may underemphasize the impact of winter rains. A cumulative curve (Fig. 10.35), with a relatively flat slope in summer and an increased slope in winter, can clearly illustrate the onset of the rainy season. Starting it with a base of zero in early summer further emphasizes the impact of winter rains after the limited rainfall of summer.

**Stepped cumulative curves.**    Sometimes cumulative curves should be stepped to reflect reality. For example, Fig. 10.36 compares planned billings with actual costs as two line curves, indicating substantial delays (8 months) and increased costs (from $1.7 to $2.2 million). Billings and revenues, however, are shown as stepped curves which reflect submittals and payments. The gray shaded area shows the date that invoices were submitted and the dark shaded area shows when payment was received. On this project, payments were prompt, as indicated by the small gray shaded area. On some projects, an owner's late payments will be painfully apparent. The vertical gap between the actual cost curve and the dark shaded payment stepped curve approximate a contractor's financing needs.

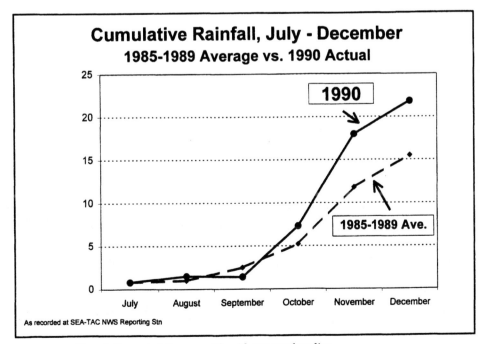

**Figure 10.35**  Cumulative rainfall curve with mid-summer baseline.

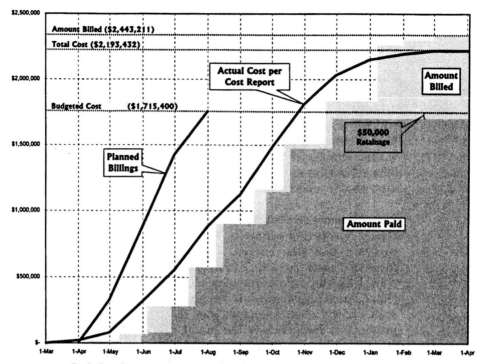

**Figure 10.36**  Cash flow forecast versus actual with progress payment requests and payment.

## 5. Histograms (periodic charts)

Histograms are used to show planned and actual resource use (i.e., the number of workers). The planned staffing of a project (ideally a steady increase of crew size followed by a constant crew size and then a slow decline to project completion) can be compared to the actual impacted staffing (with smaller initial crew size during good weather, excessive peaks during inclement weather, and wildly fluctuating levels through to completion). This comparison is especially effective if the start and finish of the impact are noted on the diagram and correspond to reduced staffing and then excess staffing to make up for lost time.

**Labor histograms.**   The same labor data as displayed in the ELIPSE schedule can be displayed graphically, giving a better "feel" for the level of effort, fluctuations, correlation with weather conditions, impact of changes, etc. To fully appreciate the level of effort, plot the histogram at the same scale as the schedule. Comparison labor histograms of the as-planned versus as-built labor hours or crew size are very helpful in analyzing impact. Although a total labor histogram can show overall impacts, histograms by craft are generally most instructive for demonstrating impact. Histograms by basic work code can also be informative.

Labor histograms generally work best when showing weekly averages. Daily histograms have excessive nonsignificant fluctuations due to personnel changes, illness, etc. Monthly histograms don't contain enough detail to ascertain the impacts of changes.

Instead of showing labor hours, show equivalent crew size, which reviewers can better visualize. Divide the weekly labor hours by 40 (or some other average weekly hours worked) to get the equivalent crew size. If acceleration is an issue, differentiate the overtime hours with shading or color.

Generating an as-planned labor histogram is often more difficult than expected. Few contractors resource-load their schedules and you will have to either allocate labor hours to activities or schedule the work by cost code. Either approach takes considerable time and judgment and is often inaccurate. It is even more difficult when using unit cost or unit price estimates. Then you must allocate part of the unit cost to labor cost, convert that to labor hours based on the average labor rate, and then make a forecast.

Keep the graphics simple (Fig. 10.37). Two-dimensional graphics are usually better than three-dimensional diagrams. Pay attention to the readability of the result and try several different formats.

When examining resource use, display both the total crew size and its composition, as in Fig. 10.38. The actual crew size histogram shows the carpentry crew shifted from concrete forming to rough framing and then to finish work. A common baseline, as shown, is often more effective than stacked resource bars. A line drawing for the total resource use is also more clear than a second vertical bar.

Figure 10.39 is for the same project but compares the as-planned crew size (as a line diagram without shading) to the actual crew size (a line diagram with shading). The difference is apparent and supported the contention that the contractor was unable to start work until much later than planned, subsequently forced to accelerate, worked far more hours owing to the piecemeal

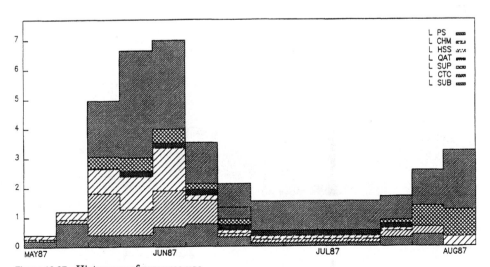

**Figure 10.37**  Histogram of resource use.

**Actual Crew Size**

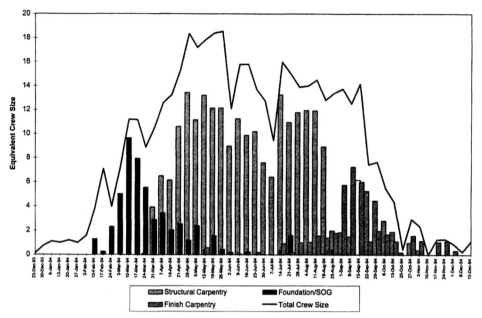

**Figure 10.38** Actual crew size with breakdown by function.

**Planned vs Actual Crew Size**

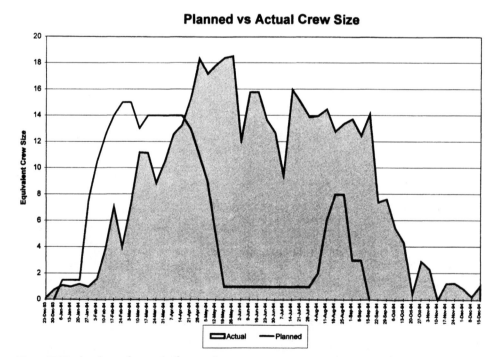

**Figure 10.39** As-planned vs. actual crew size.

release of drawings and continuous revisions, and finished later than planned. Be ready to refute the owner's allegation that the reason for delay was contractor understaffing. Note that both the as-planned and as-built resource use are shown as line diagrams instead of vertical bars.

**Weather histograms.** For weather-sensitive operations, the daily rainfall, temperatures, and wind can be displayed at the bottom of the schedule on either the detailed as-built schedule or the ELIPSE schedule.

## 6. Plan drawings

Multiple plan drawings, as "snapshots" of the project status at various points in time, can be used to analyze and explain construction progress and impacts. Create them manually on multiple drawings or with computerized overlays.

## 7. Tabular schedule reports

Tabular data can help support testimony and other exhibits and are used extensively in claim documents or written reports. Such data are seldom used as demonstrative evidence except to lay a foundation, since understanding them requires careful study and considerable experience. The presentation of specific dates or tabular data is best done in conjunction with graphics.

Tabular reports are also needed to help decipher most network diagrams, as the logic constraints on the diagrams are seldom clear (except for timescale arrow diagrams). When printing tabular reports for analyzing schedules, include the following information (see Fig. 10.40):

- Activity numbers and descriptions.
- Prior and successor activity numbers, or prior and successor activity descriptions.
- Early/late start and early/late finish dates (actual start and finish dates for completed activities).

PMS80 Update: 05 Page: 2
CRITICAL PATH SCHEDULE - REPORT NO 1 Update date: 31MAY96
PROJECT FILE:C1516 RECORD: 5 SORT:Float,EarlySt PINNELL/BUSCH, INC. -- PORTLAND, OREGON USA Print: 12:14 04JUN96

| NUM | ACTIVITY DESCRIPTION | DEP | DAY DUR | PRIOR ACTIVITIES | SUCCESSOR ACTIVITIES | EARLY START | EARLY FINISH | LATE START | LATE FINISH | FL-OAT | DAY REM | % COM |
|---|---|---|---|---|---|---|---|---|---|---|---|---|
| 162 | CERAMIC TILE | 09 | 15 | 161S6 | 171 180 175 177 | 5AUG96 | 23AUG96 | 12AUG96 | 30AUG96 | 5 | | |
| 151 | CONTINUE ELEC PNLS WIRE/TERMS | 16 | 20 | 147 | 155 | 30JUL96 | 26AUG96 | 7AUG96 | 4SEP96 | 6 | | |
| 155 | ELEC GENERATOR SYSTEM | 16 | 8 | 151 | 213 | 27AUG96 | 6SEP96 | 5SEP96 | 16SEP96 | 6 | | |
| 103 | MASONRY VENEER | 04 | 5 | 101-5 | 107 | 27JUN96 | 3JUL96 | 9JUL96 | 15JUL96 | 7 | | |
| 107 | ROOFING SYS | 07 | 15 | 101 103 | 161 121 109 110 | 5JUL96 | 25JUL96 | 16JUL96 | 5AUG96 | 7 | | |
| 175 | TRIM PLBG FIXTURES/TEST | 15 | 10 | 162 | 177 | 26AUG96 | 9SEP96 | 5SEP96 | 18SEP96 | 7 | | |
| 177 | TLT PARTS & ACCYS | 10 | 8 | 162 175 | 191 | 10SEP96 | 19SEP96 | 19SEP96 | 30SEP96 | 7 | | |
| 50 | SITE UTILITIES AND FINISHES+ | | 112 | 51H 89 | H | A09MAY96 | 16OCT96 | 17JUN96 | 28OCT96 | 8 | 97 | 13 |
| 13 | FABRICATE HVAC DUCTWORK | 15 | 20 | 9 | 131 | 31MAY96 | 27JUN96 | 12JUN96 | 10JUL96 | 8 | | |
| 52 | UG ELEC MAIN FEED | 16 | 12 | 51S2 | 54 | A31MAY96 | 13JUN96 | 12JUN96 | 25JUN96 | 8 | 10 | 10 |
| 54 | RETAINING WALL FTGS | 03 | 10 | 27S2 53 52 10 51 | 55 | 14JUN96 | 27JUN96 | 26JUN96 | 10JUL96 | 8 | | |
| 131 | DUCTWORK INSTALLATION | 15 | 30 | 31-4 13-4 | 163 171 133 135 145 132 | 24JUN96 | 5AUG96 | 5JUL96 | 15AUG96 | 8 | | |
| 55 | MASONRY RETAINING WALLS | 04 | 10 | 54 | 57 | 28JUN96 | 12JUL96 | 11JUL96 | 24JUL96 | 8 | | |

**Figure 10.40** Typical tabular schedule report with key data fields for schedule analysis.

- Total float for each uncompleted activity.
- Responsibility and work area.
- Sorted by float and early start to facilitate tracing the critical path and then near-critical activity chains.

If still unable to decipher the critical path, plot it with the timescale arrow diagraming method described in Sec. D. Generating a timescale arrow diagram is usually faster than expected, especially if using the recommended grid paper and plotting only the critical and near-critical paths.

Proof of entitlement does little good without proof of damages. Unfortunately, inadequate cost records often require proving damages based on expert opinion and inexact estimates, rather than quantification and analysis. Claim reviewers are reluctant to accept the numbers generated, making settlement difficult. If forced to arbitrate or litigate, the calculations are usually contested by another expert using similar methods. Compensation, if any, is often late and insufficient.

This chapter enables claim preparers to move beyond conventional practices and use more accurate methods of computing damages. These methods are relatively easy and provide reliable damage estimates that are more acceptable to claim reviewers, arbitrators, and the courts.

The topics covered in this chapter include:

If you are reading this chapter from start to finish, break your review into related sections (A–C, D, and E–K). For guidelines on estimating claim damages, proceed directly to Sec. D; then read Secs. E through K for specifics on estimating each type of damages. Sections A–C provide background material to help avoid problems.

The legal term *damages* results from a breach of contract. Only if a dispute proceeds to arbitration or litigation are damages awarded. This book uses the term damages for any amount due for a change order request or claim, in accordance with industry practice. The claim itself should normally be characterized as a *request for equitable adjustment* rather than a *claim*. Most change orders are paid as either contract adjustments (based on procedures specified in the contract) or equitable adjustments (based on negotiation of an equitable settlement within the contract terms).

## A.  General Comments and Suggestions

The following comments and recommendations will make your damage calculations more acceptable and will increase the likelihood of recovering costs.

### 1.  Emphasize damage calculations

Proving damages by calculating the cost is the ultimate objective of a claim—the bottom line. Yet usually only 10 to 20 percent of claim documents are devoted to proving damages, with the rest focusing on entitlement.[15] This is not a successful strategy, and the emphasis needs to change.

1. *Provide more detail.* Detail persuades claim reviewers and mind-numbing detail overwhelms their resistance. Give them a lot of information—to take apart, verify, and absorb. Simplified methods of estimating and substantiating costs may suffice for bidding or for in-house cost control but are insufficient for convincing claim reviewers unless the reviewer is an experienced cost estimator. A six-page cost analysis substantiating $1 million in damages doesn't contain enough information for most claim reviewers to analyze.

Break cost items into multiple subsidiary items and avoid combining calculations. This provides more detail for analysis and assures having a claim even if some elements are rejected on the basis of entitlement. Psychologically, twenty subtotals from multiple complex calculations appear less costly than a simple calculation for a million dollars, even though the totals are identical.

2. *Use cost estimates that are understandable to the claim reviewer.* A contractor's unit price estimate is often inadequate for convincing a claim reviewer of the extent of damages. Although accurate (because of a contractor's experience and historical cost records from prior projects), a designer or owner's representative is seldom familiar with current unit costs and cannot verify the numbers used. They prefer an estimate that can be analyzed.

Estimates based on crew composition, hourly unit rates, and productivity best meet their needs. Owner's claim reviewers, even without much field experience, can confirm hourly labor and equipment rates, evaluate the proposed crew size, verify that the composite hourly rate is computed correctly, examine the productivity rate for reasonableness, and check if the total costs are correctly calculated. They can apply their analytical skills. In contrast, a unit price estimate cannot be analyzed—there is nothing to analyze. It must be evaluated, and most reviewers lack sufficient costing knowledge to evaluate it.

   3. *Use more reliable, verifiable methods of proving damages.* Using more reliable costing methods (e.g., the measured mile approach instead of a total cost claim for inefficiency) provides more convincing damage calculations. A systematic approach using a checklist (summarized in Fig. 11.1, on page 347), referencing data to source documents and providing a clear audit path from the grand total to each detailed element of cost, allows the claim reviewer to verify the cost. The result is a higher degree of acceptance.

   4. *Use actual costs to help prove entitlement.* Claim reviewers are reluctant to accept a claim unless an actual loss is proved. Proof of a loss reduces the possibility of an inflated claim and indicates the contractor was damaged.

   For example, cost overruns for impacted activities with cost underruns for nonimpacted activities also support entitlement. On the other hand, increased material quantities concurrent with labor overruns indicate excess labor costs were probably caused by quantity takeoff errors or extra work that wasn't added to the budget, instead of labor inefficiency.

## 2.  Use preliminary damage estimates to guide and prioritize entitlement focus

Preliminary damage calculations should precede analysis of entitlement. It is senseless to analyze entitlement for issues with negligible damages. A brief review of cost overruns guides the claim preparer in prioritizing issues to analyze for entitlement and helps focus on those with the largest dollar return.

## 3.  Integrate damage calculations with entitlement and causation analysis

Analyze damages concurrently with entitlement, not after entitlement. An early identification of damages allows further scrutiny of entitlement issues. Claims often have multiple theories of entitlement that fit the facts, and the theory with the best recovery should be selected.

   Causation is often time-variable. For example, an increase followed by a decrease in costs should correlate with the start and finish of an alleged impact. The lack of a parallel change indicates a lack of causation (linkage) between the two. Therefore, it is important to review not only the final cost reports but also interim cost reports to determine if the trends and intermediate costs correspond to the alleged impacts.

   Damage calculations should be done concurrently with entitlement analysis. However, they should not be presented concurrently. It is easier to first convince an owner that the contractor is entitled to extra payment and then introduce the rather sensitive topic of just how much. Make an effort to "disconnect" the two decisions in the mind of the claim reviewer by placing the damage calculations at the end of the claim, after the reviewer is convinced of entitlement.

## 4.  Estimate, record, and analyze costs by cost code

Contractors break down project costs by cost codes (also called cost phases) in order to estimate, record, and analyze costs. All contractors should use the 16 Construction Specifications Institute (CSI), two-digit division codes. If applicable, they should use the primary five-digit codes, which facilitates comparison to the owner's budget, the use of estimating databases, etc.

## 5.  Analyze by cost category

Damage analysis *must* consider variances by cost category (labor, equipment, materials, subcontract, small tools and expendables) for each cost code. Analysis at the aggregate cost level (for each cost code) may be misleading. For example:

- Labor cost overruns should parallel increased equipment usage. Otherwise the problem may be a bid error.
- A material cost overrun accompanying a labor overrun indicates a probable quantity takeoff error, not inefficiency.
- A decrease in labor costs with increased subcontract or equipment costs indicates the work was subcontracted or changed to an equipment-intensive operation after the bid.
- Higher than estimated jobsite overhead labor supports a claim of impact from multiple change orders and RFIs.
- Small tool cost variances should correlate with labor cost variances.
- Fuel, oil and grease, or maintenance cost variances should correlate with an increase in equipment costs. Otherwise the reason for increased equipment costs may not be inefficiency.

Lack of correlation in cost variances between cost categories may result from bid errors, inaccurate cost records, or failure to revise budgets for change order work. Claim preparers should check for correlation when preparing a claim based on the bid estimate, and adjust the analysis as necessary. They can also use correlations as supplemental evidence of impact and inefficiency.

## 6.  Analyze labor and equipment hours and material quantities

Claim preparers should compare not just the variance in cost but also the variance in labor hours, material quantities, and equipment use. Analyzing labor dollars instead of hours will miss a bid error due to using a low labor rate or a crew composition with too many low-paid workers. Likewise a labor cost overrun may be due to overtime premium, not to extra labor hours. An increase in lumber prices can cause material costs to balloon without an increase in material quantities, and the substitution of rented for owned equipment can cause an overrun in equipment costs.

### 7.  Magnification of cost by productive work day limits

When computing direct cost or efficiency costs for a crew, increase either the hourly rate or the number of hours for the effect of paid travel time, setup and close-down time, and other work day limits on the productive time. For example, it might take the crew on a sewer construction project 30 minutes to set up barricades, divert traffic, remove manhole covers, and move equipment from the storage lot to the work area. Shutdown may take another $\frac{1}{2}$ hour. If the crew works an 8-hour day, only 7 hours are productive, and they have already lost 12.5 percent of the work day, so estimated costs should be increased by 12.5 percent.

### 8.  Record work quantities accomplished, productivity, and inefficiency due to impact

Proving impact damages is difficult unless the contractor records weekly quantities of work accomplished.

Contractors *should* record weekly production for each cost code so that:

1.  Labor costs can be controlled before they get out of hand.

2.  Productivity variances can be tracked for substantiating impact claims.

If monthly progress payment requests are accurate (or can be adjusted to be accurate), monthly productivity rates can be calculated from the pay requests. If some months were impacted and some not, the measured mile approach can be used to compute inefficiency.

### 9.  Maintain credibility and avoid misrepresentations

Abandon small marginal claims unless they are numerous or support an overall contention (e.g., unreasonable inspection). List them, however, without a dollar value, as "Costs Not Included." This will indicate your reasonableness, that there were more impacts than those claimed, and that they may be more significant than expected. It will also allow reopening negotiations for those items if they should prove to have significant impact.

One of the biggest obstacles to recovering damages is the tendency of the owner's negotiator, arbitrator, or judge to suspect that the contractor caused some of the problems. It is difficult to prove a negative (i.e., that the contractor made no errors). It is better to admit to some minor contractor errors, even if the need to do so is questionable.

This book is not a legal document and may recommend requesting payments that are not compensable under the strict terms of a specific contract but which should be due as a matter of equity. Be careful when preparing or certifying claims, especially when subject to federal regulations such as the False

Claims Act and the Contract Disputes Act, to avoid violating the letter of the law. If there are questions, consult an attorney. *Never submit inflated or unjustified change order requests or claims.*

## B.  Bid Estimating Procedures and Their Effect on Damage Calculations

Comparing the bid estimate (or cost accounting budget) with actual costs indicates whether a potential claim exists and is often used to "prove" a claim as a total cost claim. Therefore, the completeness and accuracy of the bid estimate may be essential for a successful claim.

Many claim preparers and reviewers have false expectations about bid estimates. They assume accuracy at the finest level of detail and use estimates without adjustment. Claims preparers must recognize the shortcomings of bid estimates and correct them before utilizing the estimate for claims analysis.

### 1.  The purpose and shortcomings of a bid estimate

The purpose of an estimate is to determine the probable total cost of constructing a specific project. It is not necessary to estimate the exact cost of each project element, as long as the total estimate exceeds or equals the actual total cost.

Bid estimating is typically accomplished quickly, with insufficient time for a careful review. Consequently, minor errors are frequent but usually balance out. Actual construction costs may vary widely from the estimate—differing over time, by cost code, between work areas or supervisors, with variable weather, and with different working conditions. Variations also occur when widely variable tasks are lumped together in the same cost code for estimating. Only in the aggregate do actual costs tend to equal estimated costs, partly because the estimate is a target for field personnel. If over budget, they try to reduce costs. If under budget, they focus their efforts elsewhere and the cost of under-budget items tends to increase.

Another reason that the estimate and actual costs vary is that some small contractors lack a good cost accounting system and do not use historical costs to estimate new projects. Consequently, their estimators use past personal experience and rely on the fact that the company's projects usually end up profitable.

Estimates for negotiated contracts often vary from actual cost, especially if based on preliminary plans and specifications for fast track or CM/GMP (construction management/guaranteed maximum price) contracts. The estimate of each cost code is what a typical building subsystem or feature *should* or *could* cost, based on past experience on similar projects and the expectation that the designer will design the project within the budget. The final design may differ considerably, with cost control maintained at an overall project level by reducing the scope (and therefore cost) of entirely different portions of the project to meet the budget. Contractors should prepare detailed estimates for fast track and CM/GMP contracts with work quantities, assumptions, and descriptions to protect themselves from unreasonable owner expectations.

## 2.  Quantity takeoffs

The accuracy of a cost estimate depends upon the quantity takeoff. Some bidders on unit price contracts use estimated quantities from the contract documents instead of doing an independent quantity takeoff. Although acceptable for minor bid items, this can result in severe cost problems if the quantities for major work items differ widely from the estimate. If the accuracy is uncertain, claim preparers should verify the quantity takeoff before examining pricing.

## 3.  Bid estimating methods

There are five basic estimating methods; any may be used for bid estimating or computing claim costs. Their accuracy varies widely, and claim preparers need to adjust accordingly. The methods are:

1. *Unit price estimates.* Unit price estimates are for a dollar amount per unit (e.g., $64.50/cubic yard) including markup for overhead and profit. The unit price times the estimated quantity of work is the bid amount.

Unit price estimates are used for unit price contracts, minor bid items, or quoting subcontract work on lump sum contracts (i.e., a framer may bid $3.50/square foot for rough carpentry to avoid taking off the quantities when bidding). Unit price estimates are also used by general contractors to check a subcontractor's bid.

Unit price bids may be a function of market conditions (what it takes to win the bid) rather than the expected costs. Contractors increase their prices when there is less competition or if they already have enough work. Claim preparers need to adjust unit prices accordingly.

Some contractors unbalance unit price bids for early cash flow or to account for expected quantity variations. When preparing claims on a unit price contract, adjust for unbalanced bids, back out overhead and profit to find the direct cost, and break down direct costs into labor, materials, etc.

Unit price estimates are also used for budget or conceptual estimates based on preliminary plans (e.g., schematics or design development level drawings).

2. *Unit cost estimates.* Unit cost estimates are for the direct costs (without markup) per unit of work and are more accurate than unit price estimates, as each cost category (labor, materials, equipment, etc.) is estimated separately. Most construction estimates are unit cost estimates.

Labor should be estimated and tracked by labor hour instead of labor cost. This eliminates the need to adjust historical costs for inflation and differences in local labor rates.

Unit cost estimates can be based on multiple data sources:

a. *Historical record–based unit cost estimates.* Unit cost estimates based on cost accounting records are more likely to be accurate, if:

- The contractor does similar types of work on multiple projects over a period of years.
- The contractor's procedures and personnel are consistent from project to project.
- Field personnel accurately record actual costs.

- Costs are adjusted for inflation, project size and complexity, and working conditions.
- The estimators use the historical costs for estimating new work.

*b. Personal experience–based costing.* Estimates based on the estimator's personal experience and judgment can be reasonably accurate, if:

- The estimator has extensive, fairly recent experience in that type of work.
- The cost factors for the work remain substantially the same in the intervening period.
- Costs are adjusted for inflation.
- The estimator has feedback of actual costs on more recent projects.

*c. Market-based costing.* On some minor bid items, contractors use unit prices bid by other contractors on other projects. These may need to be adjusted.

*d. Plug numbers.* Plug numbers are not usable for claims analysis. A "plug number" is simply a guess based on past experience. It works well for minor bid items, as it saves time and enables bidders to focus on the important bid items. Although the numbers can be grossly inaccurate, they have little effect on the total contract amount.

*e. Estimating manual–based costing.* Using an estimating manual (either printed or on CD-Rom) is similar to using historical company data, except that the data have been compiled by others from a wide range of companies. The result may not be applicable to a specific project. In addition, the costs must be adjusted for inflation if not updated recently, for varying labor rates and productivity, and for different working conditions.

Estimating manual costs are usually somewhat above actual costs for contractors regularly performing that type of work. However, if the work is unusual and the crews are unskilled or poorly equipped, a contractor's actual costs can be higher than the manual prices.

The best-known estimating manuals are those by R.S. Means Co. and Richardson Engineering Services for building construction and civil works.

*f. Catalog estimates.* Estimating specialty subcontract work from industry-published catalogs is used by plumbing, electrical, HVAC, and other specialty trades. Estimators prepare a detailed quantity takeoff, apply unit labor hours from the catalog, and adjust by a factor to arrive at a bid amount. The factor reflects their company's productivity, the conservative rates of the catalogs, project working conditions, and current market conditions.

Catalog cost estimates may have little relationship to a contractor's actual costs. Therefore, specialty contractors should keep thorough records of actual labor hours, compare their actual hours to the catalog estimates, and modify their adjustment factors accordingly.

*g. Systems unit cost estimating.* Contractors (especially mechanical and electrical subcontractors) use unit cost estimates by building system when bidding from preliminary drawings (e.g., for a fast-track or CM/GMP contract).

3. *Computer estimating.* A number of software vendors offer computerized estimating programs with a unit cost database and production rates for a composite crew that can be modified by the user. They include:

- WinEst
- Timberline Precision (they also have accounting/job costing software)
- AgTec (for earthwork and utilities)
- Bid Tek

4. *Subcontractor and vendor quotes.* Subcontractors perform 75 to 95 percent of building construction and 10 to 50 percent of civil work construction. General contractors have limited knowledge of these costs and must rely on subcontractors' estimates. To ensure competitive pricing, general contractors should solicit at least three quotes or make an independent estimate of specialty work.

5. *Detailed composite crew and productivity estimates.* The most detailed, and generally the most accurate, method of cost estimating is to:

a. Develop a composite crew that will perform a specific work item.
b. Determine the hourly cost of that crew.
c. Estimate the crew's productivity.
d. Compute the unit cost.
e. Multiply the unit cost times the estimated quantity of work.
f. Add markup for overhead and profit.

Detailed productivity estimates (similar to the example in Sec. E.8) are the easiest type of estimate to review and are therefore favored by claim reviewers. Unlike unit price estimates, they can be broken down, analyzed, and verified. Use this method whenever possible.

## 4. Common bid errors

When relying on estimated costs to analyze and compute damages, be aware of possible bid errors that might affect the analysis. These errors can include:

- Underestimated quantities due to mistakes in the quantity takeoff or summations.

- Underestimated difficulty of the work or time required to do the work.

- Incorrect assumptions about the manner and method (and therefore the cost) of work.

- Underestimated unit costs of labor, equipment, or materials.

- Omitted work items.

- Prices based on unacceptable substitutes for "as equal" materials.

- Failure to obtain quotes for materials and equipment.

A claim analysis based on an inaccurate bid estimate will lead to incorrect conclusions. A variance between the estimate and the actual cost may not indicate impact or changed conditions. Therefore, carefully review the estimate and modify it if necessary.

## C.  Cost Accounting Procedures and Their Effect on Damage Calculations

The bid estimate should be corrected for bid errors when converting it into a job cost accounting budget. Otherwise the budget is unusable for cost control and unreliable for claims analysis. The budget must also be adjusted for change orders and quantity variances during construction. Actual costs and work quantities must be accurately collected and posted to the accounting system.

Unquestioning reliance on cost accounting reports for claims analysis may lead to serious errors. In addition, the level of detail and the breakdown of cost codes may not be conducive for analyzing costs. Therefore, claim preparers should verify the accuracy of the cost accounting reports, correct the data if necessary, and be able to break down the budget into more detail or restructure it, if required.

### 1.  Purpose and shortcomings of cost accounting data

The basic purposes of job cost accounting are to (1) control costs during construction and (2) provide data for estimating future projects. The use of cost accounting data to prove claim damages is seldom anticipated and the chart of accounts rarely segregates the costs of extra work, delay, and impact from the cost of original bid work. The accounting data needed for claim analysis may therefore not be readily available and claim preparers will need to adjust the data using other records such as the superintendent's daily reports.

**Reasons for inaccurate cost allocation.**  Recorded costs may be inaccurate because some field personnel do not realize the importance of accurate data and may charge time to any cost code to save time. Or they may be unsure which account to charge their time to, or charge the wrong code to hide a cost overrun in other work.

**Failure to establish cost codes for extra work.**  Some contractors don't establish separate cost accounts for extra work or impacted original bid work, or don't record work quantities accomplished except for monthly pay estimates. As a result, variations in productivity due to impact cannot be computed with the measured mile approach.

**Reasons for cost variances.**  Actual costs vary from the estimate for reasons other than extra work and impact, owing to:

- The learning curve effect as the crew becomes familiar with the work and more productive.
- Differing working conditions of access, crowding, etc., between different work areas.
- Variations between different field supervisors in planning, directing, and motivating crews.

- Weather variations that make construction more difficult.
- The work for some cost codes being widely variable but grouped together to simplify estimating and accounting.
- Carelessness, inadvertent errors in charging costs, lack of awareness of the need for accurate records, lack of information on what codes to use, or attempts to deliberately hide cost overruns in other areas.
- Changes in crew morale and teamwork.
- Differences in union work rules and jurisdictional issues.
- The natural variation in productivity due to happenstance.

## 2.  Types of cost accounting reports

There are many job cost accounting reports and different names for each depending upon the accounting software. The basic, most important reports are:

**Weekly labor reports.**  Weekly labor reports track labor costs and hours per cost code for weekly cost control and *must* report work quantities accomplished and productivity. If not, weekly productivity cannot be determined and only the final labor cost report (with the final work quantities) can be used to determine productivity.

**Monthly summary report.**  Monthly cost reports can be either at a summary level or with details of each vendor and subcontractor payment. They record *actual costs to date* for each cost code by cost category (labor, equipment, subcontractor, etc.). The reports also list the *budgeted cost,* for comparison of planned cost to actual. However, this comparison cannot be made unless the *budgeted work quantities* and *actual work quantities accomplished to date* are recorded (for earned value analysis) or until the project is complete.

**Transaction reports.**  Transaction reports are used for detailed information on specific costs by vendor or subcontractor, for specific materials supplied, etc. The reports provide the name of the vendor/subcontractor, the material or service provided, and the date of service or the date the invoice was received.

**Equipment reports.**  Equipment utilization reports are needed to determine equipment usage and costs but often don't exist. If available, they will substantiate standby and other equipment claims. If not available, the data can be reconstructed from the superintendent's daily reports if equipment use is recorded.

## 3.  Adjusting cash-based cost accounting data

Many job cost accounting reports record costs on a cash basis—based on when payment is made (or authorized)—not on an accrual basis when funds are committed, the work accomplished, or an invoice received. Therefore, the monthly cost reports may not accurately reflect the current cost status.

Labor costs are paid and recorded the week after being incurred and are therefore current. Subcontract costs are recorded at the end of the month the work was performed and billed. However, subcontractors may overbill or owner's representatives may reduce payment for minor deficiencies. Material invoices are recorded after being reviewed and authorized but may be received late and may include material stored on site. Equipment costs may be incorrectly charged or not charged at all if the project is over budget. In addition, project managers may hold some invoices before approving them for payment.

When examining accounting reports, verify the backlog of received but unprocessed invoices and outstanding commitments for work performed but not yet invoiced and entered into the system.

Better accounting systems use an accrual/deferral basis that books purchase orders and subcontracts when issued and carries both commitments and actual expenditures.

## 4. Activity-based accounting and other means of allocating overhead

Contractors can change their accounting practices to recover a fairer share of overhead on extra work.

**Allocation of overhead.** Job cost accounting categorizes all expenses at a project site as job costs, to track and control costs and to report gross profit by job. Nonproject general and administrative costs (home office overhead) are allocated to projects on some rational basis, generally proportional to job costs. Most field activities are classified as direct costs, and the balance is jobsite overhead (also called general conditions), including all functions performed in the job trailer. Project support functions at the home office are often charged as home office overhead, which is not recommended. Some contractors charge equipment costs to a company overhead account instead of directly to projects, which is also not recommended.

Projects with numerous changes incur more overhead support than projects without problems, and change order work requires more overhead support than original bid work. Conventional accounting practice, which distributes overhead equally, therefore distorts reported job profitability and weakens cost control of home office functions. It also contributes to inadequate compensation for extra work.

**Activity-based accounting.** Recovery of change order costs can be increased with activity-based accounting, which allocates the different elements of overhead cost separately instead of allocating all overhead as a single percentage. This is a recent innovation for manufacturing companies and significantly improves cost control and recovery of overhead costs for contract changes. Construction overhead costs are dramatically higher on change order work and problem projects than on original bid work and normal jobs. Therefore, *contractors should either use activity-based accounting or reallocate some conventional overhead costs to direct job support cost categories* to improve cost control and to recover more extra work costs.

**Generally accepted accounting principles and change order accounting.**  The more costs categorized as direct costs and the more cost codes available (within reason), the more accurate the system. However, the allocation must be consistent over time and between different projects. Overhead cannot be allocated to change order work at a higher percentage than the normal allocation of overhead unless supported by documented facts.

**Recommended reallocation of overhead costs.**  The easiest step of reallocating overhead costs is to *reallocate project support functions from home office overhead to job costs*. This requires project managers and other project support personnel to code their time cards by projects, an easy task. Some steps are even easier; long-distance telephone providers can bill by a four-digit code for each project. Cell phone use assigned to project personnel can be allocated to their project, as can many other charges. In addition, equipment use should *always* be charged to projects. The administrative effort and cost are minimal in comparison to the cost recovery on a large claim.

Another step is to *reallocate some job site overhead functions to direct costs or direct job support costs* instead of to jobsite overhead, making it easier to recover those costs for change order work. Specific recommendations are:

- Establish a cost code for preparation and resolution of change order requests, including estimating, scheduling, document preparation, and negotiation; establish a separate cost code for each large change. Include these costs in change order requests and claims. Tracking these costs will probably reveal that contract-specified markup for change order overhead is low and will help justify a higher markup or reimbursement of actual costs.

- Estimate and track the costs of clarification and plan review, preparation of shop drawings, estimating, procurement and material control, surveying/layout, testing, etc., for major work items and change order work. Include these costs in change order requests and claims.

- Charge some general conditions items to those *direct job* cost codes that they support instead of charging them as general jobsite overhead. For example, crane use can be charged to setting and stripping forms, placing rebar, individual change orders, etc., with the balance to general conditions.

- Track small tool costs on all projects and develop historical records of costs for different types of projects as a percentage of labor cost (or as a percentage of some other factor) for estimating and change order allocation.

Some of these steps will meet resistance from owners' representatives but are well worth the effort. Consult your accountant and an experienced construction attorney before changing your accounting system and procedures.

**Reimbursement of change order preparation costs.**   On federal work, the cost for preparing change order requests (but not claim preparation costs) can be recorded separately and charged against the change order. This includes the costs for rescheduling to account for the change, researching the most efficient way to perform the work, etc. Such costs are essential to the progress of the work, as change order work cannot be performed without these efforts. This should also be true for other project owners. Instead of charging the costs for preparing change order requests to a general jobsite or home office overhead cost code, account for them individually and include them on change order requests.

The cost of preparing and negotiating a claim is generally not reimbursable as part of a change order but can sometimes be recovered in the disputes process. These costs should be recorded for better cost control of the claim process and included in claims. At the least, the recorded costs will support an argument based on equity and are valuable bargaining chips. Whenever possible, characterize claims as change order requests to increase the probability of recovering preparation costs.

## D.   How to Compute Damages and Equitable Adjustments

Computing damages is different from bid estimating. Many estimators are not aware of the most convincing type of damage estimate and use the same estimating methods for computing damages as used for bidding or negotiating new contracts. The goal of damage computations is different. The goal of a bid estimate is accuracy and economy of effort. Damage estimates need to be reasonably accurate but *must* be convincing.

Forward pricing, (estimating the cost of changes before performing the work) is recommended. The second choice is force account pricing, which is used when entitlement is acknowledged, but either the scope of work cannot be defined or an agreement cannot be reached on the estimated costs. The least desirable choice is after-the-fact pricing, which must be used when the others aren't.

**Flowchart of the process.**   The process of computing damages for claims, or an equitable adjustment for change order requests, is summarized in Fig. 11.1.

**Sequential steps.**   Damage calculations can be based on forward-pricing estimates, force account records, or after-the-fact pricing. These calculations are described as discrete sequential steps but in reality tend to overlap and repeat. This section describes the first three steps for each of the approaches. The remaining steps, which are common to all three approaches, are described in other sections of this chapter and other chapters as noted on the figure.

## 1.   Forward pricing of acknowledged change order work

Forward pricing is possible only when the owner's representative agrees to entitlement, the contractor prepares an estimate and a change order request, and the parties negotiate a final price before the work is done.

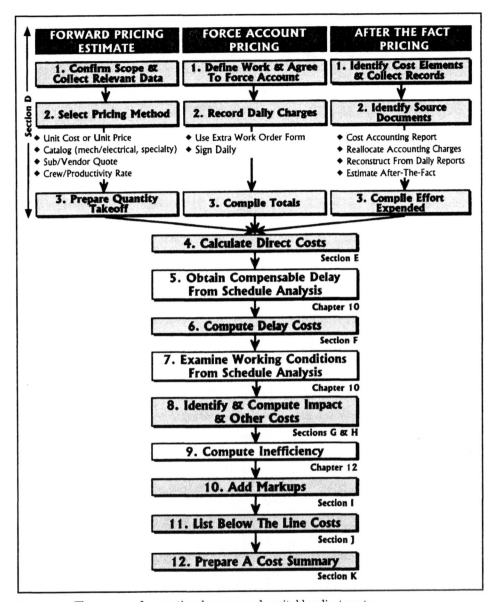

**Figure 11.1**  The process of computing damages and equitable adjustments.

*Forward pricing is preferred, as the owner is assured of a fixed price for the work before deciding whether to proceed and the contractor has an opportunity to beat that price and increase profits.* Forward pricing reduces the need for some recordkeeping, although the costs should be identified separately in the cost accounting system.

**Sequential steps when forward-pricing changes.**  The following three tasks are performed before calculating costs (Fig. 11.1):

1. *Confirm the scope of the change and collect all relevant data.* Owners may initiate a change without defining the objectives, specifying the work to be done, or determining the conditions affecting the work. Contractors, in an effort to serve their client or in a rush to start work and avoid delay, sometimes price the work and start construction without collecting needed information or fully defining the scope. The scope of work is then found to be greater than expected, and more costly, but the contractor is obligated to perform for the agreed price.

2. *Select an estimating method.* Determine what estimating method to use (Sec. B.3). This can be a unit price or unit cost estimate, a catalog estimate, a price based on subcontractor and vendor quotes, or a detailed crew composition/unit rate/productivity estimate. The estimating method can vary for different elements of the claim.

3. *Prepare a quantity takeoff.* Make a quantity takeoff compatible with the selected estimating method and available data.

**Preparing change order requests.** Forward-priced change order estimates can be prepared on the recommended estimating forms and incorporated into change order requests (Chaps. 4 and 5, respectively).

**Contingencies.** One problem in preparing a forward-pricing estimate is accounting for unknowns. This can be resolved by being somewhat high in quantity takeoffs and pricing or by providing a separate contingency against a possible event or condition, based on an expected value analysis. (Expected value is the estimated cost of the event or condition occurring times the probability of occurrence.) If the reviewer objects, exclude the event or condition from the estimate and agree that it would constitute a changed condition if it occurs.

When establishing a contingency, consider the probability of the following:

- Unseasonable inclement weather delaying work and making it more expensive.
- The work being more costly or taking longer than estimated.
- Materials not available or not delivered when needed.
- Suppliers or subcontractors not meeting their quality, cost, or schedule commitments.
- Unexpected material waste, spoilage, excess quantities, and pilferage.
- Overbreak, overexcavation, expansion/compaction factors, and stockpile loss for earthwork.
- Excessive small tool and expendables costs for labor-intensive work.
- Failure of the installed work to pass inspection, requiring rework.
- Overlooked miscellaneous costs, including engineering and shop drawing preparation, surveying, testing, shipping and special handling, rehandling and storage, permit and fees, and additional management or supervisory time.

- Delay and impact to other work (either original bid work or previously agreed change order work).

**Include all elements of cost.**   *When estimating extra work, look beyond the direct costs to determine if there will be delay, acceleration, impact, or increased overhead costs.* Consider all the cost elements listed in Sec. D.6.

**Sensitivity of estimated costs to assumptions.**   When estimating claim costs, quickly check the sensitivity of your computations to variations in the assumptions. If the owner's representative can show that a small variation in the assumptions eliminates the cost overrun, it will irreparably damage your claim position. On the other hand, if a small variation results in significantly higher costs, increase the contingency.

**General recommendations.**   Recommended procedures for estimating forward-priced change orders are as follows:

- *Propose and promptly negotiate changes* to avoid after-the-fact change and more difficult negotiations.

- *Negotiate more conservative quantities and productivity* if contract markups are inadequate.

- *Agree on a unit price change order if concerned about the quantity of work.* Then ensure the unit price will be adequate for the probable quantity of work; if necessary, include a separate payment for mobilization and other fixed costs.

- *Price with contingencies,* or exclude the conditions.

- *Price for no delay* by including overtime costs, acceleration, mobilizing a separate crew, etc. (to avoid delay), or include delay costs in the cost estimate and ask for a time extension.

- *Verify there are no impacts, or exclude impact from the agreement.* Alert all affected subcontractors and suppliers and obtain their sign-off before agreeing to the change.

- *Use two-part change orders to get some immediate financial relief for major changes* when entitlement is acknowledged but agreement on costs is difficult and partial payment is needed. The owner issues a unilateral change order for partial payment based on their estimate, while continuing negotiations to reach a final figure.

## 2.  Force account pricing—extra work orders

Force account may be the best way to proceed if unable to agree on a price before starting work or if the scope of work is too indefinite to price without a large contingency. Both parties keep detailed daily records of the actual effort expended, which are consolidated and signed daily by a representative of each party, with any disagreement noted. Payment is based on the units of effort expended (labor and equipment hours plus materials used) times the unit rate of each. Contract-specified or negotiated markup is added to determine the total cost.

The advantages of force account are that the owner pays no more than the work costs and the contractor is reimbursed for all costs. The disadvantages include the lack of an incentive to work more efficiently (crews tend to "dog it" on force account work), the inadequate markup normally allowed (which often doesn't cover actual overhead costs and profit), and the owner's lack of assurance of what the costs will be until the work is completed.

**Force account recordkeeping for disputed work.**  Force account work records should be maintained for disputed extra work, even if the owner denies entitlement, to facilitate preparation of after-the-fact pricing for a claim. Ask the owner to acknowledge and sign that the number of hours recorded is correct but that entitlement is disputed. This eliminates one item of dispute from subsequent negotiations, reducing costs and conflict.

**Sequential steps when pricing force account work.**  The following three tasks are performed before calculating costs (Fig. 11.1):

1. *Reach an agreement with the owner's representative.* Agree on the work to be done and the procedures for recordkeeping as noted above.
2. *Record daily force account efforts.* Record daily force account work using a standard form, similar to Fig. 4.7, for the daily labor, materials, equipment, and subcontract efforts. Input the data to a spreadsheet to automatically generate daily costs (Sec. C.14 of Chap. 4).
3. *Compile the totals.* Compile the total hours worked and materials used.

**Include all cost elements.**  Determine if there was delay, impact, or increased overhead costs and include in the change order request. Do not submit only force account costs unless you specifically exclude delay and impact.

**General recommendations.**  Recommended actions when performing force account work are:

- *Record costs and get signatures from the owner's representative every day* (even if there is no agreement on entitlement). Note any disagreements and promptly resolve them.
- *Manage force account work with foremen instead of superintendents,* if the contract doesn't adequately compensate for supervisory time and performance doesn't suffer.
- *Assign capable personnel, but not your key supervisors and workers* to force account work.
- *Maximize owned equipment use in lieu of labor,* as Blue Book equipment rental rates are reasonable to generous, whereas contract-allowed markup

on labor is generally inadequate. That may not be true on federal contracts using government-developed equipment rates.

- *Use owned instead of rented equipment* unless it would impact bid work, as little markup is normally allowed on rental equipment. If using rental equipment, include fuel, oil, grease, mobilization, and maintenance costs.

- *Minimize estimating, supervision, engineering, and administrative support,* as these costs are inadequately compensated, but do not skimp to the extent that costs increase and the owner can claim mismanagement. Or record and claim support costs specifically incurred for each change.

- *Use special crews to avoid impacting bid work and to simplify recordkeeping.*

## 3. After-the-fact pricing of claims

The most common method of pricing extra work is after-the-fact. Price the work based on actual time and materials used if special cost accounting codes were established. If those records were not maintained, it is usually possible to price the work from other documents.

**Sequential steps when pricing costs after the fact.** The following three tasks are performed before calculating costs (Fig. 11.1):

1. *Identify all cost elements and collect pertinent job records.* Identify all of the cost elements incurred and then collect all relevant records on those cost elements.
2. *Identify cost data sources.* Identify the source documents with the cost data, which may include one or more of the following:
   a. *Use cost accounting data.* If accurate and complete, cost accounting reports are the best documents for proving actual cost. They should be supported by material and subcontract invoices.
   b. *Reallocate accounting charges.* If cost codes were not established for extra work, charges not correctly allocated, or adjustment is needed for impact (without doing a full-scale inefficiency analysis), supervisors can reallocate charges from one cost code to another. Someone familiar with the work should do the reallocation, while referring to contemporaneous records (e.g., the superintendent's daily reports) to refresh their memory. Reallocation can be highly subjective, subjecting it to challenge and possible rejection. Therefore, make every effort to quantify the reallocation and record the reasons in order to defend them.
   c. *Reconstruct costs from other records.* The superintendent's diary report or a subcontractor's diary may contain references to extra work or impacts and the labor and equipment used. These or other project records can all be used to reconstruct the costs. The most frequently used records are:
      - *Time cards.* Time cards are sometimes annotated with comments and descriptions of work accomplished. These notes are useful for correcting computer input errors, identifying time spent on extra work, and

estimating impact. Unfortunately, the information recovered may not justify the effort required to obtain it.

- *Superintendents' daily reports.* Most daily reports include crew sizes, which may be the only available information on actual labor, subcontractor labor, and equipment use.
- *Equipment use logs.* Equipment use logs may provide the hours of use (often by cost code), from which standby time and other equipment costs can be determined.
- *Certified payroll reports.* On most public works contracts, contractors must file certified payroll reports affirming that employees were paid the specified wage rates. These reports can be used to reconstruct labor costs, if they are accurate.

  d. *Estimating after-the-fact costs.* Some costs will not be adequately identified. Reconstruct these costs by estimating the probable cost of individual items and reconcile the total to match the recorded cost. There is little difference between a forward-pricing cost estimate and an after-the-fact pricing estimate. If the owner can accept one, they should accept the other.

3. *Compile efforts expended.* The last task is to compile the effort expended—labor hours by craft and wage rate, equipment use hours, material quantities, and subcontract costs. This may require large nested spreadsheets linking each individual cost record to an overall claim summary.

Attach actual invoices, quotes, and price lists to substantiate subcontract and material costs. If not too voluminous, attach time cards. The extra volume is usually beneficial, as it lends weight to the claim. Voluminous supporting documents should not appear in the claim document itself but be referred in an appendix, a separately bound document, or a box of files.

When attaching source documents, organize them for review and index them to summary sheets, providing a clear audit trail. Highlight pertinent portions of the data for easy identification. This helps reviewers verify the costs, gives them more confidence in the accuracy and fairness of the claim, and results in a quicker, more generous settlement.

Some contractors are reluctant to provide much documentation, partly owing to the fear that anything provided may elicit more questions and provide ammunition for counterarguments. This can be true, and care should be exercised. Avoid providing information that leads to a challenge or the opening of other issues. However, a reasonable level of detail is expected; providing less leads to questions and a reluctance to settle.

**General recommendations.** Recommended actions for pricing after-the-fact changes are:

- *Give prompt notice of change,* in accordance with the contract, to prevent denial due to inadequate notice and to enable the owner's representative to maintain independent records.

- *Try to convert extra work to force account* if unable to forward-price. Submit daily records on a standard work order form and ask the owner's represen-

tative to sign the records agreeing the work was done but without acknowledging entitlement.

- *Keep extensive records* and increase recordkeeping efforts when changes occur.

- *Adjust inadequate contract-specified markups* by adding actual direct support costs (estimating, supervision, expediting, etc.) to the claim.

- Document administrative and supervisory time spent in preparing, negotiating, and performing extra work, to prove the inadequacy of contract-specified markups.

- *Negotiate and use actual jobsite and home office overhead* for major changes if the specified markup is too low.

- *Characterize the claim as a change order request,* and include change order preparation costs.

- *Start negotiations immediately,* using a win/win approach, and press for timely resolution.

## 4. Unit price payment for change orders and disputed quantities of bid work

Owner's representatives may offer to pay contract unit prices for extra work, in addition to using them for variations in quantities if they believe the prices are favorable. If the working conditions are more difficult for the extra work than the bid work (due to smaller quantities of work, inadequate notice, etc.), however, contractors can refuse to use the contract unit prices and insist on actual costs plus markup. Conversely, owner's representative are sometimes reluctant to pay contract unit prices if they believe them to be high. Contractors should insist on using those prices if the working conditions are the same.

Disputes sometimes arise over measurement of quantities installed on unit price contracts. To minimize these disputes and ensure payment:

- Videotape all project areas before starting construction to record existing conditions and photograph key areas of potential dispute before, during, and after construction.

- Survey existing site conditions for earthwork, if the information on the plans appear inadequate or incorrect.

- Install settlement plates over existing soils subject to significant consolidation, if being paid for in-place earthwork fill yardage.

- Carefully read the contract terms regarding neat line versus overbreak measures and payment and act to avoid excessive overbreak.

- Check for unusual measurement provisions and alert the field crews before work starts, if compensation might be affected by means and methods or if special recordkeeping is required.

- Agree with the owner's representative on truck yardage and other simplified measurements before the fact instead of afterward.

### 5. Total cost, modified total cost, and actual cost methods

Contractors sometimes use the total cost approach if they can't quantify costs by specific issues or separate added costs from original contract costs. To compute, subtract the bid cost from the actual cost and claim the difference, plus the as-bid overhead and profit. The advantages of the total cost approach are simplicity and ease of preparation. The disadvantages include dependency on:

- The accuracy of the original estimate.
- The assumption that none of the costs is due to contractor errors.
- The assumption that other factors (e.g., weather) didn't cause the increased costs.

The courts and most owners reject total cost claims (except for individual cost codes) but will accept a modified total cost claim if no other method to compute the costs is applicable. The modified total cost method adjusts for bid errors, contractor mistakes, and other problems not the owner's responsibility.

For a cost-plus claim, add up the actual costs, add overhead and profit, and claim the total. This is used for a "cardinal change," when the change results in work that is fundamentally different from what was bid. The change can be in either the final product or *how the work was performed*. Cost-plus claims are sometimes called total cost claims to facilitate acceptance by the claim reviewer. Most reviewers have heard of total cost claims and may be more willing to accept a modified total cost claim than "time and materials."

Although conventional practice is to use total cost and cost plus claims for the total project, they can be used for individual claim issues, while using more exact methods for other claim issues. It depends upon whether and how the work was impacted and the adequacy of the cost records.

### 6. Elements of damages

Claim preparers should use a complete, clearly defined, and logically classified list of damages to:

- Help identify all reimbursable costs so that they are included in the claim.
- Facilitate preparation of damage estimates.
- Conform to how the contract and contract law treat entitlement for different types of damages.
- Generate more detail for review and analysis.
- Aid the claim reviewer in understanding the claim.
- Help convince the claim reviewer that the amount claimed is correct.

The following classification of damages is generally recommended:

1. *Direct costs.* Direct costs are those costs specifically attributable to building the project and are necessary either to accomplish the project as bid or to change the contract scope. Direct costs are organized by claim issue and broken down by cost category (labor, equipment, materials, etc.).

2. *Delay.* Delay causes the project, or parts of it, to take longer or be done later, resulting in:

- Extended overhead—of jobsite and home office overhead.
- Escalation—in the unit costs of labor, materials, equipment use, or subcontract costs.
- Labor and equipment standby—while waiting for a problem to be resolved.
- Demobilization and remobilization—due to delay and the need to return later.
- Extended warranty—and other costs due to delay.
- Changes in working conditions with impact—inefficiency, additional tasks, and materials.
- Lost profits—for projects that could not be bid, owing to delay.

3. *Impact from disruption and change in working conditions.* Impact can result in:

- Inefficiency—of labor and equipment use in performing original bid work.
- Additional tasks—not changing the scope of the project, such as access roads.
- Additional materials—such as extra base rock to allow working on wet subgrades.

4. *Acceleration.* Acceleration results from having to complete faster than anticipated. The resulting impacts are:

- Changes in working conditions—resulting in inefficiency due to overtime, crowding, etc.
- Additional tasks—including preparatory and support work, engineering, and supervision.
- Additional materials—to expedite progress.
- Mobilization—of additional equipment.
- Expediting—for early material delivery.

5. *Other costs.* Other costs may include:

- Additional overhead—jobsite or home office for many claims or a specific claim.
- Change order preparation and negotiation costs—separate from claim costs.
- Future costs—such as higher insurance or workers' compensation premiums.

6. *Markup.* Markup for indirect costs is added to cover:

- Jobsite overhead—expressed as a percentage of direct costs.
- Home office overhead—expressed as a percentage of jobsite costs.
- Profit—expressed as a percentage of total job costs.
- Bond, insurance premium, and taxes—computed as a percent of job cost plus profit.

7. *Below the line costs.* Costs added at the end, without markup, include:

- Retainage and unpaid contract balances
- Interest—for late payment of retainage, claimed amounts, or progress payments.
- Attorney fees and claim preparation cost—for costs incurred after the change order request is rejected.
- Credit for nonconforming and unsatisfactory work—if the work will not be corrected.

**7. Organize claims by change order approval status**

Provide the owner's representative a status log of change order requests and claims, with submission dates and review durations to pressure them to negotiate. Propose a schedule for negotiations and insist on meeting as agreed. Classify change order requests and claims by approval status, using the following categories:

- Accepted and waiting to be incorporated into a change order.
- Acknowledged entitlement, but disputed amount of cost and/or time extension.
- Disputed entitlement, and therefore also for cost and time.

**8. Organize claims by origin**

Claims may also be classified by design issue. This is not generally recommended unless you want to directly blame the designer for some issues.

- Scope change—a change requested by the owner.
- Design omission—a needed item not included in the plans and specifications.
- Design ambiguity or interpretation—a needed item not clearly defined and therefore not bid.
- Design error—a mistake in the design that must be corrected with more cost or time required.
- As-built discrepancy—a change required because as-built conditions are different than indicated.
- Operating requirement—a change necessary for the continuing use of an existing facility.
- Contract administration—a problem caused by poor contract administration.
- Differing site condition—an existing site condition different than indicated or normally encountered.

**E.  Calculating Direct Costs**

Direct costs include extra work that increases the scope of the finished project or ancillary work necessary to accomplish the finished project (e.g., access roads).

Direct costs are normally broken down into the following cost categories:

- Labor
- Materials
- Equipment
- Subcontract
- Expendables (materials)
- Small tools (equipment)

## 1. Labor costs

Labor is the most variable, the most difficult to estimate and control, and often the largest direct cost.

1. *Wage rates.* Labor rates vary from project to project. Federal projects and state-funded projects subject to state Davis Bacon acts require workers to be paid a specified rate. Union agreements may require different wages and fringes depending on location, the type of project, or special agreement. Be careful when using labor cost rates, and determine if they are correct and whether all burdens and related costs are included.

   To minimize disputes with the owner's representative, negotiate labor rates early so that they can be used for all change order requests and claims without further argument. Include the following costs if applicable:

   a. *Base wage rate.* The base wage is determined by the individual's trade, experience, and work performed, in accordance with union scale or the individual's hiring agreement or as required by federal or state law for public works contracts.

   b. *Labor burden.* Labor burden normally includes the following costs:

   - Payroll taxes which have a limit resulting in a lower average rate if some employees' compensation reaches the limit:
     - Social security (FICA)—currently 6.13 percent.
     - State and federal unemployment—generally between 0.5 and 3 percent.
     - Local taxes—variable.
   - Workers' compensation insurance premiums—depends on the trade and the company's experience rating (from 1 to over 10 percent).
   - Health benefits—depends upon union agreements or company practice.
   - Union or company fringe benefits—varies from zero to over $3 per hour for vacation, holiday, sick pay, health benefits, and union dues.
   - Industry support or training fund checkoff.

   Labor burden may also include the following special costs:

   - Hiring and termination costs—including relocation expenses for the worker and family, travel expenses, training, skill-level certification (e.g., welders), etc. These costs are spread over the average hours worked per employee and included in the burden.

- Bonus and company benefits—including picnics, perks, and use of company vehicle for supervisory and salaried employees.
- Additional overtime or other fringe benefits and bonus—if needed to obtain a reliable, skilled labor force, which happens when the local construction market is booming.

c. *Travel and subsistence expenses.* Travel and subsistence costs may be included in the labor burden or accounted for separately.

d. *Other costs.* The following costs may be categorized either as labor burden or as other costs:

- Insurance premiums for liability and property damage are often tied to labor costs and are generally 0.3 to 0.6 percent of gross labor costs. These costs are normally added as markup.
- Small tools and safety supplies are estimated as a percentage of labor costs (they may be categorized as materials) and range from 2.5 to over 5 percent.
- Supervision costs can be included in the labor burden, added in the overhead markup, listed separately, or included in a composite crew rate.

e. *Overtime and shift work premium.* In the United States, overtime premium is $1\frac{1}{2}$ times the base rate for work in excess of 40 hours a week for nonexempt (i.e., nonmanagerial or nonprofessional) employees, and for some employees working over 8 hours a day. Union agreements may dictate a higher rate for some overtime (e.g., on holidays). Shift premium is normally $\frac{1}{2}$ hour per shift on union contracts. Shift workers may overlap $\frac{1}{2}$ hour to maintain operations, which add further costs.

f. *Escalation.* When forward pricing, determine if the work extends into a period of higher labor rates. If so, determine the amount of work to be accomplished at the higher labor rate, the hours of work, the increase in labor rate, and the total cost increase.

g. *Composite labor rates.* Most contractors establish composite crew rates for specific crews with multiple trades and differing skill/pay levels. These rates are used to calculate costs for that composite crew or are converted to an average hourly labor rate to simplify estimating labor costs.

2. *Productivity.* The factors affecting labor productivity are discussed in Sec. A of Chap. 12.

3. *Labor availability.* Check for the availability of skilled workers within the work force, but not working on the critical path, or available to hire. This requires a detailed labor-loaded schedule and a resource forecast (discussed in Chap. 10). Do not price extra work at straight time or an average labor rate if the crew must work any overtime. Price it at the overtime rate and add a fatigue factor to both the change order work and the original bid work as described in Sec. G.2 of Chap. 12. Alternately, bring in a separate crew to do the work or add delay and impact costs.

4. *Crowding or trade stacking.* Check if there will be too many workers in one area or an overlap of incompatible trades. Create a resource forecast or plot a timescale diagram to determine what activities are concurrent for each

area. Note the number of workers assigned to each activity during the period of extra work, determine the number in limited work spaces, and check for work conflicts. If crowding or trade stacking is forecasted, reschedule or add inefficiency cost.

5. *Management, administration, engineering, and supervision labor.* Jobsite and home office management, administration, engineering, and supervisory labor costs are included in overhead rates. However, extra work and problem projects consume an inordinate amount of these resources. Therefore, contractors should claim direct support labor costs if they are significantly greater than those included in the markup (see Sec. C.4 above and Sec. B.6 of Chap. 5).

*Overhead costs for small extra work items can exceed the direct costs.* Superintendents sometimes spend several hours responding to a foreman's request for help and determining a solution with the designer. The superintendent then picks up the materials, lays out the new work, instructs the foreman, and supervises installation. Little of this time is covered in normal overhead.

Examine your contract to determine what is reimbursed. For example, the standard AGC contract allows extra payment of jobsite costs but not home office estimating and engineering (e.g., shop drawing) costs. If using the AGC contract, send the estimator to the jobsite to estimate changes.

## 2. Material costs

Material costs are easier to estimate and vary less than labor costs but are also a large portion of total costs.

1. *Quantity takeoffs. The most important step in estimating material cost is an accurate quantity takeoff.* This requires reasonably complete plans and specifications and careful determination of the needed quantities. When preparing a takeoff, use conservative conversion factors (e.g., bank versus loose versus compacted cubic yards, cubic yards to tons) and order more than may be needed. The cost of being $\frac{1}{4}$ cubic yard short on a concrete order or having a few feet less pipe than needed greatly exceeds the cost of ordering extra materials.

2. *Elements of material costs.* The elements of materials cost include:
   - Purchase costs as invoiced, less discounts, plus late payment fees.
   - Taxes and insurance.
   - Shipping, if not F.O.B. jobsite, and expedited delivery costs if needed quickly.
   - Receipt, unloading, inspection, testing, materials handling, and temporary storage.
   - A credit for salvage.
   - Special insurance coverage, if not F.O.B. jobsite.
   
   Additionally, material costs may also include the following:
   - An allowance for loss (e.g., stockpile loss for rock aggregate, deterioration, wastage, pilferage) and overage quantities. These may be covered in the quantity takeoff.

- Procurement costs such as purchasing, expediting, and small order costs are normally included in overhead costs. Include extraordinary procurement costs in the material estimate.
- Ancillary materials, such as fasteners or pipe hangers, which may be estimated separately or factored into the cost of the primary material. Unanticipated ancillary materials, such as seismic constraints, can cause cost overruns that may not be compensable.

3. *Pricing methods. Whenever possible, base material costs on firm, fixed-price quotes by reliable suppliers.* Other methods of pricing can be risky. These methods include using historical records, estimating handbooks such as Means, vendor price lists (with an escalator factor if not current), and the estimator's experience and judgment.

4. *Risks and recommended practices.* One of the greatest risks in pricing materials for extra work is that the quote may expire and costs increase by the time a change order is negotiated. To prevent this, schedule the work and include a required authorization date with your change order request or claim.

Another risk is that the supplier may not be able to meet the agreed delivery date (which should always be specified). Or the material supplied may be defective, not meet specifications, or cause additional costs from difficult installation, attempts to correct the problem, removal, replacement, and delay. Responsibility for the problems may be unclear, and the contractor may ultimately absorb the costs.

Avoid using questionable suppliers for critical materials on change order work. A small difference in price does not justify the risk of delays or unsatisfactory performance. If forced to use unknown or unreliable vendors, quantify the risk and include it in the estimate or shift the risk to the owner.

Credit early payment discounts only if the contractor can use those funds for the early payment. It makes no sense for a contractor to incur interest costs or forgo investment interest by making early payment if the owner receives all the benefits.

If some materials must be returned, give a credit less restock charges and handling costs. Salvage credits should be the minimum recoverable under "close out" sale conditions, less the cost of sale, unless the contractor has an immediate use for them. Many equipment yards are full of salvaged material that has been quietly rusting and taking up space for years.

If an alternate source is used under an "or equal" clause, there are risks that the material may not truly be equal or the designer may not consider it so. This is more of a problem when bidding, when material prices are quoted at the last minute before bid opening, but can also occur with change order work.

5. *Services.* Charges for services (e.g., testing) and consultants (attorneys, surveyors, claims consultants, etc.) are usually listed under material costs but can be included under subcontract costs or have their own category. Do not overlook them.

### 3. Equipment costs

Equipment costs include construction plant (e.g., concrete or asphalt plants), general conditions equipment such as cranes and forklifts and operating/field equipment (e.g., scrapers) that are part of direct costs. General conditions equipment costs vary depending on the project duration; operating equipment costs vary with the amount of work done; and plant costs vary depending upon the quantity of material processed and the duration of operation.

1. *Elements of equipment cost.* The elements of equipment cost include:

- *Ownership costs* that depend upon the age and annual hours of use and includes
  - Interest or the cost of capital
  - Depreciation
  - Property taxes
  - Storage
  - Insurance
  - Attachments (dozer blade, ripper, winch, etc.)
- *Operating costs* that vary with working conditions, use, and age, and include
  - Maintenance
  - Major repairs
  - FOG (fuel, oil, grease)
  - Tires or track maintenance and replacement
  - Attachments
- *Mobilization/demobilization costs* that are spread over the period of use at the project. Include hauling equipment and labor, assembly and disassembly, modification, testing, and other costs. Multiple mobilizations or extended standby costs can be factored into the unit rates.

2. *Equipment rates.* Contractors should use actual costs documented by their cost accounting system. If they use "equipment cost pools" or have highly depreciated equipment, this is not possible and they should use listed equipment rates.

The Rental Rate Bluebook for Construction Equipment, published by Data Quest, Inc. (the Blue Book), is the most widely used guide for equipment rates. The AGC *Contractor's Equipment Cost Manual* published annually by Data Quest, Inc. is also widely accepted, including the U.S. Court of Claims, U.S. Department of Transportation Board of Contract Appeals, and the Armed Services Board of Contract Appeals. The Board of Contract Appeals reduces the printed rate by $1/_2$ for standby.[15]

Other sources of equipment rates are:

- Associated Equipment Dealers (AED) Rental Rates Compilation rates. These rates include overhead and profit, tend to be higher than AGC, and are generally higher than the internal cost of owning equipment.
- NECA Tool and Equipment Rental Schedule.
- Means cost estimating manuals.

- State Department of Transportation (DOT) equipment rates.
- Construction Equipment Ownership and Operating Expense Schedule (EP 1110-1-8), published by the U.S. Army Corps of Engineers.

The listed equipment rates are based on typical annual hours of use. If the actual annual hours of use are less, the rates can be increased. The rates can also be increased for severe working conditions.

3. *Estimated hours or period of use.* Equipment use for change order work is normally estimated by the hour but may be based on daily/weekly/monthly rates and generally parallels labor hours unless the equipment is operated intermittently.

Do not use equipment committed to critical path activities for extra work unless the extra work is (1) done on overtime, (2) performed by a separate shift crew, or (3) costed to include delay costs.

4. *Standby and mobilization/demobilization costs.* Idle equipment costs include only ownership costs, not operating costs. Include opportunity costs only if the contractor has to buy equipment for other jobs. If the duration of a delay is known and the mobilization/remobilization costs are reasonable, contractors may demobilize equipment to use elsewhere and remobilize it later.

5. *Rental equipment versus contractor-owned equipment.* If using rented equipment, include mobilization/demobilization costs, fuel, oil, grease, and any required maintenance costs in the rate.

## 4. Subcontract costs

Subcontract costs are normally based on firm, fixed price quotes and can be lump sum or unit price. Even then, costs may exceed the estimate.

1. *Obtaining quotes and awarding subcontracts.* Change order work not performed by the general contractor's forces is assigned to existing subcontractors without competitive pricing, unless the general contractor prepares a check estimate. Extra work not assignable to an existing subcontractor is normally bid to several subcontractors. They may quote slightly different parts of the work, making it difficult to compare prices and ensure that all work items are included.

Subcontractors should provide the same level of price breakdown for change order work as the general contractor. This may be difficult to enforce, as many do not have standardized estimating procedures or good historical pricing data. Many subcontractors quote unit prices or a lump sum amount without a breakdown, which is difficult for the claim reviewer to evaluate.

2. *Risks.* The risks of subcontracting extra work include:

- Subcontractor financial failure, since many are not bonded. Even if bonded, the delays and impact costs are seldom recoverable and can be extensive.
- Misunderstanding what isn't covered in the subcontractors' bids, leaving gaps in the estimate and some work not included.

- Late performance that impacts other work or delays the project and causes further impacts or liquidated damages.
- Poor-quality work that requires rework, impacts other work, or delays the project.

## 5. Costs of expendables

Expendables, also called consumables or supplies, are materials consumed in the course of construction and not incorporated into the work. Examples include form lumber, nails, welding rod, etc. Expendables are often estimated as a percentage of labor costs or as a percentage of certain types of material costs.

## 6. Small tool costs

Tools and equipment costing under $1000 (some use $250) are considered to be small tools and are "consumed" on the job. The useful life of small tools is limited because they are frequently broken, lost, or stolen. The cost of small tools, usually estimated as a percentage of direct labor costs, ranges from 2 to over 5 percent of direct labor costs (including burden). Small tool costs vary with the type and work conditions. For example, working over water results in higher than normal small tool costs. Work in a high crime area has a greater theft factor. Small tool costs also depend on morale, jobsite security, and the contractor's control over issuance and return.

Small tool costs vary by construction trade, as follows:

- Equipment operators—very little to none (survey equipment is jobsite overhead).
- Carpenters—high (saws, ladders, etc.).
- Laborers—moderate to high, depending upon the project and their duties.
- Plumbers, pipefitters, and sheetmetal workers—high.
- Electricians—high.
- Teamsters—moderate (chains, binders, jacks).

## 7. Sample direct cost calculations

A typical crew/productivity rate calculation is shown on pages 364 and 365.

Compare this to a contractor's unit price estimate that may use $150 per cubic yard for a quick estimate:

| | Unit Price Estimate | |
|---|:---:|---:|
| Concrete: | 100 cy @ $150/cy= | $15,000.00 |

Claim reviewers would have a difficult time evaluating the quick estimate shown above unless they were experienced construction cost estimators.

| Crew and Productivity Rate Estimate | | | | |
|---|---|---|---|---|
| | No. | Hours | Hourly base rate | Subtotal Am't |
| **1. Labor daily rate.** | | | | |
| Carpenter foreman | 1 | 8 | $18.13 | $145.00 |
| Carpenter journeymen | 2 | 8 | 16.99 | 271.84 |
| Carpenter apprentice 3 | 1 | 8 | 16.70 | 133.60 |
| Laborers | 4 | 8 | 9.45 | 302.40 |
| Operating engr. (crane) | 1 | 8 | 16.25 | 130.00 |
| | 9 workers | | | $982.84 |

**2. Labor Burden (see Exhibit A)**

| | |
|---|---|
| FICA | 6.20% |
| Medicare | 1.45 |
| Federal unemployment | 0.80 |
| Workers' compensation | 7.40 |
| State unemployment | 1.35 |
| Insurance | 0.25 |
| Health & welfare & pension | 1.50 |
| Labor Burden | 18.95 / 100 × $982.84     $186.25 |

**3. Subtotal Labor Daily Rate**

$1,169.09 per day

**4. Equipment Daily Rate**

Crane, Bucyrus-Erie 61-B 1 day                    $725.00*     $725.00 per day
30 ton w/60' boom
(* OCE Equip Ownership Manual, see attached page xx)

**5. Small Tools**
@ 5% of labor costs        $1,169.09 × 0.05                    $58.45 per day
**6. Consumables**
@ 3% of labor costs        $1,169.09 × 0.03                    $35.07 per day
**7. Safety Supplies**
@ 1% of labor costs        $1,169.09 × 0.01                    $11.96 per day
                           DAILY COMPOSITE CREW RATE     $1,999.57 per day

**8. Estimated Production Rate**
Form and pour 100 cy wall estimated @ 2 days                    ×2 days
                                                               $3,999.14

**9. Materials (see Exhibit B)**

| | |
|---|---|
| Form lumber 20 MFBM @ $240/M (invoice attached) | $2,400.00 |
| Form Hardware, lump sum estim. @ 10% of lumber cost | 240.00 |
| Concrete 100 cy $45.00 (quote attached) | 4,500.00 |
| MATERIALS COST | $7,140.00 |

**10. Subcontract**

| | |
|---|---|
| P&J Concrete quote to place and finish concrete (attached) | $2,000.00 |
| R&R Steel quote to furnish and place rebar (attached) | 1,275.00 |
| | $3,275.00 |

*(Continued)*

| 11. Recap by Cost Category | |
|---|---|
| Labor (#3×2 days) | $2,338.18 |
| Equipment (#4 & 5×2 days) | 1,566.90 |
| Materials (#9, #6×2, & #7×2) | 7,234.06 |
| Subcontract (#10) | 3,275.00 |
| Subtotal Direct Costs | $14,414.14 |

## F.  Computing Compensable Delay Costs

Delay costs are based on the total compensable delay and often result from several claim issues.

### 1.  Determine the amount of compensable delay

The schedule analysis explained in Chap. 10 determines the amount of delay. The excusable but noncompensable delay is included in the would-have-been schedule. *The compensable delay is the difference between the completion dates for the would-have-been schedule and the as-built schedule.* It is used to compute delay costs, as noted below.

### 2.  Extended jobsite overhead

Extended jobsite overhead costs include:

- Jobsite overhead labor (supervision, engineering, estimating, and administration) costs

- Plant and equipment costs (trailers, cranes, pickups, concrete mix plants, etc.)

- Expenses (telephone, garbage, office supplies, etc.)

Use either estimated or actual jobsite overhead costs, or a combination if it is more convenient and more accurate. Many contractors use estimated jobsite overhead rates if they haven't maintained adequate job cost accounting records. Since extension of the project duration doesn't affect non-time-dependent costs (mobilization, purchase of office equipment, etc.), delete them from the actual overhead when computing extended jobsite overhead rates. Divide the balance by the duration that the actual costs were incurred for the daily rate and multiply the result by the number of days delayed. The average overhead is commonly used because jobsite overhead varies during the course of a project. In some cases, peak monthly overhead rates may be more appropriate.

Some contractors stop charging internal equipment rental against projects that are over budget. Add these unrecorded costs to the claim. Also add project managers and other direct support costs incurred in the home office, even if not included in the accounting reports (if their time can be documented or reasonably estimated).

| Extended Jobsite Overhead | |
| --- | --- |
| General conditions item description | Cost per work day |
| * Field Superintendent | $340 |
| * Carpenter foreman | 311 |
| * Job trailer | 16 |
| * Phone/pager | 17 |
| * Fax | 8 |
| * Copier | 10 |
| * Weather service | 5 |
| * Travel | 18 |
| * Storage trailer | 5 |
| * 35 ft 4wd reach lift | 105 |
| * Two 60-ft boom lifts (Ivy HiLift) | 210 |
| * 20-ton hydro crane | 720 |
| * Forming material (Mason's Supply) | 429 |
| * Planking material (Ivy HiLift) | 29 |
| * Safety equipment (harness, lanyards,...) | 50 |
| * Miscellaneous tools and equipment | 599 |
| **Total daily cost** | $2,872 |

The extended jobsite overhead cost is therefore:
33 days delay at $2,872/day =    $94,776

Many daily costs must be calculated from monthly costs or on some other basis. A typical jobsite overhead calculation would be as follows. The superintendent's daily cost would be monthly salary plus fringe benefits and taxes, bonus, pickup, etc. The total is divided by 30.5 calendar days or 21 working days per month.

## 3. Extended home office overhead

Project delays cause less revenue to be received than if the project had been completed on time and other work undertaken. Therefore, some of the home office overhead will be unabsorbed and must be reallocated to other projects. Unabsorbed overhead and extended home office overhead are the same concepts; the term extended is normally used for construction, while unabsorbed is used for manufacturing.

1. *Eichleay formula.* The generally accepted method of computing extended home office overhead (HOOH) in the United States is the Eichleay formula. It is easy to use, easy to understand, and accepted by most owners. The U.S. Federal Circuit Court has stated that the Eichleay formula is the only proper method of calculating unabsorbed home office overhead. Yet some federal agencies have continued to resist it, and some state courts have disallowed it. While flawed, it is a rational method and better than the alternatives. A modified Eichleay formula eliminates the shortcomings and fairly computes extended home office overhead.

---

**Original Eichleay Formula**

[actual Project Billings / Total Company Billings During Project] ×
[total Home Office OverHead During Project] = Overhead Allocated to Project

[overhead Allocated To Project] / [actual Project Duration] =
Daily Home Office Overhead Allocated to Project

[daily Project Home Office Overhead] × [days of Delay] =
Extended Home Office Overhead Amount

---

To use the Eichleay formula, contractors must show that:

- The owner delayed the project.
- Contract revenues were reduced and some home office overhead costs were unabsorbed.
- It was not practical to find additional work to cover the unabsorbed overhead.
- It was not practical to reduce home office overhead costs for the period of delay.

Total company billings and home office overhead are taken from the financial statements (the profit and loss and G&A reports).

The shortcomings of the original Eichleay formula are:[22]

- It allocates overhead based on revenues (billings) instead of on direct costs, which are more closely related to overhead expenditures.
- It understates the daily overhead rate by including days of delay in the average rate (corrected with the modified Eichleay).
- It doesn't fairly account for disputed change orders and delay costs not included in billings. This is also corrected with the modified Eichleay. Or, contractors can interpret "billings" to include what they billed, not what was finally accepted.

2. *The modified Eichleay formula.* The Eichleay formula can be modified to correct its shortcomings. The most common and equitable modification substitutes the original contract price for the actual billings and the original contract period for the actual contract period. These changes give a greater overhead value than the original formula. Since the modified formula is based on the original cost and duration, it can be used while the project is still ongoing. It has, however, been rejected by some courts.

The modified Eichleay formula is as follows:

---

**Modified Eichleay Formula**

[Original Contract Amount / Total Company Billings During Original Contract Period] ×
[Total Home Office Overhead During Original Contract Period] =
Overhead Allocated to Project

[Overhead Allocated to Project] / [Original Contract Duration] =
Daily Home Office Overhead Allocated to Project

[Daily Project Home Office Overhead] × [Days of Delay] =
Extended Home Office Overhead Amount

---

3. *Disallowed overhead costs.* Federal regulations disallow interest, advertising except for recruitment, charitable contributions, entertainment, bid preparation costs, and bad debts in the overhead calculations for federal projects. This is unfair and discourages an essential community service (charitable giving). These costs can be included on private construction or state and municipal government projects unless state law dictates similar deductions. Excessively high "perks" such as a boat (unless used for business entertainment) and distribution to shareholders in the form of excessive bonuses should be excluded.

4. *Inability to obtain other work.* Although most contractors simply compute and submit the amount computed by the Eichleay formula, they may be required to prove that the delay prevented them from obtaining other work. These reasons may include limited bonding capacity, lack of additional management personnel or equipment, or the short or uncertain duration of the delay. Evidence of bidding other work indicates, but does not prove, that the delay on the subject contract did not prevent them from obtaining other work.

5. *Credit home office overhead included in change orders during delay.* Contractors should not claim both extended home office overhead and home office overhead included in the markup for change orders performed during the delay. To avoid double dipping, determine the percentage of home office overhead in the contract-specified markup for change order work and provide a credit for change order work performed during the period of delay. If the contract-specified markup does not cover all overhead costs and profit, a credit is not necessary.

6. *Adjust home office overhead charges.* To recover a more equitable share of overhead costs on change orders, *contractors should charge all costs directly attributed to projects to those projects* (described in Sec. C.4).[15]

7. *Other methods for computing extended overhead.* Alternative methods of computing extended home office overhead are:

- Allegheny (burden fluctuation method).
- Carteret—which computes excess overhead.
- Hudson formula—which is used in England.
- Direct markup percentage—which is unfair if the dollar value of the extra work causing the delay is minor or if the cause of delay is work stoppage without added work.

## 4. Escalation

Costs subject to escalation include:

- Labor wages and fringe benefits.
- Material and subcontract prices.
- Equipment operating costs—fuel, lubricants, and repairs.

**Escalation factors.**    Escalation is not limited to the consumer price index (CPI), the ENR (*Engineering News Record*) indexes, or even *Dodge Report* numbers. Escalation can increase dramatically (5 percent or more) in a few months. For example, missing the bidding season results in higher subcontractor markup if they become busy and no longer need the work. Some will not bid, while others substantially increase their quotes. The increases are compounded if suppliers and lower tier subcontractors also increase their margins.

**Labor.** Wage escalation occurs when delay pushes work into a period of higher labor rates. To compute damages for labor cost escalation, determine the hours that are paid at the higher rates, identify the difference in wages, and compute the total.

**Subcontract and materials purchases.** If quotes were received prior to and after the delay, computations are easy. However, purchase of expendables and some permanent materials may occur when needed, at the then-current price. Compute the escalation from either an estimate of local conditions based on telephone calls to local vendors or a construction materials price index, such as ENR's.

**Equipment costs.** Compute the escalation of these costs on a percentage basis based on published increases or a survey.

5. **Equipment standby or demobilization/remobilization**

When work is delayed, equipment is idle or underutilized. Compensation is due whether the equipment is at the jobsite or at the contractor's yard but committed to the project and not available for use elsewhere. Computation of standby costs is discussed in Sec. E.4.

6. **Labor underutilization or layoff and rehire costs**

During periods of delay, labor may be idle for a period of time and then underutilized or inefficient. Or, if experienced workers are laid off and find other work before the delay is resolved, they will be replaced with new workers who may not be as skilled. Delay also causes learning curve inefficiencies as the newly hired or rehired workers get up to speed.

7. **Materials storage, financing costs, and deterioration**

Delay often requires additional handling and temporary storage of delivered materials, postponing delivery of some materials (with extra costs), additional financing costs (unless paid for as materials on site), and possible deterioration of stored materials (e.g., stockpile deterioration of crushed rock).

8. **Subcontractor mobilization/remobilization and other delay costs**

Subcontractors are subject to the same delay costs as general contractors, as listed above. In addition, they are often required to demobilize and remobilize, incurring further costs.

### 9. Extended warranty

Delay may extend the period a contractor or vendor is required to repair installed equipment, and the manufacturer's warranty may expire before the contract warranty period expires. Either exclude these costs or estimate and include them in the claim.

### 10. Change in working conditions and resulting impact

Delay often results in work being performed under different working conditions than originally anticipated, which may increase costs and cause subsequent delay.

The largest problem is winter weather. Many projects are delayed during design and are bid with minimal time to complete weather-sensitive work before the onset of winter. The delay of weather-sensitive work into winter causes impact and subsequent delays that may extend work into the next construction season. Or a delay to bridge piledriving can push the activity into an environmental fish window, causing more delays. Delays to resource-constrained work can force it into a period of peak resource demand when the resource is unavailable, again causing delay.

See Sec. G for a description of impact costs and Chap. 12 for a discussion of impact inefficiency.

### 11. Financing costs

Delays may increase financing costs from either increased borrowing or increased use of equity funds. Contractors are due those costs plus interest on retainage for the period of delay. For compensation, many courts require linking a specific loan directly to the delay or, as a minimum, proving that borrowing was directly attributable to the delay. The use of equity or general corporate borrowing is accepted by some courts, but linkage to the delay may be difficult to prove.

### 12. Lost profits and loss of operating business

Although difficult to prove, lost profits from delay can be substantial and should be claimed. To recover lost profits, a contractor must prove:

- Work was available.
- There was a reasonable expectation of being low bidder on some work.
- There is reasonable certainty the work would have been profitable, based on company history.
- They were prevented from bidding more work owing to limited bonding capacity (most common), productive capacity (key equipment), key personnel, or the inability of company management to supervise additional work during the period of delay. Written notification from the bonding company is the best proof of limited bonding capacity.

Proving lost profits *should not* be difficult. Profit is the basic objective of business. Lost profits is a well-accepted owner-suffered damage, when a contractor is late and liquidated damages are not in the contract. It should be equally applicable to contractors. To demonstrate the reasonable certainty of lost profits, show a history of profits prior to the project, reduced profits during the delay, and a return to profitability after the delay.

The courts have often ruled that loss of future profits was beyond the contemplation of the parties when forming the contract and that future profits are "...*conjectural and speculative at best....*" To counter the contemplation argument, give prompt notice of delay and state the magnitude of consequential damages. If the owner could have ended the delay and mitigated the damages but did not, you may be able to use the notice to overcome the contemplation argument. Use examples of owner lost profits that have been compensated to support your argument.

To prove damages from the destruction of an operating business, claim either the present worth of the lost income or the sale value of the business (less salvage). Add expenses and the owner's lost salary during liquidation. A professional appraisal and, if possible, evidence of a potential buyer helps prove loss of business value.

Many claims for lost profits are traded off during negotiation. These costs should not be abandoned without significant concessions by the owner's negotiator. Lost profits are recoverable with a carefully prepared argument and adequate documentation. Consultation with an attorney is recommended if pursuing a claim for lost profits or destruction of a business.

### 13. Quantify delay costs

Quantifying delay costs prior to starting work helps prevent delay. The contractor's project team, the owner's representative, and key subcontractors need to know the cost of delay—for the total project team—before starting construction. Discuss the costs, and the parties' responsibility to mitigate damages, at the partnering workshop and/or preconstruction conference. It is psychologically more difficult for the owner or a subcontractor to procrastinate if they are aware of the cost. All parties should know their duty to mitigate damages and the possibility of legal action if they ignore the impact of their actions on others. Contractors should estimate the cost of delay to weather-sensitive work scheduled for just before the expected onset of inclement weather. Then they should perform a cost-time tradeoff analysis of the optimum acceleration (described in Sec. A.2 of Chap. 10). Contractors can also determine the optimum project duration and submit a value engineering change proposal to accelerate and minimize the owner's total project costs.

## G. Identifying and Computing Impact Cost

A change in working conditions frequently causes impact, which increases the cost of doing the work without changing the end product. It is sometimes called the "ripple effect" and can include:

- Inefficiency in doing the work, which is the largest contributor to impact, the most difficult to compute, and the focus of Chap. 12.

- Increased unit rates of labor (from using more expensive labor), materials (from expediting or more expensive materials), and equipment (required due to different working conditions).

- Additional tasks or materials not originally required to accomplish the original contract work.

## 1. Determining the change in working conditions

*Impact is caused by different working conditions that make the work more costly.* The difference in working conditions is determined by comparing the as-built schedule with the would-have-been schedule and determining the differences in working conditions for each activity. For example, a footing excavation activity may have been scheduled for 5 work days during good September weather under the would-have-been schedule but actually took 12 work days during heavy rains in late October on the as-built schedule. A 45 percent increase in actual costs over estimated would appear reasonable under those conditions, especially if 15 percent of the increase was for overexcavation of wet soil and drainage of storm water.

See Sec. A.3 of Chap. 12 for a detailed list of impacts from different working conditions.

## 2. Inefficiency

Refer to Chap. 12 for a detailed discussion of inefficiency costs from impact.

## 3. Increased unit rates

Increased labor rates can result from overtime paid to attract additional workers in a booming construction market. It can also include travel or relocation expenses, per diem, and camp expenses for a remote site.

Increased equipment rates can include higher unit operating costs or maintenance costs due to adverse conditions (difficult soil or rock, mud, excessive moisture, etc.). It can also include the use of larger equipment, rental equipment instead of owned, and purchase of new equipment instead of used.

Increased materials costs can result from expediting expenses, premium for early delivery, use of more expensive vendors, a higher grade of materials that is available, shipping longer distances, air freight instead of truck or rail, etc.

## 4. Additional tasks and materials

Impact frequently creates a need for additional tasks and materials that don't change the finished project but are necessary because of different working conditions. To document actual impact, check the cost records and daily reports, and question project personnel about additional tasks or materials resulting from impact, as they may not consider them to be extra work. To estimate future impact, compute these costs the same as direct costs and categorize them as either direct costs or impact. Include:

- Weather protection and temporary heat.
- Wet weather access roads.
- Scaffolding or other means to access the work.
- Additional base rock to allow working on a wet subgrade.
- Additional temporary protection of materials.
- Rehandling materials and temporary storage.

### 5. Increased errors and rework

Impact often overloads management and supervisory personnel, leading to inadequate planning and insufficient oversight of the work. This, in turn, leads to contractor errors and more defective work requiring correction. If these problems can be traced back to supervisory overload, the resulting contractor errors and rework should be included in the claim.

### 6. Ripple effect of consequential delays and impact

Although impact is sometimes called the ripple effect, the term can be more appropriately applied to consequential impact from an initial impact. For example, an owner impact causing inefficiency may delay work into winter, experiencing additional impact in a ripple effect.

## H.  Computing Other Types of Damages

The other costs related to changes are primarily increased direct support costs.

### 1. Additional jobsite and home office overhead

Changes and impacts often cause a contractor to increase jobsite overhead personnel and expend more home office support. This results from increased reporting requirements, preparation or review and control of additional shop drawings, expediting procurement, more complex scheduling, etc. If excessive, document these costs and claim them separately as part of the change.

### 2. Change order preparation and negotiation costs

Include the costs of preparing and negotiating change order requests. Keep change order request preparation costs separate from claim preparation costs and request them as a separate line items. Claim preparation and negotiation costs should be reimbursable but are more difficult to recover. If not recovered, they may be used as a negotiation chip. To improve the probability of collecting these costs:

- Characterize most claims as change order requests and submit them during construction.
- Obtain a written request from the owner's representative asking for a change order proposal.
- If working on a U.S. government contract, note that the FARs allow recovery of expert costs for claim/change order preparation.

## I. Adding Markup

Markup is computed as a percentage of the sum of direct, delay, and impact costs. It includes:

- Jobsite overhead
- Home office overhead
- Profit
- Bond, insurance, and taxes

Industry practice is to compute each markup separately, add it to the subtotal, and compute the next markup on the revised subtotal. This is fair, as the percentages are developed on the same basis.

## 1. Overhead

Jobsite overhead is computed as a percentage of direct costs (direct field costs, delay and impact) and home office overhead is a percentage of jobsite costs. Divide the jobsite (or home office) overhead by the actual direct field (or jobsite) costs and multiply by 100.

Overhead is either fixed or variable. Variable costs change in relation to the level of effort. For example, jobsite utility costs increase with activity and home office overhead eventually decreases with a reduced volume of work. Federal agencies and other project owners argue that fixed costs are already included in the original contract and pay only for variable overhead costs. However, many so-called fixed overhead costs such as a superintendent's time are not free for the owner's use in doing change order work. Change order work should actually carry a higher overhead rate than the average, as it is disruptive and requires more management and supervision than original contract work.

Many contracts specify a fixed percentage for markup that supposedly includes jobsite overhead, home office overhead, and profit. These markups are often 15 percent for labor and materials and 8 percent for subcontract costs, which is usually less than actual costs. Although not grossly unfair for anticipated change order work, these rates aren't adequate for major changes. Use actual overhead costs if the value of the contract changes exceeds 5 or 6 percent of the original contract.

Owners may demand a credit for the markup on deleted contract work. The contractor's overhead, however, does not decrease in proportion to the reduction in the contract amount, and in some cases increases due to the cost of estimating and negotiating the change. Contractors should insist on retaining the overhead markup for deleted work unless specific overhead cost savings are identified and the cost of estimating and negotiating the reduction is credited.

Contractors should request compensation for excessive change order estimating. Compensation should be requested for excessive estimating costs for (1) deductive and additive change orders resulting in minimal or no change to the contract amount, (2) evaluation of contract alternatives that are not implemented, and (3) no cost change orders. Both cases result in no compensation for the additional cost of estimating and negotiating the change order or implementing the change, again without compensation for the contractor's estimating services.

## 2.  Jobsite overhead

Mark up direct costs for the additional jobsite overhead required to estimate and process the changes, plan the work, procure materials, supervise installation, etc.

Extra work, delay, impact, and other factors which increase the contractor's direct field costs also increase jobsite overhead costs and should carry those costs. Even if the jobsite overhead costs don't increase substantially, the added work should carry a portion of the jobsite overhead burden, in accordance with standard accounting procedures. In some cases, the added work should carry more overhead than original contract work, as when it prevents jobsite and home office personnel from spending adequate time controlling the cost of original contract work.

Contractors staff projects with sufficient management, administrative, engineering, and supervisory personnel for constructing the project as designed plus handling the normal level of extra work. The extra staffing isn't for the owner's benefit, however, but is provided in anticipation of the added work and should be paid for by that work.

## 3.  Home office overhead

All jobsite costs need to be marked up for home office overhead, using a percentage of job costs. To determine the home office overhead percentage, divide the annual G&A (general and administrative) costs from the contractor's annual profit and loss report by the annual job costs for all projects. Adjust the G&A costs for unallowed items and for shareholder distributions characterized as bonus.

The markup on extra work for home office overhead is separate from the extended home office overhead computed by the Eichleay formula. If an extended overhead claim is made, credit the home office overhead portion of the markup for change orders completed during a period of delay. Otherwise the same costs are claimed twice.

Contractors with a recent history of low income or losses, with subsequent low salaries for company owners, can use imputed salaries instead of actual to establish a fair value for their management.

## 4. Profit

Profit is added as a percentage of the total cost (jobsite cost plus home office overhead), based on:

- Industry practice—10 percent is usually requested and accepted.
- A reasonable profit for the risk and effort required, based on judgment and experience.
- Historical records for the company (widely variable).
- Industry averages, which are usually 5 percent or less and are not applicable to individual changes.
- The Corps of Engineers guidelines from their negotiating manual.[23]

A "reasonable" profit isn't necessarily what was used in the bid; it can be more or less. Some owners assert that contractors should receive less profit for after-the-fact change orders, as there is no risk. To the contrary, collecting for extra work is always difficult, with a substantial risk of inadequate payment. Typical profits allowed by government agencies vary from 6 percent on very large changes to 12 percent on small changes or changes that include a lot of risk and special skills. On most contracts, the normal profit markup on changes should be 10 percent.

On federal contracts, profit is disallowed for a suspension of work but included for changes pursuant to the changes clause (e.g., for differing site conditions). Therefore, base the claim for an equitable adjustment on the changes clause instead of the suspension of work clause.

## 5. Insurance, bond premium, and tax

Payment and performance bond premiums, some insurance premiums, and sales and use taxes are computed as a percentage of contract revenues. Add them to the total cost, after adding overhead and profit markups.

## J. Listing Below-the-Line Costs (Other Sums Due)

Other sums due, generally presented below the total line for direct costs plus markup, are here termed "below-the-line costs." They are not marked up, but interest is added if not paid promptly.

## 1. Retainage and other contract balances due

When a claim is submitted before the end of a project, retainage and possibly other contact balances are due. List them in the claim document so they aren't overlooked in the final settlement, but don't include them in the claim amount. Larger amounts are more difficult to settle.

## 2. Interest

Interest can be claimed on the following:

- Slow monthly progress payments if later than specified in the contract, or applicable state law or federal regulation.

- Late release of retainage at the end of construction.

- Interest on retainage, if the project is delayed by the owner.

- Interest on disputed work from the date it should have been paid until actually paid.

- Borrowings to fund extra work or delay costs (as noted in Sec. F.11 above).

In some states, prejudgment interest is recoverable if the amount due is determined at the time due, even if the amount is disputed. The interest rate is prescribed by either contract or the legal rate of interest. Postjudgment interest accumulates on total amount due, including attorney fees.

Some states have enacted prompt payment acts governing how soon public agencies must pay contractors and specifying an interest rate for late payment. The Oregon prompt payment act also requires prompt payment of subcontractors by general contractors and sub-subcontractors by subcontractors, with significant interest penalties for late payment. Contractors should verify state law to determine the requirements and specified interest rates.

Formerly on federal contracts contractors needed to prove they borrowed specifically for the subject project. Now these costs may not be allowable. However, federal acquisition regulations allow interest after the contracting officer's decision on a claim if you eventually prevail—an excellent reason for expediting the submission and attempted settlement of claims.

Whether interest is due, and how much, depends on the contract, the owner (federal, state/municipal, or private), and applicable law. Contractors may need legal advice regarding their specific circumstances.

### 3. Claim preparation costs and attorney fees

Claim preparation costs and attorney fees may be awarded to the prevailing party in arbitration if allowed by the arbitration agreement. Even if not allowed, claim them to have something to trade in negotiation and to show the arbitrator what the change really cost. When presenting a claim, segregate costs for change order preparation and negotiation from claim preparation costs, since the former is usually recoverable.

On federal contracts, claim preparation costs are not reimbursable because they were not incurred in the performance of the work, but change order preparation costs are reimbursable if incurred during construction, if the request was obviously valid, and if the matter wasn't yet a claim. Change order request (COR) preparation costs *should* be compensable on all projects, as work cannot be performed without compensation and only by preparing a COR can the contractor be compensated. COR preparation is therefore part of performance. Contractors prevailing under the Equal Access to Justice Act may recover attorney and expert witness fees for claims if they are a small business and if the government's position is not substantially justified.[17]

Prevailing parties in litigation are generally not entitled to attorney fees in the United States unless statutory or contractually authorized. The only exception is when the losing party acted fraudulently or in bad faith in bringing or maintaining the litigation, in which case attorney fees are recoverable as damages. Some states allow payment of attorney fees in their prompt payment, mechanics'/construction lien, or statutory payment and performance bond laws.[24] Other states disallow recovery of attorney and consultant fees altogether.

### 4. Credit for nonconforming and unsatisfactory work

If some of the work is satisfactory for use but not conforming to the specifications, reasonable credit must be given to reflect the difference in value between the work done and the work specified. If the work is unsatisfactory for use, repairs must be made or a credit given for the cost of repairs.

Owners sometimes refuse reasonable credit for satisfactory but nonconforming work and insist on strict performance, which requires removing the installed work and replacing it exactly as specified in the contract. Case law, based on the doctrine of economic waste, has enabled some contractors to successfully challenge such actions when noncompliance was unintentional, the work was performed satisfactorily, and the work was generally as specified. The case law was established in a dispute in which the contract gave the owner discretionary authority to accept nonconforming materials with an adjustment in contract price and the owner inspected installation of the nonconforming work but didn't note the failure to conform until all work was complete.

### 5. Contingency

A contingency is an additional amount in an estimate for a possible future event or condition. It may be definable and have a reasonably certain probability of occurrence, or it may be completely unknown. Normally, in construction, each line item has a built-in contingency, but in some cases the risks are so great that they should be identified as a separate line item. This may clarify why an extra work estimate is substantially higher than the owner's estimate and may result in excluding the condition from the agreement.

## K. Preparing a Cost Summary

A tabular cost summary brings all the costs together on one page and clarifies the issues and their relative importance. The summary is often used as a worksheet in negotiations to track the current status of negotiations. Organize and word it carefully to best present your case. Supplement the tabular summary with a simple graphic, showing the amounts of each major claim element and their relative magnitude.

### 1. Tabular cost summary

A typical claim summary follows.

**Claim Cost Summary**

| Construction Change Directives | Amount | |
|---|---|---|
| 1. Outstanding CCDs #045 - #072 (Exhibit A) | $102,000 | |
| | | |
| **Direct Cost** | | |
| 2. Issue #1(Exhibit B) | $155,300 | |
| 3. Issue #2 (Exhibit C) | 123,100 | |
| 4. Issue #3 (Exhibit D) | 35,500 | |
| Subtotal | $313,900 | |
| | | |
| **Delay** (Exhibit E — 12 days from 14Dec97 to 22Feb98) | | |
| 5. Extended Jobsite Overhead | $29,300 | |
| 6. Extended Home Office Overhead | $14,300 | |
| 7. Escalation | 2,500 | |
| Subtotal | $46,100 | |
| | | |
| **Impact** | | |
| 8. Excess Concrete Forming Costs (Exhibit F) | $150,100 | |
| 9. Effect of Issue #1: Acceleration 23Feb-1May98 (Exhibit G) | 103,300 | |
| Subtotal | $253,400 | |
| | | |
| **Other** | | |
| 10. Additional Overhead to Handle Issue #1 (Exhibit H) | $4,500 | |
| | | |
| **Subcontractor Claims** | | |
| 11. XYZ Mechanical (Exhibit I) | $51,500 | |
| 12. A&Z Electrical (Exhibit J) | 8,500 | |
| Subtotal | $60,000 | |
| | | |
| **Total Direct Field Costs** | | **$779,900** |
| | | |
| **Markup** (Exhibit K) | | |
| 13. Jobsite Overhead (except on #5-6 & #10) @ 7.0% | +$51,226 | |
| 14. Home Office Overhead (except on #6) @ 4.5% | +$36,757 | |
| Subtotal | $867,883 | |
| 15. Profit @ 10%   +$86,788 | | |
| Subtotal | $954,671 | |
| 16. Bond, Insurance, and Taxes @ 1.45% | $13,843 | |
| Subtotal   $968,514 | | |
| | | |
| **Total Amount of Claim** | | **$968,514** |
| | | |
| **Below the-Line Costs** | | |
| 17. Retainage Due (10%) (Exhibit L) | $336,201 | |
| 18. Approved but Unpaid Change Orders #14 - #19 (Exhibit M) | 45,415 | |
| 19. Interest on Late Payment (Exhibit N) | 4,895 | |
| 20. Interest Due on Unpaid Claims (Exhibit O) | 6,372 | |
| 21. Claim Preparation Costs (Exhibit P) | 31,200 | |
| | | |
| **Subtotal of Below the Line Costs** | +$424,083 | |
| **TOTAL AMOUNT DUE*** | | **$1,392,597** |

*Additional interest is due if not paid by 15 June.

The Construction Change Directives are separated from the direct cost claims because entitlement is acknowledged and only the amount is being negotiated. This makes it easier to treat them as a foregone conclusion and press for full payment. Below-the-line costs are shown as a separate item for the same reason and because they are not marked up. The claim preparation costs can be included in each claim instead of being listed separately.

All summarized costs are referenced to a separate exhibit with detailed backup. This facilitates review by the owner's representative and lends credibility to the document.

## 2.   Graphical cost summary

A simple graphic communicates the relative magnitude of the cost elements and complements the tabular cost summary. The image is retained much longer than a tabular listing. A well-designed graphic can also persuade the claim reviewer, at a gut level, that the costs are reasonable. The ancillary notations on the left of Fig. 11.2 show the direct cost of extra work as 25 percent of

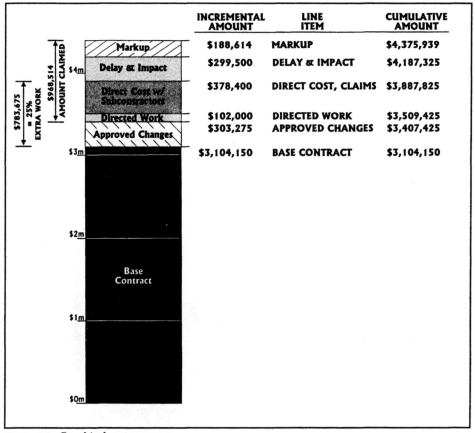

| | INCREMENTAL AMOUNT | LINE ITEM | CUMULATIVE AMOUNT |
|---|---|---|---|
| Markup | $188,614 | MARKUP | $4,375,939 |
| Delay & Impact | $299,500 | DELAY & IMPACT | $4,187,325 |
| Direct Cost w/ Subcontractors | $378,400 | DIRECT COST, CLAIMS | $3,887,825 |
| Directed Work | $102,000 | DIRECTED WORK | $3,509,425 |
| Approved Changes | $303,275 | APPROVED CHANGES | $3,407,425 |
| Base Contract | $3,104,150 | BASE CONTRACT | $3,104,150 |

Figure 11.2   Graphical cost summary

the base contract amount. This is an extraordinarily high percentage, as a normal contingency for new building construction is 5 percent, with 10 percent considered high. It supports the contractor's allegation of massive disruption and places the impact and delay cost in perspective as being only 8 percent of the total direct cost.

The owner's representative may assert that the claimed amount is inflated and the statistics unreliable. To respond, compute the average percentage reduction to date, from the change order request amount to contract modification amount, and adjust the amount currently claimed to the "expected settlement amount." For example, if your initial proposals for all change order requests included in the $303,275 of approved changes totaled $433,150, your average reduction to date is 30 percent. Applying this to the $102,000 plus $378,400 in disputed extra work gives an "expected settlement amount" of $336,280. The expected total direct cost of all changes would therefore be $639,555, or 21 percent of the base contract—still an extraordinary amount. Keep in mind, however, that the 30 percent reduction to date will probably increase for the balance of the disputed work.

Figure 11.2 shows markup as a separate element and illustrates how to display the cost summary in a graphical form. Normally, to avoid emphasizing the amount of overhead and profit being claimed, include the markup in each individual cost element.

Proving inefficiency is difficult, but it is essential if you hope to recapture your impact losses. Unfortunately, job records often lack sufficient information, and few estimators know the best method of computing loss of productivity. Conventional methods of estimating inefficiency (e.g., total cost, expert opinion, or industry studies) are thought to be inaccurate and are not always applicable. Therefore, many inefficiency claims are settled at cents on the dollar or are abandoned during negotiations.

This chapter describes a systematic rational approach for proving inefficiency by analyzing the available data and then selecting the best method(s) to compute inefficiency. A combination of new methods and innovative variations on existing methods is next applied to accurately compute and convincingly communicate the amount of inefficiency. This approach is easy for experienced construction personnel to apply if they follow the procedures in the sections below.

To prepare an inefficiency claim using a specific method, proceed to the applicable section (D through H). To better understand the concepts of productivity and inefficiency, start with Secs. A and B. Then read Sec. C to learn how to select the best method to analyze inefficiency. First-time readers are advised to read this chapter from start to finish and to break their review into related sections (A–B, C, and D–H).

## A.  Assessing Productivity and Inefficiency

Claim preparers must understand the factors determining productivity, the basic elements of inefficiency, and the causes of impact. The causes of impact, listed in Sec. A.3, indicate a problem and point to its cause. They are *indicators*.

### 1.  Factors determining productivity

Productivity is the measure of output, in labor hours per unit of work (or units of work per labor hour). It is a computed value, based on two measured values: labor hours and work quantities (or percent) completed. Productivity varies widely from project to project and over time on a given project, owing to the following factors.[25]

#### Factors intrinsic to the work and site

- Difficulty of the work due to complexity, tolerances, constructibility, and working conditions.
- The quantity of work performed, with higher productivity for larger work quantities.
- Environmental conditions including temperature, lighting, humidity, wind or rain, etc.
- Quality control, inspection, testing, startup procedures, and owner acceptance.
- Travel time to and from the work area, setup and shutdown/cleanup time.

#### Management and supervision

- The extent and quality of project scheduling and the coordination of different crews.
- Prejob planning—layout, task sequencing, equipment selection, crew balance, etc.
- Management and supervisors' skills at reading and interpreting plans and specifications, scheduling and organizing work, providing materials and tools, and directing and motivating workers.

#### Organization

- Work continuity at the same or similar task and logical sequencing to the next task.
- Availability of materials in the form and at the location and time needed.
- Adequate tools and equipment.
- Access, layout, and work space.

**Workers**

- Availability of skilled workers.
- Worker education, training, experience, and skill at the task being performed.
- Individual work ethic, morale, and motivation.
- Trade practice and union work practices, if applicable.

**Absence of impact**

- The effect of other work and ancillary operations such as traffic control and scaffold erection.
- Accidents, equipment breakdown, and other unanticipated problems.
- Owner disruption, interference, delay, acceleration, or change.
- Worker fatigue, poor morale, etc.

Although cost accounting systems can accurately measure productivity (e.g., 0.15 labor hour per cubic yard), the loss of productivity due to impact is difficult to segregate from productivity losses due to other causes. These other causes include changes in the difficulty of the work, inclement weather, less efficient management and supervision, and differences in worker skills. In addition, variations from the estimate may be due to an overly optimistic estimate, inaccurate charging of time, or an inadequate budget breakdown.

2. **Basic elements of inefficiency**

Inefficiency is loss of productivity, usually expressed as a percentage of the actual or the optimum productivity. The basic elements of labor inefficiency are:[26]

1. *Ineffective work.* Working ineffectively (painting large surfaces with brushes) leads to project completion, but less efficiently than other methods (spraying).

2. *Excessive material handling.* Excessive material handling should be identified separately from other ineffective work, as it occurs frequently, is easily identifiable, and represents a large component of construction costs.

3. *Working slowly.* Working slowly is relative to the worker's normal speed for the task performed and varies with the worker. Speed is not the only factor; some workers with a steady pace outperform others who work frantically or carelessly.

4. *Waiting.* "Excessive" idle time is certainly inefficient, but short breaks are necessary for optimum performance and some idle time is inherent in even the most efficient crew operations.

5. *Moving outside the work area or inefficient traveling between tasks.* Moving within the immediate area of the task being performed or to the

next task is productive. Moving back and forth between tasks or moving too far is inefficient.

6. *Rework.*   Rework is always inefficient.

Inefficiency is the difference between what was accomplished and what "would have been" accomplished in the absence of the impact. There is no way of directly measuring inefficiency (except for a time and motion study analysis, explained in Sec. H). Fortunately, the courts and most owners recognize this difficulty and accept a lesser degree of proof than is required for extra work or overhead cost damages.

## 3.   Reasons for loss of efficiency—the causes (and effects) of impact

Loss of efficiency is the end result of one or more causes. There may be a long chain of cause-effect relationships between the initiating action, event, or condition and the ultimate effect on productivity. The causes of inefficiency may be initiating events/conditions or intermediate causes/effects. These causes/effects are indicators of inefficiency. When preparing an inefficiency claim, you should search for and identify these indicators and trace them back to the initiator and forward to the ultimate inefficiency.

The indicators are listed below by general category, to help in identifying them and ensuring their inclusion in a claim. Many overlap, occur in multiple categories, and are both causes and intermediate effects. They are:

1. *Adverse or abnormal environmental conditions.*   Poor environmental conditions may be an initiating event or condition, or an intermediate cause/effect. For example, an unexpected summer rainstorm may impact rough framing. Or the initiating event may be a stop-work order that delays framing into winter; the intermediate effect/cause would be winter rains; the ultimate effect would be reduced framing productivity.

Environmental conditions vary from adverse, which does not justify a time extension, to abnormal, which does justify a time extension, if the project is delayed. They include any of the following:

- Rain—makes surfaces slippery, requiring extra care and a slower pace; wet weather gear restricts movement; wet surfaces require wiping; equipment is slowed; and moisture-sensitive work must be protected.
- Cold—requires heavier clothing (which restricts movement and reduces speed), makes exposed hands or fingers numb (reducing productivity and increasing the risk of accidents), and requires heating or protecting work from low temperatures.
- Wind—can make operations slower or even dangerous. For example, roofing can be shut down by a moderate wind when combined with rain. Wind and cold together can result in a subzero wind chill factor that slows work and can cause frostbite.
- Snow—can increase cold weather impacts, especially in conjunction with wind. Snow can melt on contact, making workers cold and uncomfortable, and become a safety hazard If it refreezes. Snow may require protective covers or removal.

- Heat—has a direct physiological impact on workers, reduces productivity, and can cause heat exhaustion. It can also require more expensive methods, such as cooling concrete.
- Humidity—intensifies the effect of excess heat and slow productivity and extends drying time for paint or plaster.
- Noise—interferes with communication and damages hearing if it's too loud and adequate protection isn't provided.
- Dust or odors—reduces productivity (especially painting or chemical odors).
- Poor lighting or darkness—reduces productivity and requires labor and expense for lighting on multiple-shift operations or winter operations in northern regions.

A typical inefficiency rate for working outside in rain, cold, and other winter conditions is 20 to 30 percent, but it can range from negligible to over 100 percent. Inefficiency rates are very task-specific; correspond to the severity of the conditions; depend upon the individuals involved, their motivation, and training or experience working under adverse conditions; and are affected by local industry practice and expectations. Work in Alaska, for example, continues at a reasonable pace (but inefficiently) under conditions which in the Lower 48 would completely shut down the job.

2. *Delay.*   Delay can result from inclement weather, owner actions, or some other event or condition. Delay can cause the following impacts:

- *Idle labor* while waiting for direction to continue, for a supervisor to check the plans and obtain tools or materials for a new work area, or for the receipt of different materials. Idle labor can be sent home early if the delay is short (but show-up time must be paid) or can be laid off. However, layoffs damage morale and may result in the loss of productive personnel if they find other work.
- *Ineffective work* on nonessential tasks or ineffective operations in order to keep personnel busy and partially productive.
- *Working at a reduced pace* due to smaller crews, worker slowdown, insufficient equipment, etc.
- *Equipment standby* for equipment that cannot be demobilized but is not currently needed on the job and may possibly be needed later.
- *Performing work in different conditions* than would have occurred without the delay. For example, boiler installation may be delayed until the mechanical room is crowded with other crafts, or hoisting rebar to the fifth floor may be delayed until the crane is overcommitted and available only intermittently.

3. *Acceleration.*   Acceleration is a result of contractor management action and is an intermediate effect/cause. The initiating event/condition may be an owner directive to accelerate, owner refusal to grant a time extension with the threat of liquidated damages, or an effort to avoid pending inclement weather. Acceleration can cause the following impacts:

- *Overtime* resulting in premium labor rates and inefficiency due to worker fatigue if the overtime is protracted.
- *Fatigue* is caused by excessive hours of work (e.g., overtime) or excessive exertion. Fatigue lessens physical ability and decreases the ability to move, think, and respond quickly, leading to accidents and errors.

- *Mental fatigue* occurs after 60 to 75 minutes of repetitious work demanding careful attention. Brief rest periods (or a change in activity) every 30 to 50 minutes greatly reduce mental[27] and physical fatigue and should not be labeled idle (inefficient) time.

- *Boredom* may accompany mental fatigue and cause reduced productivity but is less frequent in construction than in manufacturing because of the variety of activities. It is due to a dull job or to the worker's lack of interest, is highly variable from one individual to another, and is difficult to measure.

- *Work pace inertia, absenteeism, turnover, accidents, and poor morale* result from extended overtime and dissatisfaction with conditions. Workers tend to adjust their pace and accomplish the same amount of work during overtime as during a normal work week. They may be absent without notice due to exhaustion or the need to handle personal business. Turnover increases as the additional earnings allow them to quit working; accidents are more frequent, and morale suffers.

- *Multiple-shift operation* causes inefficiency due to (1) overlap between the two crews, (2) poor lighting, (3) circadian rhythm inefficiencies of swing and graveyard crews, (4) the costs of illuminating the work and other features necessary for multiple shift operation, (5) shutdown for equipment servicing, maintenance or breakdown, and (6) additional costs for shift premium.

- *Mobilization and demobilization of additional labor and equipment* results in direct costs and learning curve inefficiencies. It may also result in less skilled workers, crowding, etc.

- *Overworked supervisors* may be too busy and unable to handle the faster pace, larger workforce, and more frequent changes. They will make more mistakes, plan inadequately, fail to correct obvious errors, and may become ill or quit.

- *Other impacts* may result from poorly planned and implemented acceleration or from excessive acceleration. The result can be poor layout and instruction, inadequate materials and tools, inadequate or insufficient equipment, inadequate access or work areas, start-and-stop operations, out-of-sequence construction, concurrent activities, learning curve losses, frequent or unplanned changes, overstaffing, crowding, trade stacking, lower-quality workforce, and poor-quality work requiring rework.

4. *Design ambiguity, errors, and omissions.*  Design ambiguities, errors, and omissions take supervisor time to identify, clarify, decide on alternative methods, obtain tools and materials, etc. In the meantime, the crews may be idle or underutilized. The result is delay, disruption, changes, unconstructible work, etc.

5. *Poor contract administration.*  Poor contract administration by the owner's representative may impact productivity:

- Failure to obtain right of way, permits, or licenses not provided by the contractor.

- Late response to requests for information, submittals, or value engineering proposals.

- Late decisions on problem resolution, material or color selection, etc.
- Late surveying or untimely inspection of completed work.
- Failure to inspect or improper inspection.
- Failure to coordinate multiple general contractors or other parties.
- Failure to accept reasonable substitution requests.
- Late delivery of owner-furnished materials and equipment.
- Untimely payment or refusal to pay for legitimate changes.

6. *Interference and other disruptions.* Disruption is any interruption to the orderly course of work that makes it inefficient. Interference is a special type of disruption—an action by the owner or a third party that directly disrupts construction and causes inefficient work. Disruption and interference include:

- *Stop-work orders or design holds* that delay work, cause start-and-stop operations, or require out-of-sequence work.
- *Start-stop operations* cause idle time, interrupt learning curve productivity gains, and result in workforce fluctuations. If the crew is assigned to other work, additional inefficiencies result from the relocation, closeout and cleanup at the current site, and setup at the new site. Obtaining different materials and tools, moving tools and materials between work areas, and moving back to the original work area also cause inefficiencies.
- *Out-of-sequence work* when the normal, logical, and most efficient order of progress is changed. Out-of-sequence work results in start-and-stop operations, additional supervision time to check the plans and lay out the work, late layout of the new work area, wasted time moving between work areas, and time to set up different tools and materials or to wait for materials and tool delivery. Additional inefficiencies result from retraining or reorienting the crew, learning curve inefficiencies as the crew gets back into operation, crew imbalance, crowding, damage to installed work, inefficient operation, additional cleanup, and additional administrative costs to purchase materials out of order.
- *Demobilization and remobilization* by subcontractors who aren't able to continue work and must either move to other projects or remain idle. If the period of delay is significant and certain, demobilization and remobilization is an alternative to equipment standby.
- *Crew imbalance* when the work output of sequential crews is different and some crews catch up with the preceding crew and therefore have to slow down.
- *Limited flexibility* from multiple work areas on hold. This prevents efficiently coping with normal disruptions by shifting crews to another noncritical fill-in operation in another work area with the same type of work, tools, and materials.
- *Overinspection* when quality requirements are enforced too strictly, unnecessarily, and beyond industry practice.
- *Directed work* where the owner's representative interferes by directing or pressuring a crew into performing work in a specified manner or sequence.

7. *Changes and uncertainty.*  Changes in the work can alter the working conditions, which can disrupt the crew's rhythm and productivity, damage morale, require reassignment of workers, and waste materials. Changes require more administrative and supervisory time than original bid work and may result in errors and rework if administrators and supervisors are overloaded. The types of changes include:

- *Constructive change* when the owner forces the contractor to perform extra work by misinterpreting the contract or fails to grant a time extension which forces the contractor to accelerate.
- *Untimely change* which causes standby, last minute schedule changes, start-stop operation, higher procurement costs, etc.
- *Unclear change* which causes uncertainty and require more supervisory time.
- *Uncertainty* from numerous continuing and last-minute changes damages morale and causes more inefficiency.

8. *Multiple changes.*  Multiple changes have a synergistic cumulative impact on morale and on the effectiveness of planning, scheduling, and efficiency that exceeds the sum of the individual impacts from each change. The impact is difficult to recognize since it develops when the number of changes increases beyond some unknown point.

Cumulative impact occurs on projects that have become a moving target with numerous continuing changes that overwhelm the contractor's ability to quantify, estimate, schedule, negotiate, and implement. One change melds into the next, and no one individual has a clear picture of the overall project status. Changes are not promptly authorized, working conditions change, bid work and previously authorized change order work are impacted by subsequent changes, and the schedule becomes outdated. The owner's representative cannot promptly resolve pending change order requests, creating a financial burden from the unpaid extra work.

9. *Inadequate access, work space, and site conditions.*  Poor access, inadequate work space, and unfavorable site conditions can result from late access, acceleration, other impacts, or poor contractor management. The results can include:

- *Excessive travel time* from an assembly area to the work area.
- *Crowding* due to acceleration or restricted work areas, which results in conflicts, damage to installed work, and safety problems. Inefficiencies are also caused by an oversized crew; by insufficient space for efficient layout, materials storage, and work; and by unhealthy or unsafe conditions.
- *Trade stacking* is crowding, but with the added complexity of different crafts working in the same space.
- *Limited access* with resulting delays, double handling of materials, and excess use of labor instead of equipment. For example, costs on multistory buildings will greatly increase if manlifts are unavailable, too far from the work area, or overcrowded. Blockage of corridors by a crew can also restrict access.

- *Inadequate work areas* for storage, laydown, fabrication, staging, or circulation.
- *Poor site conditions* include muddy, snow-covered, or poorly laid-out conditions for access and storage. The conditions may be caused by weather, poor planning, or owner changes or may be inherent in the project scope.
- *Differing site conditions* with conditions different from those indicated in the contract documents or normally encountered.

10. *Materials and resource problems.*   Acceleration, delay, disruption, or changes can result in resource problems including:

- *Shortage of qualified workers,* resulting in insufficient crew size to complete on time or the use of inadequately skilled workers with resulting delay, cost overrun, and quality problems.
- *Undersized or insufficient tools or equipment* for the conditions encountered.
- *Excessive fluctuations in resource needs* causing excessive under- or over-staffing or multiple equipment mobilization/demobilization, learning curve inefficiencies, overload or underutilization of supervision, and extra labor hiring/layoff costs.
- *Material unavailability* due to lack of cranes or other equipment for transport, insufficient labor, material not on site, incorrect material, or supervisory error.
- *Additional materials handling,* storage, and rehandling. Materials handling is a major cost of construction, and impact can easily double it.

11. *Supervision and management problems.*   Impact frequently affects supervision and management efforts, resulting in:

- *Additional overhead* due to excessive change orders, requests for information, or other administrative requirements to replan, reschedule, track and control progress, etc.
- *Dilution of supervision* due to crews being too large or too spread out for effective control.
- *Overworked supervisors* who make mistakes or lack time to properly plan, lay out, instruct the crews, and supervise.
- *Excessive supervision* due to the crews being smaller than supervisors normally direct, resulting in higher than normal supervision ratios and costs.
- *Increased supervision* due to working supervisors being converted to nonworking supervision.
- *Changes in supervision or management* with resulting confusion and reorganization.
- *Layout errors* due to last-minute changes, supervisor fatigue, or overwork.

12. *Poor morale.*   The causes of poor morale are as follows:

- *Lack of clear direction and adequate tools* prevents workers from performing their duties and causes frustration and resentment.
- *Uncertainty and job insecurity* occur when workers and supervisors are uncertain about the next task or whether they will be working next week. They spend more time seeking information, and their morale suffers. If concerned that work will be delayed and that they might be laid off, they will slow production to ensure continued employment.

- *Lack of job satisfaction* arises if the work is not being done well or if it is frequently changed and completed work is torn out and replaced.
- *Fatigue, unsafe conditions, or poor environmental conditions* create workers who are exhausted, cold, wet, concerned about safety, or otherwise dissatisfied with the working environment. They will resent the hours and working conditions, and productivity and morale will suffer no matter how much overtime pay they're making,
- *Feelings of injustice* can be caused by overinspection or unsatisfactory owner-furnished facilities.
- *Mismanagement or poor supervision* such as an abusive or authoritarian supervisor, poor coordination by supervisors, start-and-stop operation, or poor scheduling also damages morale.

The effects of poor morale include:

- *Lowered productivity.* Workers don't work as hard or as efficiently. In some cases, they will deliberately slow the pace of work.
- *Reduced quality and more rework.* Work must be repaired or removed and reconstructed.
- *Theft and loss of small tools.* Dissatisfied employees are more likely to steal tools and materials, or fail to secure and care for company-owned small tools.
- *Absenteeism.* Workers fail to show up, resulting in unbalanced crews, lowered production, and delay.
- *High turnover.* This is prevalent at the end of a project, owing to poor morale and uncertainty about the future. The result is learning curve inefficiency and lowered productivity.
- *Difficulty in hiring skilled workers.* Good workers have their choice of jobs and are reluctant to work on a troubled project.

13. *Safety constraints.* Impact may cause increased safety constraints. These include:

- *Required safety measures* due to change in conditions, including belting off when exposed to falling, waiting for an overhead crane to clear, and shoring trenches. For example, an unmarked telephone duct required a waterline contractor to shore as trenching under it exceeded 5 feet deep. The crew waited for shoring delivery and they worked inefficiently because of interference by the shoring.
- *Caution and slowdown* when conditions are unsafe or perceived to be unsafe.
- *Disruption and slowdown* due to an accident or near accident.
- *Increased accidents* with loss of productivity by the injured worker, those who come to the worker's aid, and everyone else in the area. Continuing costs include payments to the injured worker and increased workers' compensation rates.
- *Hygiene procedures* on hazardous waste work (asbestos, lead, confined sites, etc.).

14. *Errors and poor workmanship.* Errors and poor workmanship frequently result from disruption, change, acceleration, or poor morale, which in turn, lead to:

- *Increased wastage* when work is accelerated, as everyone is too busy to avoid waste. If morale is poor, no one cares.
- *Increased cleanup* caused by poor working conditions, multiple work areas, stop-and-start operation, acceleration, crowding, poor morale, etc.
- *Increased close-out and punch list costs* due to being rushed to finish or due to early owner move-in. Excessive punch list work is an indicator of poor quality control during construction, which frequently occurs on an accelerated or impacted project.
- *Rework* when workmanship is inadequate, which may result from adverse weather, delay or acceleration, disruption, poor morale, supervision problems, etc.

15. *Other impacts.* There are many other ways that impact can occur. Approach each analysis of inefficiency with an open, inquiring mind and accept nothing at face value.

## 4. Need for more research, understanding, and emphasis on productivity

Contractors can achieve higher profits by giving more attention to productivity, not just to meeting budgets. Cost reduction, below the budget, must be a goal and productivity must be tracked. Tracking productivity also ensures that impacts and other changes will be promptly identified, documented, and addressed. It practically guarantees the recovery of extra costs.

Emphasizing productivity requires looking at jobsite overhead differently. Management, supervisory, and engineering staffing levels are often insufficient to *effectively* handle the normal project workload, let alone to respond to the needs of an impacted project. Temporarily staffing up at the beginning of a project will establish control of the work and will enable the permanent project team to maintain more effective control without incurring continued high overhead.

## B.  Proving Causation

Causation is the linkage from the initiating event or condition through a chain of intermediate cause/effects to the ultimate impact on efficiency. Proving causation is part of proving entitlement (discussed in Chap. 9), but it is also discussed here because it is so important to proving inefficiency.

### 1.  Link entitlement to damages through causation

For a successful inefficiency claim, you first prove entitlement—that the owner is responsible for the initiating event or condition that increased costs—as discussed in Chaps. 3 and 9. Then prove causation. Causation may be simple and direct, or complex and only indirectly provable.

For example, the initiating action, event, or condition could be differing site conditions, which results in the work (say, excavation) being more difficult than anticipated. This, in turn, causes inefficiency, which is compensable and justifies a time extension if the contract has a differing site conditions clause. If the owner fails to promptly grant a time extension, the contractor will be forced to accelerate to avoid liquidated damages. This will result in entitlement for the impact costs of the acceleration. The impact depends upon the contractor's expediting actions, which may include a combination of overtime (resulting in fatigue), adding crews or special equipment (resulting in learning curve inefficiencies), increasing crew size (resulting in crowding, trade stacking, or overstaffing), rescheduling (resulting in disruption), and other actions (e.g., using more expensive materials to speed installation).

In addition, there may be a "ripple effect" of multiple related impacts. For example, the initial delay may push the work into winter weather, which causes additional impacts and inefficiencies, which then delays work into the next construction season.

To make the most effective case and to classify causation, you can follow the model described below.

#### 2. Use an action-response model to help prove causation

Halligan, Demsetz, and Brown have established an action-response model that helps clarify the often complex chain of cause and effect that results in inefficiency.[28] The model lists three basic action-responses which interact and lead to inefficiency. These and an additional five action-responses that frequently occur are:

1. Initiating action, event, or condition:
   - Owner action (untimely design changes, slow response to RFIs, etc.).
   - Force *majeure* and third party actions (strikes, regulatory changes, etc.).
   - Environmental conditions resulting in difficult working conditions (muddy, cold, wet, etc.).
   - Contractor actions (poor site management, inadequate subcontractor coordination, etc.).

2. Change in working conditions.

3. Initial response.
   - Physical response in terms of output or productivity.
   - Changes in morale and motivation, which lead to changes in performance.

4. Management and supervisory action (notice to the owner, schedule or operational changes, etc.).

5. Owner response (acknowledging or denying responsibility, directing action, etc.).

6. Subsequent management and supervisory action (additional schedule, operational, or recordkeeping changes).

7. Subsequent possible change in working conditions due to management and supervisory action.

8. Subsequent crew response. This may initiate another round of action-response.

All eight action-responses need to be examined by claim preparers. They can be described in a narrative, a fragnet, or some other graphic that traces the linkage from the initiating event or condition to the resulting inefficiency.

3.   **Graphically correlate impact and inefficiency over time**

One of the best methods of proving causation is by graphically displaying labor use or productivity and the initiating impacts on a timescale diagram. A correlation between initiating impact and resulting inefficiency is strong evidence of linkage, although the owner may allege other reasons for the inefficiency.

*Productivity charts.*    If weekly or monthly productivity data are available, plot them on a timescale and note the beginning and end of the initiating impact. A correlation between a drop in productivity and the beginning of impact, with an increase in productivity when the impact condition ends, provides strong circumstantial evidence of causation. You will need to include a carefully reasoned analysis and adjust for learning curve effects, but the productivity chart is often more effective than the narrative in convincing a claim reviewer. Productivity histograms are a graphical form of measured mile analysis. They provide a strong evidence of causation when correlated with impact conditions.

*Labor histograms and periodic line graphs.*    If productivity data aren't available, use a secondary effect, such as crew size. Although not as conclusive as productivity, labor hours or crew size histograms are effective, especially if the as-planned labor forecast can be compared with actual labor use.

The steps in creating a comparison labor histogram (or periodic line graph) are as follows:

1. *Convert estimated cost to labor hours.*    First, if the estimate (or a subcontractor's quote) is lump-sum, break it down by cost code for the various elements of the work. This requires knowledge of trade practices and costs, or a discussion with the estimator and superintendent.

   Second, if the estimate is not broken down by cost category for each cost code, separate the labor costs. In the absence of a detailed estimate: (1) select a reasonable percentage of cost for labor (often around 30 percent, but varying with the trade and type of activity), or (2) deduct other costs such as materials and equipment, with the balance being labor cost, or (3) reestimate the labor element.

   Third, develop an average hourly labor rate based on an assumed composite crew, their individual rates, and payroll taxes, fringes, etc. Then divide the labor cost by the average hourly labor rate to obtain the estimated hours of labor.

2. *Allocate the estimated labor hours to the construction activities.* Starting with the original project schedule, either (1) allocate the estimated labor hours from the cost codes to the schedule's activities, or (2) create a new schedule that corresponds to the cost codes with the estimated labor hours assigned to each cost code and spread over time.

Enter the estimated labor hours into a spreadsheet or scheduling software and generate the forecasted labor requirements—by day, week, or month. It is possible, but tedious, to generate the same information by hand. Separate labor hours by trade or subcontractor, and by major function (cost code or activity) for the more important trades.

3. *Plot the results on a histogram.* Plot labor hours or crew size by the weekly, not the daily, average. Daily values often fluctuate wildly, confusing the picture and masking underlying trends. The goal is to identify patterns, not specific details. Optionally, plot a 4-week moving average, which reveals trends.

It is usually more informative to plot equivalent crew size instead of labor hours. To convert labor hours to equivalent crew size, divide by the standard hours of work per day, or per week if working with weekly subtotals.

Labor can be shown as a single bar, or the crews may be differentiated. Showing the crews separately is helpful, but can be confusing. It may be best to focus on a single crew working on the critical path.

4. *Use and advantages of histograms and line graphs.* Labor forecasts are normally presented as histograms, but line graphs can also be used. Line graphs are sometimes better for comparing two sets of numbers, as in Fig. 10.39, as the differences are easily identified even if they are close to the same value. Alternately, you could use a solid histogram for one and a line graph for the other, as in Fig. 10.38. Annotate histograms and line graphs with comments on important events, impacted conditions, and other issues.

Labor histograms can be impressive and convincing, even if they don't provide conclusive proof, as they do provide supporting evidence of inefficiency. They can provide proof of crowding and acceleration.

## 4.  Use a process of elimination

If the linkage between an initiating event or condition and an inefficiency is weak, eliminate other factors to strengthen your case. It is not necessary to use the process of elimination in the claim document; that can be saved for negotiation or for testimony to counter expected owner arguments. For example, if claiming impact from excessive inspection during inclement weather, an analysis proving that the weather had little actual effect on productivity would greatly strengthen your case and counter a probable owner argument.

## C.  Using Rational Analysis to Prove Inefficiency

Rational Analysis is a systematized approach to identifying, computing, and proving inefficiency. It consists of seven basic steps and several intermediate tasks that generally occur in the following order (Fig. 12.1).

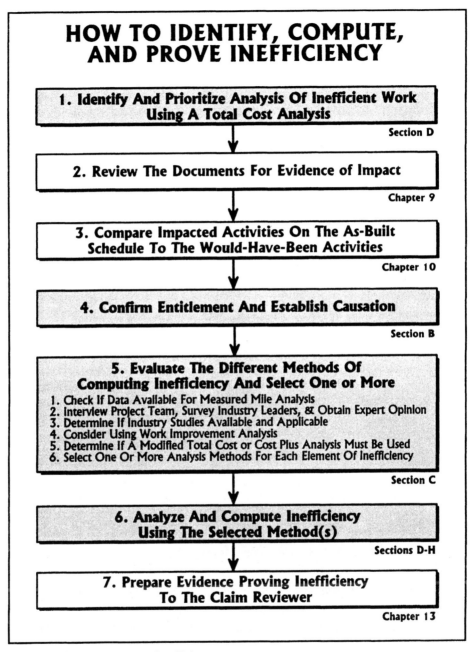

**Figure 12.1**  Rational analysis of inefficiency.

## 1. Identify and prioritize analysis of inefficient work

First, identify the inefficient work. Examine the cost accounting reports and compare the estimated labor costs with actual costs for a total cost analysis (of each impacted cost code). Confirm that other factors did not cause the overrun. For example, labor costs may exceed the estimate because of changing from equipment-intensive operations to labor-intensive, or because of performing work in-house rather than subcontracting as planned. In either example, there is no impact and the increased labor costs will be largely balanced by decreased equipment or subcontract costs.

Next select the cost codes with significant overruns for further analysis and prioritize them by dollar amount and expected ease of analysis.

## 2.    Review the narrative text and other documents to determine if the work was impacted

Review the narrative text (described in Sec. B of Chap. 9) and other documents for evidence of impact, the amount of inefficiency, possible reasons, and responsibility. Review the superintendent's daily report, annotated timecards, photographs, the project manager's diary, and other documents.

If necessary, obtain additional information to establish the pertinent facts. Examine the inspection reports, subcontractors' records, special accounting reports, and other records. Consider preparing a detailed as-built schedule, described in Sec. G of Chap. 10, to determine what happened and why.

## 3.    Compare impacted activities on the would-have-been schedule to the as-built activities

Compare the as-built schedule to the would-have-been schedule to establish the difference in working conditions under which each impacted activity was, and would have been, performed. This may include differences in weather, crowding, competition for scarce resources such as crane time, and longer activity durations. For details, see Sec. E.6 of Chap. 10.

## 4.    Confirm entitlement and establish causation

If you have not already established entitlement for the initiating action, event, or condition (Chaps. 3 and 9), do so. Then trace and document causation as described in Sec. B above and Secs. E, G, and H of Chap. 10.

## 5.    Evaluate the different methods of computing inefficiency and select one or more

Next compare the data available with the data needed for the different methods of computing inefficiency so that you can select the best method(s). The steps you will follow are:

1. *Check if data are available for a measured mile analysis.*   The favored method is the measured mile. If there was a period of unimpacted production to compare with the impacted productivity and if adequate cost accounting data exist for the measured mile approach, use the measured mile method as described in Sec. E. If cost accounting data aren't available, it may be possible to use other records for a measured mile analysis. If that isn't possible, proceed with the next step of the evaluation.

2. *Interview the project team, survey industry leaders, and develop expert opinion.* If proceeding with the evaluation, question the project team. If potentially helpful, informally survey other contractors to determine their experience with similar conditions and consult with an outside expert. At this time, the claim preparer should formulate a tentative opinion on the extent of inefficiency, which can be used to adjust priorities, guide further analysis, and serve as a check on the results from the method eventually selected.

3. *Determine if industry studies are available and applicable.* Then examine Sec. G below and other trade literature to determine if industry studies are available and applicable, and if adequate data exist. These methods rely on graphs of the percentage of inefficiency relative to some factor and are relatively simple to use.

4. *Consider using work improvement analysis.* If none of the above three methods can be used, or if the amount involved is large and a more extensive investigation is warranted, consider using work improvement techniques to compute inefficiency. This method is difficult to apply and requires specific detailed information described in Sec. H.

5. *Determine if a modified total cost or cost plus analysis must be used.* If none of the other methods is suitable or affordable, the alternative is a modified total cost analysis or a cost plus claim as described in Sec. D and Sec. E.13 of Chap. 3.

6. *Select one or more methods of analysis for each element of inefficiency.* Select the most applicable method based on the available data, the type of work and impact, and the expected resistance by the claim reviewer. A combination of methods for the various parts of the claim may be best, especially on a large difficult project with multiple impacts. However, always use a second method (such as a modified total cost analysis or expert opinion) for corroboration.

## 6.  Analyze and compute inefficiency

Apply the selected method and compute the inefficiency. For details, see Secs. D through H.

The computed percentage of inefficienciy is often based on the original contract labor hours. If an impact affects previously negotiated change order work, the estimated or actual labor hours for those change orders should be included with the original contract labor hours when the loss of productive hours is computed.

## 7.  Prepare documentation for proving inefficiency to the claim reviewer

Last, prepare documentation and other material, as described in Chap. 13, for inclusion in the claim document and for oral and visual presentation to the claim reviewer. This may include the claim document and exhibits plus the anecdotal testimony of the project team members, presentations by experts, photographs, videos, and possibly a site visit.

"Proving" inefficiency means *convincing* and normally involves the following factors:

- Demonstrating there was inefficiency and that the contractor had a cost overrun.
- Identifying the initiating actions, events, or conditions.
- Proving entitlement—that the owner was responsible and that the results are compensable.
- Proving causation by linking the initiating event or condition to the ultimate impact.
- Computing damages.

### D.   Total Cost, Modified Total Cost, and Cost Plus Methods

The total cost, modified total cost, and cost plus methods are normally applied to the total project and are used when no other method is applicable. Each method can also be applied to individual tasks or cost codes.

### 1.   Total cost method

The total cost approach subtracts the estimated costs (plus change order amounts) from the actual costs and claims the difference plus markup. This method is usually rejected by the courts and arbitrators, and is not recommended for claims. However, it is useful for identifying potential inefficiency claims and prioritizing analysis efforts, as described in Sec. C.1.

### 2.   Modified total cost method

If impacts are so numerous and diffuse that determining the loss of productivity from each is impossible, the total cost approach can be modified and used. In many cases, this is the only method that can be used. Give a credit for bid errors, contractor errors during construction, and unexpected events or conditions that are not the owner's responsibility (e.g., weather).

The modified total cost method also can be applied to individual cost codes, instead of to the entire project. This makes it easier to eliminate bid errors and to separate compensable costs from contractor error. It should be used, however, only if no other method is applicable.

### 3.   Cost plus method

Cost plus, also called time and materials or actual cost plus markup, consists of determining the cost expended, adding markup, and claiming the total. It can be used in lieu of the modified total cost method, and must be used when the work is so changed that it is materially different from what was bid. When occurring at the project level, this is called a cardinal change, and the theory of recovery is termed *quantum meruit*. A cost plus claim may be characterized as a modified total cost claim to avoid the negative connotation of the term "cost plus."

## E. Measured Mile Method and Learning Curve Adjustment

The measured mile method is the preferred method of computing efficiency. It consists of comparing the productivity during the impacted period to productivity during an unimpacted period. The difference is the amount of the claim.

### 1. Requirements for measured mile analysis

To use the measured mile analysis, weekly production quantities must be recorded. Earned value analysis compares actual productivity rates (labor hours per unit of work) to planned. Earned value is essential for controlling labor costs, the most volatile element of construction costs, and is needed for a measured mile analysis. If production quantities aren't being recorded and there is forewarning of a pending impact, detailed cost accounting records or work sampling can be instituted to record productivity before and after the impact.

A period of impacted production must be identified in order to use the measured mile method, and there must be an unimpacted period to compare with it. In addition, the work and the productive resources (labor and equipment) must be essentially the same for both the impacted and unimpacted period. Only the working conditions can differ, and only due to the asserted change.

The difference in productivity of the impacted versus unimpacted condition is expressed as a percentage of the actual (impacted) productivity. It is computed from an arithmetical average of each set of data, or from selected, representative periods, and it should be adjusted for learning curve effects. The percentage is used to compute the net inefficiency and the delay due to the inefficiency.

Charting the productivity rates as vertical bars on a timescale (on either a labor histogram or an as-built schedule) can be an effective means of demonstrating the impact and reduction in efficiency. It will help the claim reviewer understand and accept the claim. Tabular productivity data, as displayed on an ELIPSE schedule (described in Sec. H.3 of Chap. 10), can also demonstrate inefficiency.

### 2. Adjust for learning curve effects

There is always a learning curve effect as a crew gets up to speed, and variations in productivity when problems arise or when conditions change. When comparing impacted to unimpacted conditions, adjust for learning curve effects and other impacts. Otherwise the compensation may be less than the actual costs. See Sec. H.7 below for an explanation of learning curve effects.

### 3.   Computing productivity from other sources if lacking adequate accounting data

It may be impossible to do a measured mile analysis from the cost accounting data. Weekly quantities of work accomplished may not have been recorded or an unimpacted period of work may not exist. If that happens, it may be possible to make a measured mile analysis from other data.

1. *Monthly productivity from progress payment requests.*   If a task continues for several months, the work quantities in the progress payment requests may be usable in determining average monthly productivity. This requires a full month of unimpacted work (with productivity up to speed) to compare with a full month of impacted work. It also requires adjustment for overstated work quantities completed, and understated quantities if the owner didn't pay for items in dispute.

2. *Productivity from videos or timelapse photography.*   A 15-minute video of unloading a trailer of structural steel, showing the crew shaking out poorly coded and randomly loaded steel, may be sufficient to "prove" a major inefficiency claim against the fabricator. Likewise, timelapse photography or videotapes can be analyzed with time and motion study techniques to determine what the productivity was, and what it would have been absent the impacting conditions. For details, see Sec. H.5.

3. *Productivity reconstructed from other job records.*   The daily diaries and other job records may contain enough information to determine production rates. This often happens on projects involving one major type of work, such as utility construction where the daily feet of pipe laid is usually recorded and indicates overall productivity. However, verify that conditions are constant. For example, on one sewer claim, the estimated productivity varied from 33 to 75 linear feet per day for different sections of the line, depending on working conditions. Some sections included manholes, lateral connections, utilities, and other incidental work items. The variation in estimated production was accounted for by converting the incidental work items into equivalent linear feet of pipe, based on the estimated hours, and by comparing productivity at each section of the line.

4. *Productivity from historical records on other projects.*   Absent other data, contractors can use productivity data from other projects with similar conditions as evidence of unimpacted productivity. This requires a good accounting system and reliable recordkeeping. A limited amount of statistical analysis coupled with anecdotal testimony and other documentation will help convince the claim reviewer that the work was similar. Be prepared for a hard sell, as owners will always question whether the projects are comparable and are reluctant to accept data that they cannot verify.

5. *Patterns of productivity.*   A correlation between impact and work being over budget (and vice versa) can help substantiate damages. On one arbitration, showing a consistent pattern of unimpacted activities being on budget and impacted activities being over budget convinced the arbitrators to award the majority of our inefficiency claim. A spreadsheet compared the cost over- and

underruns for impacted versus unimpacted cost codes. Although not absolute proof, it was the best that could be done with the data available.

### 4. Measured mile analysis based on estimates

The cost of actual impacted work can be estimated if adequate cost accounting and productivity records do not exist. Likewise, the cost of what the work would have cost under unimpacted conditions can be estimated if an estimate doesn't exist or if the means and methods changed from the estimate. Either or both estimates can be used for a measured mile analysis. Owners usually accept contractor estimates of proposed change order work and are capable of evaluating and negotiating a reasonable cost for work based on contractor estimates. Therefore, the difference between two estimates is a reasonable estimate of inefficiency—if more exact data are not available. When using estimates for measured mile analysis, it helps if the estimates are based on historical costs of similar work.

For example, building siding was delayed on one project by owner actions until the scaffolding was removed. Consequently siding had to be installed from man-lifts. An estimate was not available for the planned installation from scaffolding, and separate records were not maintained for the actual installation. It was simple, however, to estimate the cost of both operations and to claim the difference.

## F. Interview, Survey, and Expert Opinion Methods

Quantitative analysis is essential for computing inefficiency. However, anecdotal evidence and the opinions of the project team, other contractors, and industry experts are needed to supplement more analytical methods. They add a down-to-earth emotional appeal, which can be essential for convincing the claim reviewer. In addition, formal or even informal surveys can provide a clear perspective and hard data for proving inefficiency.

### 1. Interview project team members

Interview all key team members familiar with the impacted conditions using the procedures described in Sec. D of Chap. 8. Ask their opinion on what happened and why, the extent of impact, and the inefficiency percentage they experienced. They will have invaluable insights and may provide information not otherwise obtainable. However, they may miss some issues or have misperceptions, since they were involved in the project.

Emotionally felt verbal presentations by project participants can be as important as rational analysis in convincing a claim reviewer of inefficiency. Only after convincing the reviewer at the "gut" level of the extent of impact can you convince them to accept your computation of the amount of inefficiency. Supervisors and workers are often the most effective presenters, if they are believable and reasonably articulate. A supervisor, project manager, or expert who can build a word picture of conditions can also be effective, especially if they describe firsthand experiences of being tired, cold, wet, frustrated, confused, and inefficient. Personal experience generates empathy.

For documentation, excerpt key comments from the supervisors' daily reports and reproduce or type them on a single sheet for quick review and greater visceral impact. For larger projects, a formal worker or supervisor survey can provide more specific quantitative data.

### 2. Survey other contractors

Ask other contractors about their experience under similar conditions. Recently, when unable to find reliable data on the effect of winter rains on forming suspended slabs, I informally surveyed three specialty subcontractors on their adjustment for costs in the Pacific Northwest. One made no specific adjustment for winter weather (but worked primarily in a drier, warmer area), another stated 45 minutes per day loss (9 percent) for late November through February, and the third stated 15 to 20 percent for normal winter conditions from December through mid-February. Thus the 20 percent factor developed on another project for severe winter rain was determined to be reasonable and possibly too low for the subject project.

### 3. Expert opinion

A common "proof" of inefficiency is an expert's statement starting with "In my opinion (or experience)...." Many expert opinions are based on anecdotal evidence that was never quantified or documented but relies on the expert's reputation, image, and ability to project competence.

An expert need not be a consultant, although there are advantages to using an independent third party. An owner or an executive of the construction company can be qualified as an expert if that person possesses the necessary experience and recognized judgment.

To effectively present expert opinion, describe several projects with similar conditions that support the opinion. Names and incidental facts establish credibility, making it more likely that expert conclusions will be accepted.

### G. Construction Industry Productivity Study Methods

If the measured mile approach cannot be used, most claim preparers compute inefficiency from inefficiency factors from industry productivity studies, supplemented with expert opinion and a modified total cost analysis. One good source of inefficiency data is the U.S. Army Corps of Engineers' *Modification Impact Evaluation Guide* (MIEG), EP 415-1-3, July 1979,[29] which is the source of several of the charts in this section. Another source is *Calculating Lost Labor Productivity in Construction Claims* by William Schwartzkopf.[30]

Besides those described in this section, there are numerous other industry studies, several of which are reasonably accurate and generally applicable.

The ones described below are from reliable sources or are well documented and applicable to most conditions, especially if adjusted for actual conditions and supported by corroborating evidence.

## 1. Acceleration inefficiency

When acceleration occurs, the contractor may not be able to provide adequate workers, materials, and equipment in a timely manner. Support services also may be overloaded. Inspection may be late, crews may be unbalanced, and supervisors may lack time to lay out and organize the work. Conflicts may occur from overcrowding or trade stacking. The result is idle time, unproductive traveling to another work area if reassigned, and working slowly when no other work options are available. If additional personnel are added, conflicts will result from overcrowding, physical interference, and competition for equipment and space. In some cases, less skilled employees must be hired, and productivity will further decrease.

Acceleration costs can be enormous, as noted by the MIEG:[29]

> When acceleration enters the picture, the principle of diminishing returns adversely alters the normal labor cost/productivity ratio. Situations may occur where the impact costs amount to more than the cost of accomplishing the directly changed work. It is likely that the credibility of an estimate producing such results will be questioned by those unfamiliar with the facts; it is therefore important for the estimator's work to be thoroughly documented so its rationale can be defended.

There is no direct measure of acceleration inefficiency, as the impact is indirect—through overtime, crowding, overstaffing, etc. To compute acceleration inefficiency, identify the impacts, determine the inefficiency from each, and sum the total. For details, see Secs. G.2 through G.5, below.

## 2. Overtime inefficiency

Extended overtime causes significant inefficiency that must be included in an acceleration analysis.

The effect of physical and mental fatigue from overtime is documented by several studies, including those by Procter & Gamble (published as a Business Roundtable report), the National Electrical Contractors' Association, Foster Wheeler, C.F. Braun Inc., and the Construction Industry Institute. Overtime inefficiency charts are widely available and simplify the computation of inefficiency factors. However, many of the charts are based on one of the above studies or on a study (Bulletin No. 917) by the U.S. Bureau of Labor Statistics from the 1940s of manufacturing plants under wartime conditions.

The most extensive review of the various overtime studies was performed by H. Randolph Thomas and published as "Effects of Scheduled Overtime on Labor Productivity."[31] An expansion of his figure 13 comparing the various reports is as follows:

| Study | Efficiency | | |
|---|---|---|---|
| | 50 hr/wk, % | 60 hr/wk, % | 70 hr/wk, % |
| U.S. Bureau of Labor Standards | 92 | 84 | 78 |
| Foster Wheeler | 87 | 73 | — |
| NECA Survey | 88 | 85 | 78 |
| C.F. Braun | 87 | 73 | 58 |
| Procter & Gamble—12 weeks | 84 | 64 | — |
| Procter & Gamble—4 weeks | 90 | 84 | — |
| U.S. Army MIEG—4 weeks | 96 | 79 | 63 |
| Average Value | 89% | 77% | 69% |

**Corps of Engineers methodology.** The *Modification Impact Evaluation Guide* provides a chart (Fig. 12.2) for determining overtime inefficiency for various combinations of overtime up to 4 weeks long.

Another study of scheduled overtime, "Effect of Scheduled Overtime on Construction Projects," was prepared for the Construction Users' Anti-Inflation Roundtable and published in the October 1973 issue of the American Association of Cost Engineers (AACE) Bulletin.[32] An extract of Table 4 from that report, reprinted with permission of the AACE, follows:

| Work week | 50-Hr week inefficiency, % | 60-Hr week inefficiency, % |
|---|---|---|
| 1–2 | 7.4 | 10.0 |
| 2–4 | 10.0 | 14.0 |
| 4–6 | 13.0 | 20.0 |
| 6–8 | 20.0 | 29.0 |
| 8–10 | 24.8 | 34.0 |
| 10–12+ | 25.0 | 37.0 |

## 3. Crowding and trade stacking inefficiency

Crowding too many workers in a given space causes inefficiency. It results from overlapping of crews, out-of-sequence work, interface conflicts, inadequate access or layout space, and physical interference.

Trade stacking is often used to describe crowding but is better defined as the effect of conflicts arising from the incompatible work of different crafts working in the same area. Sequencing is difficult, supervision is a problem, and there may be adverse environmental conditions such as paint fumes, dust, or noise that affect other trades. Examples are painters and carpet layers in the same corridor or pipefitters and insulators together in a mechanical room. The conflicts in their work are in addition to conflicts over the limited physical space being shared.

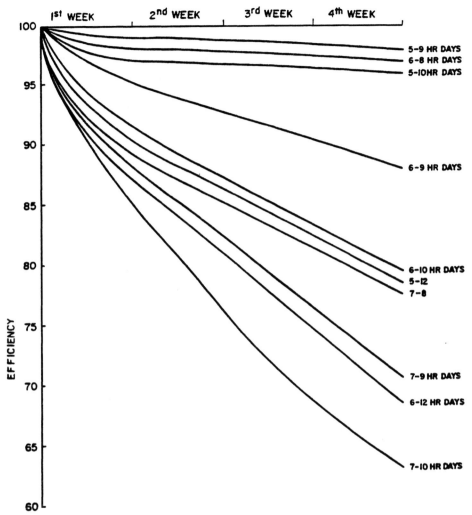

**Figure 12.2**   Effect of work schedule on efficiency. (*U.S. Army Corps of Engineers Modification Impact Evaluation Guide, EP 415-1-3, Fig. 4-4.*)

This section doesn't address the additional inefficiencies from trade stacking, which should be estimated separately from crowding inefficiencies.

Activity overlapping (concurrency) doesn't necessarily result in crowding, as there may be sufficient room for the crews to work without interference. Crowding can be avoided by rescheduling (if some of the tasks are noncritical), by a different sequence of operation, or by a time extension. Other means of avoiding crowding include overtime, multiple shifts, larger equipment, or prefabricated materials.

**Corps of Engineers methodology.**   The *Modification Impact Evaluation Guide* provides a methodology of computing efficiency loss from crowding (not including trade stacking effects), which includes the chart reproduced as Fig. 12.3. Use the chart to determine the degree of crowding at the jobsite. First determine the maximum number of workers who can efficiently work in an area. (According to Schwartzkopf, a minimum of 200 to 300 square feet per worker is needed for optimum efficiency, with 45 to 65 percent efficiency at 100 square feet.)[30] Second determine how many will be added to accelerate. Third divide the difference by the maximum efficient number and divide by 100 to obtain the percentage of crowding.

For example, assume three activities, each with five workers, are scheduled concurrently in a mechanical room and are the maximum number that can work efficiently in that space. If a three-worker crew is added for 5 days, the crowding factor is 20 percent (3 / 15 = 0.20). The chart gives an 8 percent inefficiency factor for the entire crew. Thus it will take 8 percent longer, or 0.4 day, for the four crews to accomplish 5 days' work (5 × 0.08 = 0.40). The total inefficiency is 7.2 labor days cost (0.4 crew day × 18 workers = 7.2). There also will be extended overhead costs, and a time extension of 0.4 working day will be needed. In addition, the work will now extend over a weekend. You may need to add additional time and costs for the added crew to close down for the weekend and to start up again the following Monday, since they won't finish in 1 week.

### 4.  Overstaffing inefficiency

Too many workers in a crew for effective supervision and support by the foreman causes inefficiency, regardless of and in addition to any impact from

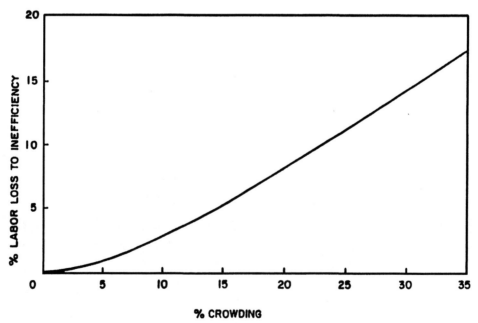

**Figure 12.3**  Effect of crowding on labor efficiency. (*U.S. Army Corps of Engineers Modification Impact Evaluation Guide, EP 415-1-3, Fig. 4-2.*)

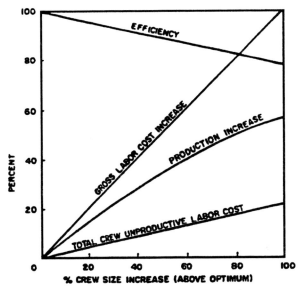

**Figure 12.4**  Composite effects of crew overloading. (*U.S. Army Corps of Engineers Modification Impact Evaluation Guide, EP 415-1-3, Fig. 4-3a.*)

crowding. The optimum crew size, according to the MIEG, is the minimum required to perform the task in the allocated duration. If work must be accelerated, one method is to add to the crew. However, adding more workers will, at some point, result in lower productivity per worker, even without crowding effects. This results from the difficulty of effectively supervising a larger crew due to "span of control" limitations, more difficult communication within the crew, difficulties of providing sufficient materials and work area, and loss of cohesiveness and teamwork.[29]

**Corps of Engineers methodology.**  The *Modification Impact Evaluation Guide* describes a five-step process to compute efficiency loss from overstaffing based on Fig. 4.3a from the manual, which is reproduced here as Fig. 12.4. For example, assume that 10 workers (the optimum crew size) are scheduled to accomplish a task in 15 days and the work must be accelerated to complete in 10 days.

1. *Step 1. Determine the rate of work.*  The rate of work is the quantity of work accomplished divided by the effort in labor-days. This is:

Original Rate of Work = (1 job) / (planned crew size × originally planned duration)
Original Rate of Work = 1 / (10 workers × 15 days) = 1/150 = 0.0067 unit/labor day
New Rate of Work = (1 job) / (planned crew size × currently required duration)
New Rate of Work = 1 / (10 workers × 10 days) = 1 / 100 = 0.0100 unit/labor day

2. *Step 2. Determine the percent acceleration (production increase).* The percent acceleration (production increase) is the new rate of work less the original rate of work divided by the original rate of work:

---
% Acceleration = (new rate − original rate) / (original rate of work) × 100
% Acceleration = (0.0100 − 0.0067 / 0.0067) × 100
% Acceleration = 50 %
---

3. *Step 3. Determine the required crew size.* Theoretically a 50 percent increase in production can be gained by a 50 percent increase in the crew size. However, as stated earlier, overstaffing inefficiencies will necessitate additional workers to achieve the increased production.

To use Fig. 12.4, start at 50 percent on the left (*y* axis), move horizontally to the PRODUCTION INCREASE curve, and drop vertically down to the *x* axis for the % CREW SIZE INCREASE (ABOVE OPTIMUM) value of 80 percent. Therefore, the crew size must be increased 80 percent—from 10 to 18 workers.

4. *Step 4. Determine the efficiency.* To determine the efficiency, divide the planned crew size (15) by the required crew size (18) to obtain the efficiency, which is 83 percent (15 / 18 = 0.83).

To use Fig. 12.4, start at 80 percent on the bottom (*x* axis), move vertically up to the EFFICIENCY line, and then move horizontally to the *y* axis value of 83 percent.

5. *Step 5. Compute the inefficiency cost.* To compute the cost of inefficiency, multiply the required crew size (18) by the inefficiency (17%/100), and divide that by the original crew size to obtain the inefficiency cost percentage.

---
Inefficiency cost = (18 workers) × 0.17 / 15 workers (original crew size)
Inefficiency cost = 20% of original cost
---

Alternately, to use Fig. 12.4, start at 80% crew size increase above optimum (on the *x* axis), go up vertically to the TOTAL CREW UNPRODUCTIVE LABOR COST line and horizontally to the PERCENT line for a value of 20 percent. Overtime premium must be added to this.

## 5. Task reassignment inefficiency

Reassigning workers to a new task, due to start-stop operation or waiting for resolution of a problem, causes inefficiency during reorientation. Moving back to the original task causes another loss of productivity. This loss of productivity results from the time it takes to become oriented to the task rather than to acquiring skills, as construction workers are trained to perform a wide variety of tasks. The amount of time depends on the worker's skill and experience and how much the tasks vary.[29]

**Corps of Engineers methodology.** The *Modification Impact Evaluation Guide* includes a chart (Fig. 12.5) for computing efficiency loss from worker reassignment, assuming that the average worker takes one day to reach full

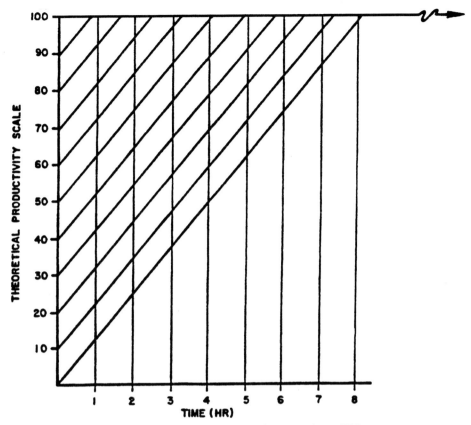

**Figure 12.5**  Construction operations orientation and learning chart. (*U.S. Army Corps of Engineers Modification Impact Evaluation Guide, EP 415-1-3, Fig. 4-3b.*)

production when shifted to a new task, with an average productivity of 50 percent. To use the chart, choose the point on the curve where productivity starts. For example, a laborer shifted from stripping forms to jackhammering concrete might take a full day to become oriented, and the loss of productivity will be 4 hours. If moved from stripping forms in building A to stripping forms in building B, the starting point might be 90 percent with only 0.8 hour to get up to speed.

The chart assumes the new work area is ready (i.e., laid out by the foreman), tools and materials are available, crew morale and motivation are reasonably high, and the crew has performed the tasks before. If not, there will be further inefficiencies. Note that the curve doesn't include relocation and setup time.

### 6.   Inefficiency from multiple changes

Extensive studies on contract changes indicate an average 5 to 10 percent of the original contract amount on federal projects and $5\frac{1}{2}$ percent on privately owned projects.[30] This is in line with the typical architectural contingency of around 5 percent for building construction.

**Leonard study.**   Based on a study of 90 cases on 47 projects, a master's thesis[33] by Charles A. Leonard at Concordia University indicates a linear relationship between inefficiency and change order labor hours in excess of 10 percent of the original contract labor hours. The coefficient of correlation was relatively high (0.88 for mechanical/electrical work and 0.82 for civil/architectural work). However, the correlation between inefficiency and the frequency of change orders (number of change orders per month) was lower. This was believed to be because some change orders include multiple changes. The study excluded the effect of large change orders at the start of a project, if they were implemented early enough to avoid impact.

The study indicates inefficiency on civil/architectural work was 12 percent when change orders were 10 percent of the original contract and increased to 26 percent when change orders were 60 percent of the original contract. On mechanical/electrical work the inefficiencies were 12 to 36 percent, respectively. A second or third major impact, such as acceleration or out-of-sequence work, increases the inefficiency substantially, as indicated in Fig. 12.6. The coefficient of correlation for a second major impact was relatively high (0.76 for mechanical/electrical and 0.74 for civil/architectural), but the coefficient for a third major impact was relatively weak (0.34 for mechanical/electrical). A third major impact could be computed with a different analysis instead of being included in the multiple changes analysis.

To apply the study, first calculate the percentage of change orders by dividing the actual change order labor hours by the actual labor hours on

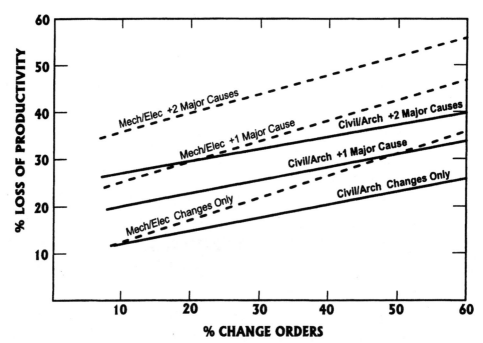

**Figure 12.6**  Effect of multiple changes on productivity. (*Source: The Effect of Change Orders on Productivity by Charles A. Leonard.*)

the original contract work. Next, find the percentage loss of productivity from Fig. 12.6 for the percentage of change orders, the type of work, and whether there are other major causes of impact. Then, multiply the percentage loss of productivity from the chart times the actual hours expended on the original contract work to determine the loss of productive hours.

**Cumulative impact from multiple changes.** The Leonard study validates the concept of cumulative impact from multiple changes, as discussed in Sec. A.3.8, and is often the most practical way to quantify the impact. The linkage from individual changes to specific inefficiencies often cannot be identified, although the ELIPSE schedule described in Sec. H.2 of Chap. 10 can show a graphical correlation.

Changes beyond the chart limits (60 percent of the original contract) probably constitute a cardinal change to the contract and should be resolved through a cost plus settlement.

## 7. Extreme temperature inefficiencies

The May 1972 issue of *CONSTRUCTOR*, the national magazine of the Associated General Contractors of America, listed the following inefficiencies from abnormal temperatures:[35]

| Effective temperature (include wind chill), degrees | Efficiency loss of gross skills, % | Loss of fine skills, % |
|:---:|:---:|:---:|
| 100 | 60 | Not estimated |
| 90 | 40 | * |
| 80 | 30 | * |
| 40 | 0 | 15 |
| 30 | 0 | 20 |
| 20 | 0 | 35 |
| 10 | 5 | 50 |
| 0 | 10 | 60 |
| −10 | 20 | 80 |
| −20 | 25 | 90+ |
| −30 | 35 | Probably can't work |

Gross skill trades include laborers, concrete handlers, cement masons, ironworkers, operating engineers, roofers, bricklayers, and glaziers. Fine skill trades include carpenters, tile setters, plumbers and pipefitters, welders, and electricians.

Published studies indicate a significant impact from humidity.[27] Published data indicate a range of 80 to 54 percent efficiency at 100 degrees F (for 25 to 95 percent humidity) and a range of 57 to 23 percent efficiency at zero degrees F (for 25 to 95 percent humidity).[35]

**8. Learning curve inefficiencies**

The use of learning curves can be part of a work improvement approach to computing inefficiency, as described in Sec. H.7, or part of an industry study approach.

**9. Inefficiencies from a combination of effects**

Adding the effects of a number of inefficiency factors together overstates their combined impact.[26] Unfortunately, there is no generally accepted method for adjusting the results of the individual calculations, although several authors have proposed methods. Therefore, claim preparers must use judgment when combining several factors.

**10. Collaborating evidence**

Industry productivity studies are frequently challenged. Therefore, it's best to provide collaborating evidence of their applicability. At a minimum, include a modified total cost analysis plus a statement by project field supervisors describing the conditions and confirming that they suffered the claimed degree of impact.

**11. Sources of additional data on productivity and inefficiency**

For more information, read *Calculating Lost Labor Productivity in Construction Claims* by William Schwartzkopf.[25] Sources for additional data include:

- Associated General Contractors of America, 1957 E. Street, N.W., Washington DC 20006-5107.
- The Business Roundtable, 200 Park Avenue, New York, NY 10166.
- National Electrical Contractors Association, 7315 Wisconsin Avenue, Bethesda, MD 20814.
- Mechanical Contractors Association of America, 1385 Picard Drive, Rockville, MD 20850.
- AACE International, 209 Prairie Ave., Suite 100, Morgantown, WV 26505.

**H. Work Improvement Methods**

One of the most powerful but least used methods for identifying and computing construction inefficiency is to use work improvement techniques. The potential benefits of using these techniques to prove inefficiency are immense, especially when the measured mile approach cannot be used. Work improvement studies are most applicable to repetitive cyclical operations and require detailed documentation of each step of the operation. This information can be obtained from (1) observation of ongoing operations, (2) timelapse photography or videotape, (3) other detailed project records, or (4) detailed "prejob" planning to identify each task.

Using work improvement techniques to analyze inefficiency claims requires understanding productivity. One of the best sources of information on productivity is *Productivity Improvement in Construction* by Oglesby, Parker, and Howell, McGraw-Hill, New York, 1989,[27] which was used as the basis of this section. Their book, based on an earlier publication, has more recent input by Greg Howell, a construction productivity consultant and cofounder of the Lean Construction Institute. The techniques are briefly described here, and readers are referred to their book for details.

## 1.   Work improvement for increased productivity—an overview

Work improvement is the analysis of a repetitive task to determine how to do it better—safer, faster, and more economically.

**What work improvement is and what it can do.**   Work improvement includes a set of planning and control techniques that are far more detailed than overall job planning and control techniques (critical path scheduling and job cost accounting). Work improvement formalizes normal detailed task planning and control. It uses a set of written procedures for preplanning tasks and processes, measuring productivity, and analyzing work to determine the most effective and efficient method of performance. The method includes a number of techniques for tabular or graphical display of production data that permit the simultaneous viewing and analyzing of hundreds of data elements, instead of the dozen or so the human mind can consider at one time.

Work improvement has been largely ignored by the construction industry—except for informal intuitive application in the field at the supervisory level—in spite of the potential for greatly improving profits. Its application is described here, however, as a tool for claim analysis.

**The work improvement process.**   The recommended process for work improvement is as follows.[27]

1. *Record how the task is being or will be accomplished.*   Observe how the work is being done and record it using stopwatch studies, work sampling, timelapse videotape, or interviews and questionnaires. If work hasn't started, preplan it using crew-balance charts, flow diagrams with process charts, models, sketches, computer simulation, or other aids.

2. *Analyze each detail.*   Evaluate the task layout; tools, equipment, and materials used; crew composition; material flow; operations; and safety. Examine each detail of the present or planned method, using the same tools that are used for preplanning. Prepare a narrative or sketches of hand work, machine work, and materials handling. When analyzing the details, ask the following questions: (1) what is its purpose, (2) why do it this way, (3) when is the best time to do it, (4) where is the best place to do it, (5) how should it be done, and (6) who is best qualified to do it.

3. *Devise a better method.*    After determining the basic objective of the task, devise a better method to accomplish the objective. To do this: (1) solicit ideas from management, supervisors, subcontractors, and workers, (2) use brainstorming and informal roundtable discussions to develop ideas, and (3) ask the following basic questions and consider the following answers:

- What? Eliminate unnecessary detail.
- Why? Reduce labor costs.
- When? Rearrange for better sequence.
- Where? Reorganize for easier access.
- How? Provide better tools, devices, jigs, materials.
- Who? Consult with others who have an inherent interest.

Then, (4) evaluate several alternative methods and select the best one, and (5) write up a detailed description and prepare a sketch of the best method.

4. *Implement the selected method.*    After selecting the best method, proceed as follows: (1) sell it to management, supervisors, and workers, (2) implement it, (3) follow up to ensure continued use and make further improvements, and (4) give credit and praise to all involved.

**Subcycle efficiency.**[36]    Many construction operations can be broken down and analyzed as cycles and subcycles. For example, an excavator has a dig/swing/unload/swing cycle, which is a subcycle of a truck load cycle, and the truck load cycle is a subcycle of a load/haul/unload/return cycle for the entire fleet.

Cycle and subcycle efficiency is a major component of productivity. It requires matching the rate of resource supply with consumption for each subcycle. This is done by either (1) tight control over the rates of supply and demand (as in careful planning and control of crane use) or (2) providing adequate buffers (queues) of intermediate work products between subcycles. Cycle efficiency also depends on crew balance, to minimize idle time for labor and equipment on subcycle operations, and on permitting workers to specialize in one task, thus becoming more efficient.

Tightly linked subcycle processes are efficient and minimize excess materials storage (as in "just-in-time" manufacturing). This is similar to overlapping schedule activities, which permits a shorter schedule. Unfortunately, tightly linked processes are risky for construction, owing to the variable and uncertain nature of the industry. Acceleration, delay, and change (especially unexpected change) can disrupt material delivery and empty the buffers, thus causing inefficiency.

When analyzing inefficiency, examine the buffers between subcycles and determine if the linkage is tight or loose. Tightly planned operations suffer far greater impacts than loosely planned ones.

**Case studies.**    Oglesby, Parker, and Howell describe a generator installation contract that illustrates the savings attainable from work improvement (Fig. 12.7). The project was estimated at 10,800 labor hours per unit by the owner's expert, bid at 7200 labor hours versus an average of 10,450 for the

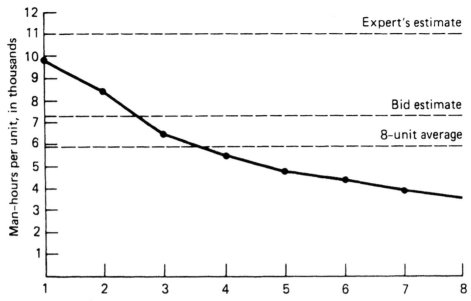

**Figure 12.7**  Arithmetic plot of labor hours per generator unit installed. (*Source: Productivity Improvement in Construction, reprinted with permission of McGraw-Hill.*)

other bidders, and completed (for the first eight units) at an average of 6000 labor hours. The project was so successful that the government instituted a suit to recover the excess profits.[29]

Work improvement can also be an informal process. For example, critical observation by a piledriving superintendent on one project revealed the crew was partly idle while piles were driven. By reorganizing the pile storage, cabling up the next pile, and dragging it into position while driving the prior pile, we drove two additional piles per day. The 8 percent productivity increase equaled our profit margin.

## 2.  Applying work improvement techniques to claim analysis

For claim analysis, work improvement techniques can be used to deconstruct an impacted operation to determine the amount of inefficiency. The procedure is as follows:

1. *Record how the task was or is being accomplished.*  If work is ongoing, record observations using stopwatch studies, timelapse videotape, work sampling, or interviews and questionnaires—just as for work improvement. Recording ongoing work with stopwatch studies and timelapse videotape is described in Secs. H.4 and H.5.

If work is completed, reconstruct the task from the records supplemented by questionnaires of crew members, interviews and surveys of supervisors, and estimates of missing data. Reconstructing the task can be difficult or impossible unless adequate records are maintained. Video or timelapse photography of an operation is invaluable if it records an entire cycle. Photographs can be helpful if enough were taken to reconstruct an operation. Superintendents' daily reports may also provide sufficient information for analysis. In many cases, however, the allocation of time to various work elements will have to be estimated.

2. *Analyze each detail of the task and break it into its fundamental elements.* Generally the best method for analyzing a task to determine inefficiency is crew-balance charts with a combined flow diagram and process chart. For details, see Sec. H.6.

3. *Create a would-have-been task (without the impacts).*  Deconstruct the impacted task by eliminating the inefficiencies caused by impact, leaving a task as it would have been performed without the impacts. This is done with a crew-balance chart.

4. *Determine the differences.*  The percent inefficiency is the difference in cycle time between the actual task and the would-have-been task, divided by the actual cycle time and multiplied by 100. Multiply the inefficiency by the actual labor cost to obtain the labor inefficiency damages. Add equipment use, increased subcontract costs, etc.

### 3.  Interviews and questionnaires for recording data

Questioning the workers and supervisors can provide invaluable insights to what happened in addition to providing workers and lower level supervisors an opportunity to contribute. Interviews of supervisors can be invaluable but should be supplemented by questionnaires of supervisors and workers, as questionnaires can be more efficient, cover more personnel, and are considerably cheaper. For details, see Chap. 7 of Oglesby, Parker, and Howell's book.[27]

### 4.  Stopwatch studies for recording data

A stopwatch study is the classic time-and-motion study effort. It requires only a digital watch with a stopwatch feature, clipboard, paper, pencil, and a careful observer. The observer records the details of a single individual, piece of equipment, or crew. More than one operation cannot be adequately recorded.

The stopwatch should be run continuously rather than stopping and restarting each time. Most stopwatches have a lap feature that stops the watch for reading while the clock continues to run. The observer breaks each task into a series of detailed steps (subcycles), recording the elapsed time from the start of recording to the start of each step, until an entire "cycle" is completed. This could be a truck loading cycle for earthwork operations, laying a standard number of bricks for masonry operations, or setting and cutting with a jig in a jobsite carpentry shop. Record several cycles to detect variances, determine the presence of nonstandard functions (i.e., shutdown for adjustment), and establish a level of reliability. Try to identify changes in the operation—from natural variations in the work, a change in conditions, or supervisory-directed improvements.

Observers need to be familiar with the operation recorded in order to break it into steps that can be beneficially analyzed. They need to be able to identify the reasons for variances (due to changes in conditions, equipment breakdowns, interference, etc.) and must annotate the record with comments on the reasons. An experienced observer can often detect major problems just by careful observation. But keep in mind that it is difficult and very tiring for one observer to correctly and consistently record production data.

Operations will be influenced because they are being observed and recorded. Minimize the effect by being unobtrusive, but do not hide, and the effect of observation will lessen over time. Workers may respond negatively. Pay attention to worker concerns, clearly communicate the purpose and need for the recording, and involve them in the process. Brief discussions with the crew or even an informal survey can avoid problems, obtain support, and elicit additional information not readily obtainable by a lone observer. Worker support will contribute to successful observation.

## 5.  Timelapse photography for recording, analyzing, and presenting data

Timelapse photography uses videotape, videodisk, or super 8-mm movie film at a slow rate and then plays it back at faster speeds. This speeds up the action, which greatly reduces the time necessary to review the film, and it eliminates the detail, which makes patterns obvious to untrained observers. The videotape or film can be analyzed for work improvement studies or inefficiency analysis. Or it can be reviewed by the project team and work crews to aid them in improving their own operations. It also can be a powerful exhibit for demonstrating impact to claim reviewers, arbitrators, or the courts.

The timelapse camera should be set up high enough so that visibility isn't blocked by action in the foreground. Recording time must include a full cycle of the operation and preferably continue for a half day or day to pick up variations. Recording-to-viewing compression ratios for analysis can vary from 1:10 for craft crews to 1:20 for equipment spreads, with quick reviews varying from 1:30 to 1:60. A 1:30 ratio allows a carpenter crew to review their 8-hour day in 15 minutes; they should view it at least twice in order to identify the patterns and develop improvements.

A timestamp in the corner of the film is needed for in-depth time-and-motion studies. It facilitates breaking the operation down into cycles and the cycles into their component parts. From these a crew-balance chart can created. As noted below, the crew-balance chart is used to create a "would-have-been" operation which is compared to the impacted operation to compute the inefficiency.

## 6.  Crew-balance charts and combined flow-process diagrams for analyzing data[27]

**General principles of analysis.**  Analysis of productivity, whether for work improvement or claims analysis, should start with the big picture and progress to the details. First evaluate the overall operation, next the task as a whole, then break it down into its basic steps, and finally analyze each step: what, why, when, how, where, who?

The cycle analyzed can be short (tightening bolts) or relatively long (a 2-week form and pour cycle for a multistory building). It can also be complex, with cycles within cycles. Or the work may be so complex or so severely impacted that a consistent cycle cannot be identified. If so, work improvement techniques cannot be used.

In analyzing productivity, keep in mind that materials handling occupies probably two-thirds of most construction operations, with some experts asserting that it occupies as much as 85 percent.

**Crew-balance chart.** A crew-balance chart is a series of vertical bar charts, one for each element of a crew. An example is provided below (Fig. 12.8), taken from a work improvement study in Oglesby, Parker, and Howell's book.[27] The same technique can be used to compare actual with would-have-been conditions. The bars are marked with each step basic to a task or work cycle, with the elapsed time as a scale to the side. The bars can be darkly shaded for productive work, lightly shaded for supporting work, and clear for nonproductive. Or the steps can be color-coded with some more complex allocation such as the following:

- Productive work (effective work categorized separately from noneffective)

- Working at less than normal capacity

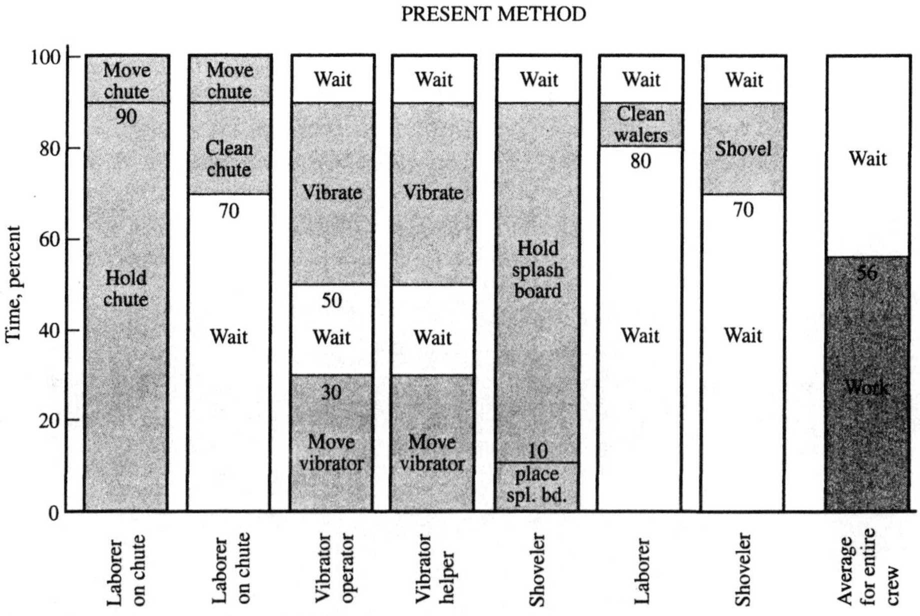

PRESENT METHOD

**Figure 12.8**  Typical crew-balance chart for concrete placement. (*Source: Productivity Improvement in Construction, reprinted with permission of McGraw-Hill.*)

- Work paced by controllable machine speed
- Work paced by uncontrollable machine speed
- Holding while someone else works
- Transporting material
- Standby (necessary but unproductive)
- Waiting for another worker to finish
- Waiting for a machine
- Waiting for material delivery
- Idle

**Flow diagram and process charts.**   A flow diagram is a plan-view sketch of work stations, equipment locations, and the flow of materials. A process chart describes the steps involved in a process and categorizes them as one of five types of functions. Both are very helpful for work improvement and could be used for impact analysis. The process chart steps are:

- Operate—to produce, change, assemble, disassemble, etc.
- Transport—to move from one location to another
- Inspect—to verify its condition
- Store—temporarily or permanently
- Delay

**Other analysis techniques.**   There are other work improvement analysis techniques that can be applied to impact analysis. They include mass diagrams, load-growth curves, operations research, and simulation. Most of these techniques are too complex for most applications but could be useful in specific instances. See Oglesby, Parker, and Howell's book and other literature on the subject for additional information.

## 7.   Learning curve adjustments

Claim preparers should be familiar with learning curves and use them when analyzing inefficiency.

**Introduction and overview of learning curves.**[27]   Learning curves are useful tools for understanding productivity and are needed for accurately computing impact. They are most applicable to complex labor-intensive tasks without external constraints.

1. *Definition of learning curves.*   A learning curve charts the improvement in productivity when an individual or crew repeats the same task. When plotted on an arithmetic scale, the curve is hyperbolic as in Fig. 12.9; when plotted on a loglog scale, it forms a straight line as in Fig. 12.10.

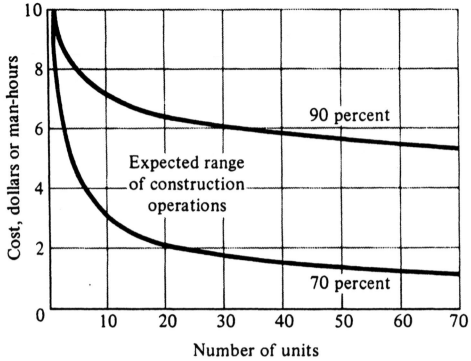

**Figure 12.9**   Expected range of construction learning curves, printed on arithmetic scale. (*Source: Productivity Improvement in Construction, reprinted with permission of McGraw-Hill.*)

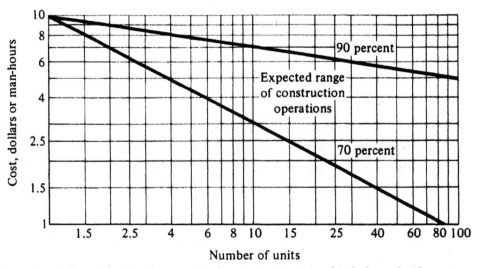

**Figure 12.10**   Expected range of construction learning curves, printed on loglog scale. (*Source: Productivity Improvement in Construction, reprinted with permission of McGraw-Hill.*)

2. *Amount of improvement.*   The amount of improvement is expressed as a percentage of the effort to accomplish a unit of work when the number of units doubles. For example, on a 90 percent learning curve, the effort (e.g., labor hours) for the second unit will be 90 percent of the first unit, the hours for the fourth unit will be 90 percent of the second, the hours for the eighth unit will be 90 percent of the fourth, etc. When expressed as a formula, the value for the current unit produced ($ax$) is:

$$a_x = a_1 x^{-n}$$

where  $a_x$ = the level of effort (e.g., labor hours) to accomplish unit number "x"
  $a_1$ = the effort required to accomplish the first unit in the series
  $x$ = the sequential number of the unit, i.e., the 20th unit completed
  $n$ = the curve percentage of the variation (i.e., 0.90 for a 90% curve, 0.70 for 70%)

3. *Reason for and rate of improvement.*   In some cases, as when production is limited by equipment capacity, no improvement is possible, and the learning curve ratio will be 100 percent. However, as noted by the authors, normal construction learning curves range from 70 to 90 percent, with the smaller values being possible only for more complex labor-intensive operations. Forming of intricate concrete structures might experience a 70 percent ratio, while placing concrete is more likely to be in the 90 percent range.[27]

The rate of improvement also varies depending upon the starting point. A carefully planned operation with skilled crews on a standard operation will have a relatively lower starting point with a "hump" at the beginning of a loglog curve (Fig. 12.11). Consequently the improvement from the first to the

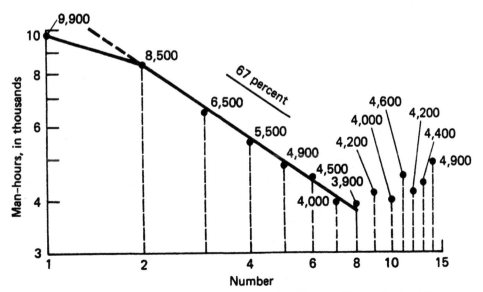

**Figure 12.11**  Impact of changed conditions on generator installation. (*Source: Productivity Improvement in Construction, reprinted with permission of McGraw-Hill.*)

second unit will be less than expected. On poorly planned, first-time, error-plagued operations, the increase in productivity for the second unit can be greater than 50 percent, with much smaller improvements for subsequent units.

Other factors influencing the rate of improvement include:

■ The level of worker skill and the degree to which they can learn on the job
■ Morale, motivation, environmental conditions, and the absence of boredom
■ The consistency of the tasks performed and the absence of delays or interference
■ The availability of improved methods, tools, and equipment

In practice, the most important factor is the skill of management and supervisors in organizing the work and their leadership skills in motivating, supporting, and guiding the workers. In practice, productivity increases eventually taper off and become flat, primarily owing to lack of supervisory attention and boredom. If a radically redesigned process is implemented, however, there can be an abrupt improvement in productivity with a changed curve for continued production. On the other hand, productivity can fall off toward the end of a job owing to boredom, complacency, or a desire to extend the work.

It should be emphasized that the learning curve concept and charts are approximations of reality and merely a useful model in predicting and understanding productivity. Human endeavors are complex and cannot be completely summarized by a chart.

**Use of learning curves in claims analysis.**  Learning curves can be used for claims analysis, especially if productivity records are maintained.

1. *Example use of learning curves for claims analysis and presentation.* One illustration of the applicability of learning curves for claims analysis and presentation is the continuation of the example in Fig. 12.7. Late material delivery, inadequate work area, and subsequent crew changes on units 9 through 14 caused a loss in momentum and a decrease in productivity as shown in Fig. 12.11.

Comparing productivity for the impacted ninth to fourteenth units (at 3900 to 5000 labor-hours per unit) from Fig. 12.11 with the average of the first eight units (6000 labor-hours) from Fig. 12.7 would have indicated that there were no damages. The impacted units took fewer hours than either the estimate or the average actual costs to date. This would have been the result of a claims analysis if the contractor had not maintained productivity records for each unit. However, the evidence of the loglog learning curve plot (Fig. 12.11) proves that compensation is due because of the even greater productivity that would have occurred without the impact.

2. *Average versus current productivity.*  Always consider the learning curve effect when preparing a measured mile analysis. The average unimpacted production from the first part of a project compared with production during an impacted later period will provide inadequate recovery. Instead, use the latest unimpacted production rate and adjust it for what production would have been

without the impact, before comparing it with production during the impacted period.

3. *Charting actual productivity.*    If actual productivity data are available, chart them on a loglog curve. Hopefully, they will follow a straight line except for the identified impacts. Correlation of the change in productivity compared to the straight-line trend preceding the condition of impact is adequate proof of causation. Computation of damages is even simpler, as you can extrapolate what production would have been.

4. *Fitting actual data to learning curves, if missing data by individual unit.* If only the total productivity quantity and labor-hours are available, fit the actual productivity (up to the time of impact) to the most likely construction learning curve. Select a curve value between 70 and 90 percent, depending upon the type of work, and use either the formula or the chart in Fig. 12.10 to determine the probable values.

For example, assume you didn't have the actual productivity for each of the first eight generators for Fig. 12.11 but have the total of 48,000 hours only for the first 8 units. Based on the type of work (labor-intensive), select the 70 percent learning curve in Fig. 12.10. Then find the values on the $y$ axis for the $x$ axis values of 1 through 8. These are 3.2 for generator 8, 3.5 for generator 7, 4.0 for generator 6,..., and 10.0 for 1, for a total of 42.6. Next divide 48,000 hours by 42.6 and multiply the result (1120) by the values from the $y$ axis. The value for generator 1 is therefore 11,200 hours and the value for generator 8 is 3600. This compares to the actual values of 9900 and 3900. The difference is due to the "hump" that often occurs between the first and second unit (which could be accounted for with additional calculations) and because the selected curve (70 percent) is slightly different from the slope of the actual curve (67 percent).

5. *Computing loss of efficiency gain when adding crews.*    If a second crew is added to an operation, it will have to start at the top of the curve and will have considerably lower efficiency than the first curve. In addition, both crews will never reach the lower end of the curve. The loss of efficiency can be readily computed after the slope of the curve is determined.

# 13

# Phase 5—Preparing Exhibits and Assembling the Claim

Analysis of entitlement and computation of damages is documented as narrative text, and becomes the draft of the claim document and exhibits. This chapter describes how to convert those drafts into a final product. It is divided into five sections:

For guidance on finalizing and polishing a claim document, proceed directly to Sec. C. To prepare exhibits for presentation go to Sec. D. If you are reading this chapter from start to finish, you might want to break your review into related sections (A–B and C–E).

The process of preparing exhibits and assembling a claim is summarized in Fig. 13.1.

## A. Introduction and Overview

The claim document and exhibits may be the only source of information supporting your position that the claim reviewer has for making a decision. In some cases, your written claim document may be the only counter to the arguments of the owner's field representative. You may not be there to explain, and even if you are, your claim document *must* be well-written and convincing with strong graphics.

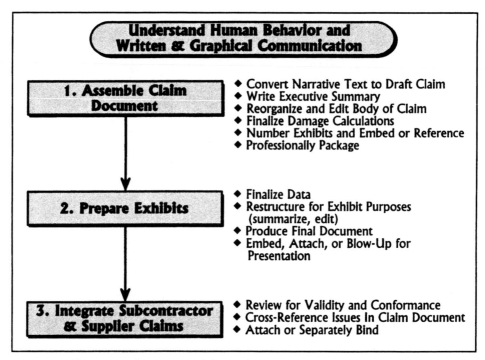

**Figure 13.1.**  The process of preparing exhibits and assembling the claim.

If negotiations fail and the dispute escalates to mediation, arbitration, or litigation, the claim document may be used to convince the mediator or trier of fact (arbitrator, judge, or jury) of the validity of your position. In that case, if you are testifying as an expert, it must be your work product and it must not be unduly influenced by your client or legal counsel. However, the exhibits, if in arbitration or litigation, should be reviewed and approved by legal counsel. Counsel may exercise considerable control over exhibit preparation and presentation but cannot change your basic conclusions.

## 1.  Requirements for a successful claim

The requirements for a successful claim reflect your goal—*to persuade* the reviewer(s) of the validity of the claim and the need to make adequate compensation. To do this, the claim must be:

- *Understandable* by clearly and concisely describing the issues and facts.

- *Convincing* by presenting a compelling argument that is realistic and contract-compliant.

- *Supported* by documented details.

*Simplicity is essential for a successful claim.* A claim must be well documented, but without excessive detail that distracts the reviewer and dilutes the impact. Write clearly and succinctly, and use only a few clear and focused exhibits. Detailed exhibits must be included, but the discussion should focus on the summary exhibits.

Prepare the final claim document well in advance of the initial presentation, allowing for detailed review by the claim support team and editing of the completed document. Be aware that editing, generating figures, printing, and binding can take up to one-fourth of the total time and effort to prepare a written report, although considerable work will have already been put into the narrative text.

## 2.   Understand and address the claim reviewer(s) and decision maker(s)

The claim must address the reviewers and decision maker(s). Consider personalities, biases, and needs, their understanding of the project and the issues in dispute, and initial position on the claim. This may require considerable research and communication with them to test your initial conclusions.

**Identify the claim reviewer(s) and decision maker(s).**  Often the reviewer is a single individual who decides whether to accept the claim. However, in some cases there are multiple reviewers and multiple decision makers, or no clearcut decision maker. Determine who makes the decision and what criteria they will consider—before preparing the claim.

**Establish ongoing communication during claim the preparation process.** Communicating with the claim reviewer(s) creates a sense of teamwork. Involve the claim reviewer(s) in the preparation of the claim by asking their opinion. They may suggest which theories will be best received and which will be rejected. Try to obtain an implicit commitment from the reviewer(s) to accept certain positions or issues. Always get a commitment for an open-minded review and fair decision. No one will refuse to give such an assurance. Later, if they are unreasonable, remind them of their commitment.

**Understand the owner's prior conduct.**   Research the owner's response to other contractors' claims. Some like extensive legal briefs and others prefer stacks of backup documents, but all want a well-written claim document that clearly explains the issues.

**Consider prior relationships.**  Consider the prior relationship between the project team and the claim reviewer(s). Evaluate any hard feelings or distrust that may exist. If it is too late to change perceptions, bring in an outsider to present the claim.

**Determine the reviewers' initial position.**   Determine the claim reviewers' initial position on the issues. Respond to their concerns with adequate documentation and refuting analysis.

**Consider the reviewers' biases and needs.**  Respect the needs of the reviewer(s). They must justify claims to their supervisors and therefore need ample documentation. For example, if they need to analyze cost estimates but lack the experience to evaluate unit price estimates, provide a crew size/ productivity estimate, as it can be broken down and analyzed.

The claim reviewers' biases have a significant effect on how the claim is received. For example, some owner's representatives are skeptical of any contractor's reliability and are unwilling to accept statements at face value. Solid proof or a formal report from a respected, independent claims consultant might help to convince them.

**Consider adverse interests.**  Realize that you may have to convince the reviewer of a fact or conclusion that is adverse to their interest. This is a difficult task. Appeal to a sense of fairness or let it be subtly known that rejection may put them in a defensive position—due to assertions about their inadequate job performance.

**Achieve understanding of the facts before attempting to convince.**  The claim reviewer(s) must understand your claim before formulating a rational opinion. You must first clarify any misunderstanding and then present:

1. The facts

2. The pertinent contract clauses and contract law

3. Your analysis

4. Your conclusions

Clarify any existing misunderstandings. Then clearly describe all the pertinent facts before presenting the analysis. If the reviewer was not involved in the project, the claim must be very clear in explaining the project and the issues in dispute. Even if the reviewer(s) were involved, they may not understand or may have forgotten important aspects of a dispute. In any case, define the issues from your viewpoint.

**Understand personality styles and how reviewers receive and process information.**  Individual personality styles determine how receptive reviewer(s) are to new information or information that conflicts with preexisting opinions. Personality styles also affect how they prefer receiving information. Some prefer written information, while others prefer oral presentations. Some reviewers like to analyze reams of detailed facts, and others rely on personal relationships when negotiating agreements. See Sec. B of Chap. 14 for more details.

Take into account obvious concerns such as possible color blindness and poor eyesight, and ensure that your charts can be read and understood by everyone.

**Meet the reviewer's expectations regarding content, format, and style.**  If submitting a claim different in substance from what the claim reviewer expected, address those expectations and explain the differences. Otherwise the reviewer may misunderstand and reject the claim.

If the claim reviewer is accustomed to a certain format and style, use the same format and style if practical. A new format may interfere with the reviewer's understanding. If changes are needed, make them all at once, explain them fully, and be consistent in format from one claim to another.

### 3. Who to consider besides the claim reviewer

In addition to the claim reviewer, consider the needs of others who may influence the claim's acceptance.

- *Owner's representative.* The owner's representative is often the claim reviewer. If not, the representative's support, acquiescence, or active resistance can influence how the claim is received. In addition, the owner's representative may substantiate or refute your assertions of fact. If possible, use partnering to gain cooperation and support in identifying acceptable recovery theories, obtaining information not available from the contractor's project team, etc.

- *The designer.* If the designer created the problem and is the reviewer/decision maker, you will need to carefully choose your words and characterize the dispute in relatively neutral terms.

### 4. How much to reveal

Generally, reveal all important information to make your case as convincing as possible. However, you may withhold a "smoking gun" or an alternate theory of recovery until a more opportune time during negotiations.

### 5. Submitting change order requests or claims on federal contracts

On federal contracts, contractors must submit under either the changes clause of the contract for change order requests (requests for equitable adjustment) or the Contract Disputes Act for claims. Change order request preparation costs are compensable and should be included in the COR, but interest is not. However, interest is payable on claims from the date of submission (even if initially rejected by the contacting officer and you eventually prevail).

If over $100,000, claims against federal agencies must be certified. When so required, comply with all certification requirements. Also remember that the Truth In Negotiations Act, which applies to claims over $500,000, can result in severe penalties for misstatements.

## B.  Writing, Graphics, and the Human Element

For a successful claim, you must understand basic principles of communication using text and graphics. You must also consider human psychology. You must *persuade.*

## 1.   Remember the human element—emotions and perception

Emotions and the perception of facts are sometimes more important in resolving disputes than rational analysis and the actual facts. You must change misperceptions and resolve emotional issues before you can convince the other party of your position.

**The importance of words.**   Choose the right words—words that give life to your writing and create an image in the reader's mind, words that communicate clearly without evoking an angry response. It is better to assert that the other party's statement is inaccurate or that events turned out differently than expected than to accuse them of misrepresentation. Avoid being fuzzy or too soft, but clearly communicate your position.

A word that has a specific meaning to you may have a very different emotional content to someone else. Use clear, concrete-specific words and simple phrases. Avoid words with negative connotations. Have several people review your claim for verbal "land mines," especially if the claim reviewer is from a different business or social culture.

**Characterization of a dispute.**   The characterization of a disputed issue is very important. For example, state you are *submitting a change order request* instead of *filing a claim,* even if in a contract-specified claims process. The visceral and procedural response is different. In addition, you can request change order preparation and negotiation costs when submitting a change order request. Likewise, a *differing site condition* assertion will be better received by the designer than a claim for *defective plans and specifications.* Be sensitive, however, to contractual and statutory requirements that may require particular language, and never misstate the facts.

**Emotions.**   Consider the emotional response to your actions and submittals. Avoid eliciting negative responses, emphasize the positive, and reinforce the feelings you want to create. Anticipate anger, outright skepticism, or caution. The claim reviewer's fear of possible embarrassment, criticism, or job loss can interfere with your arguments, regardless of the facts. Foster feelings of empathy to overcome negative factors.

**Consistency, expectations, and building trust.**   Consistency in the tone and organization of claims and negotiations and a pattern of settlement will create expectations that contribute to successful resolution. If you have a history of fair but firm proposals for extra work and impact, the claim reviewer will eventually trust your assertions. Likewise, a pattern of timely, quality work assures the owner that a contractor can be trusted.

Do not inflate a claim because you expect it to be arbitrarily cut and hope to end up with the original figure. An inflated claim engenders resistance and is counterproductive. However, include marginal issues and weak arguments, without inflation, to reserve some room for negotiation. The claim reviewer needs to achieve some compromise; otherwise there is no incentive to negotiate.

If negotiating a series of disputes, be especially reasonable in the beginning to establish a pattern that builds confidence.

**Reaping benefits and avoiding negative consequences.**  When "selling" a claim, emphasize the benefits of settling. These benefits will enhance the position of the reviewer and the organization. Benefits include the reviewer's professional reputation, as reviewers are judged on their ability to resolve disputes. The organization benefits because it acquires a reputation for fairness which results in lower bids in the future.

Tactfully suggest the negative consequences of stalemate. (Avoid threats unless you can back them up.) A settlement now saves time and frees up the reviewer for other tasks. Prompt resolution avoids conflict. Resolution avoids a protracted dispute while maintaining good partnering relationships with the contractor and encouraging cooperation on other issues. It has the potential to save money by avoiding the cost of defending against a claim in litigation or arbitration.

**General comments.**  To ensure effective communication, always keep the following in mind:

- Do not assess blame against individuals and avoid referring to individuals by name, especially when criticizing.

- Do not "preach" or attempt to practice law unless you are an attorney. Attorneys shouldn't be legalistic with a nonlawyer, as that is intimidating.

- Acknowledge contractor errors and provide appropriate credit, but don't volunteer negative information that the reviewer and owner's representative may not know.

- Never claim something is valid that is invalid.

- *Do not try to prove the other party wrong unless absolutely necessary.* Rubbing their noses in their mistakes encourages resistance.

## 2.  Written communication concepts and techniques

Written communication is an art and a craft to be learned and polished. Claims must be written skillfully to inform the reader of the facts and to convince the reader of your position, conclusions, and recommendations. Not only do words communicate information, but carefully chosen words create an image in the reader's mind, evoke a positive emotional reaction, and encourage the reader to accept your point of view.

**Narrative text.**  Convert the narrative text prepared during the analysis into a claim document by integrating your chronology and your deductions with the appropriate literary style.

Chronological writing is used for the project narrative or to describe the events as they occur and the development of a problem. It traces events over time.

Deduction is used for argument or proof and is applicable to the analysis sections of the claim. It creates a foundation with the facts and logically builds step-by-step to the conclusion.

A dramatic style is used for stirring the emotions and creating empathy. It tells a story by describing the problem, the people involved, the struggle to overcome the problem, the crisis, and the resolution. Pertinent anecdotal examples from the people involved lend credibility. Dramatic style will give life to your chronology and deductions.

The entire claim document is organized expansively. It starts with a summary, expands to the statement of facts and findings and detailed analysis, and then refers to supporting details in the appendix.

**Basic writing techniques.** Good writing techniques guarantee that your audience not only understands the information but are persuaded by your thought processes and opinions. Keep the following simple rules in mind:

- Organize each section logically, beginning with an introduction and summary, moving through analysis, and then coming to conclusions. Place more detailed analyses and backup information in appendixes. End by emphasizing the most important elements, and make a logical transition to the next section.

- Describe each section and subsection with a clearly worded header that describes the contents.

- Use paragraphs as the basic unit of thought. Begin each paragraph with a sentence that introduces the topic or makes a transition from the previous paragraph.

- Break long paragraphs into two so that the text does not appear overwhelming to the reader.

- Use active verb tense whenever possible.

- Keep your sentences simple, short, and direct. Write about the specific, not the general.

- Don't repeat yourself.

- Express coordinate ideas in similar forms.

- Be upbeat and positive. Avoid words or phrases that are abrupt, argumentative, or condescending or that send negative messages.

- Be direct and authoritative. Avoid vague "weasel" words such as possibly, could, and perhaps.

- Be accurate. Present the facts and avoid generalizations. Do not misuse words. Edit to correct your grammar, spelling, capitalization, and punctuation.

- Choose the right words. Use short, familiar, concrete-specific words. Build a word picture and keep it simple and consistent. Do not use excessive adjectives. Avoid ambiguities and abstract phrases, but use words to create empathy.

- Integrate words and graphics. Simplify some of your graphics, embed them in the text, and discuss the graphics in the text. When the reviewer reads the claim, there is no eye contact, body language, or other nonverbal communication to clarify your writing. Create a natural flow from the text to the graphics so that they work together to communicate effectively.

- Write and design your text by page and facing page so that the text and graphics explained by the text are on the same page or on two facing pages.

**Check the work for errors or omissions.**   After completing the claim, verify that you followed the plan and met your objectives. Have the claim support team check the claim document, as you may be too close to see weaknesses. Check the following:

- Are the section headers consistent with the content and are they descriptive?

- Do the sections fully cover the issues?

- Is the writing concise?

- Are the transitions between sections smooth?

- Are grammar, punctuation, and spelling correct?

- Is the writing clear?

- Are the references to source documents correct?

- Are the graphics clearly drawn, accurate, and understandable? Do they convey the point?

- Do the graphics support the text? Are more graphics needed?

- Are there duplications?

**For additional information.**   For more information on written communication, read Strunk and White's *Elements of Style*[37] before preparing a claim. It is short, to the point, and tells how to simplify writing for clarity and effectiveness.

## 3. Graphical communication concepts and techniques

Graphics visually display data as points, lines, numbers, symbols, words, shading, and color (generally) on a coordinate system. Good graphics are invaluable aids in understanding and analyzing data and in making a persuasive case. Graphics can display great quantities of data in a form that clarifies relationships and reveals patterns, trends, and specifics otherwise not discernible.

**Graphical forms.**   Graphics vary in form. The types most frequently used for claims are:

- Cumulative curves
- Histograms

- Line diagrams
- Bar charts
- Network diagrams

The blot map, sometimes called a measles chart, is another graphic form that is used less often but is potentially valuable. The blot map locates variable information (as symbols, letters, or numbers) on a map and indicates the quantity of the variable by the size of the element. It can indicate a second variable by the shape of the symbol and a third variable by color or shading. For example, blot mapping can display change order or RFI numbers spatially on a key plan of the project to indicate frequency, with number size or color indicating severity of impact.

**Lettering and figure size.**  Exhibits submitted in a claim should be $8\frac{1}{2} \times 11$ inches or $11 \times 17$ inches to fit within the claim document. If necessary, fold larger-size documents such as a detailed schedule network and insert in a pocket of the claim document.

When preparing exhibits for presentations, determine the number of viewers and the distance between them and the charts. An $8\frac{1}{2} \times 11$ inch drawing does well for hand-held charts or one shared by two people; C or D size sheets ($17 \times 22$ inches or $22 \times 34$ inches) work better at a typical arbitration hearing with three arbitrators, two attorneys, and the opposing expert viewing the exhibits from across the table. Always keep your exhibits relatively simple with large lettering. If displaying a chart before a jury from the witness box, use an E-size sheet ($34 \times 44$ inches), simple graphical forms, and large lettering. In all cases, verify that your exhibits are readable under the expected viewing conditions.

**Practical advice for creating graphics.**  When creating graphics, these few simple guidelines increase their effectiveness:

- Spell out words, avoiding codes, acronyms, and abbreviations.
- Run words from left to right, not vertically or in different directions.
- Place brief commentaries on the chart to explain and clarify the data.
- Avoid overly bright, distracting clutter or cute symbols.
- Use a readable type style with a combination of both upper- and lowercase letters and all capitalized letters, depending upon the relative importance.
- Balance line weight and the lettering size and boldness. Make coordinate system lines (e.g., month lines on timescaled diagrams and crew size lines on labor histograms) light and make the data lines darker.
- Use contrasting line weights and line styles (solid, dashed, dot-dash, etc.) to differentiate between lines (see Fig. 10.3 for an example).
- Use subtle, pleasing shading and hatching. Avoid rippling, moiré effect hatching which overpowers the chart and the information being displayed.

- Scale graphics to be wider than they are high. The optimum size is generally about $1\frac{1}{2}$ horizontal to 1 vertical.

- If emphasizing the size of the numbers displayed, draw the top of the chart just above the highest data point so that the chart looks almost full. To deemphasize the size of the numbers, leave a lot of blank space above the highest data point.

- Use simple shading, color, and line coding and label them directly instead of using a legend.

- Select colors for the color-deficient and color-blind (5 to 10 percent of viewers). For example, avoid combinations of red, green, and brown. Use blue, which contrasts with other colors.

- Use red sparingly, to create a sense of alarm, interest, or intense emphasis. Change to orange when red has been overused, to highlight an additional "alarming" fact. Use yellow as a highlight, but never as a main color, as it doesn't show up well.

- Don't use too many colors. When using several colors, look at the combination and use trial and error to avoid jarring effects. Use shadings of the same color or adjacent colors on the spectrum to show quantitative differences or changes over time.

To better understand graphical concepts and visual communication, read Edward R. Tufte's book, *The Visual Display of Quantitative Information.*[38]

## C.   Finalizing and Polishing the Claim Document

The claim document evolves from the narrative text described in Sec. B of Chap. 9. It was created for the preliminary analysis and revised continuously during document production and the detailed claim analysis. The narrative text is converted into the final claims document as described below by deleting some sections and modifying or adding other sections. The following description and procedures are offered as a guide and example only. Use them as a starting point for addressing specific needs.

### 1.   Outline of the claim document

One suggested outline for a typical claim is shown on page 438.
A common, and equally acceptable, outline includes:

- An executive summary.

- A separate section for each claim or group of claims. Each section would cover all aspects of the individual claim, including schedule impacts and costs.

- A schedule section that discussed all scheduling issues not included in the individual claims. It could also include computation of impact and delay damages.

> **Table of Contents for Claim Document**
>
> **Packaging and Front Matter**
> Cover with description of the project and claim
> Cover Letter
> Title Page
> Table of Contents and Table of Figures
> Preface (optional)
> Description of the parties (optional)
>
> **Body of the Claim**
> A. Executive Summary
> B. Statement of Facts and Findings
>   Description of the project and the disputed work
>   Analysis of applicable contract terms, and contract law, if provided by counsel
>   Description of the original strategy, schedule and expectations
>   Description of progress and problems
>   Overall analysis and conclusions
> C. Detailed Analysis of Issues
>   Issue #1 (with subsections for summary, description, analysis, conclusions, and chronology)
>   Issue #2
>   Issue #3, etc.
> D. Schedule Analysis and Summary of Delays and Schedule Impacts
> E. Computation and Summary of Damages
>
> **Appendixes**—with referenced source documents, prepared exhibits and detailed calculations

- A cost section that summarized the costs from the other sections, added markup and below the line costs, and ended with a cost summary.

- Appendixes.

## 2. Convert the narrative text into a claim document

Compare the above outline with the outline for the narrative text in Sec. B.2 of Chap. 9 and you will see how narrative text is converted into a claim document. The steps to create a claim document from narrative text are as follows:

1. *Administrative matters.* Remove section A of the narrative text from the body of the text. Save it as a working document to help you coordinate the final claim preparation effort.

2. *Executive summary—section A of the claim.* Subsection B.1 of the narrative text, Executive Summary, becomes section A of the claim after editing for clarity and making last-minute changes.

3. *Statement of facts and findings—section B of the claim.* Rewrite the balance of section B of the narrative text as the Statement of Facts and Findings. Label it section B of the claim document.

4. *Detailed analyses of the issues—section C of the claim.* Many of the issue analyses in section C will not be included in the claim document. They should be archived or deleted, along with minor issues. Rewrite the issue analyses included in the claim to remove privileged information, cautionary statements, comments, questions, etc.

5. *Schedule analysis and summary of delays and schedule impacts—section D of the claim.* Section D of the narrative analysis becomes section D of the claim, after editing and removal of privileged information, cautionary statements, comments, questions, etc.

6. *Computation and summary of damages—section E of the claim.* Section E of the narrative text becomes section E of the claim, after editing and removal of privileged information, cautionary statements, comments, questions, etc.

7. *Summary review and chronological notes.* Sections F and G of the narrative text are normally removed and archived. However, section F, the Chronology, may be included in the Appendixes to demonstrate the level of effort expended.

8. *Appendixes.* Some of the appendix material will be removed and archived; other material may be edited or rewritten and included in an appendix.

## 3. Format

Single space the claim with ample margins for easy reading and the reviewer's annotations. Double spacing is not recommended. Compactness is very important for clarity and allows the reader to view more related information per page. Use a font type size of at least 10-point and an easy-to-read typeface. Use bold, underlined, or larger-size type for section headers and subheaders. Write descriptive headers that explain the contents of each section. Place the project and claim titles, page number, and so on, in a footer or header.

Print on both sides of each page. The reader can see more information at one time and more easily refer back to earlier statements. The whole document looks slimmer and more manageable. When printing on both sides of the paper, combind the document so that it can be fully opened and laid flat.

## 4. Packaging and introduction

The packaging normally includes the following:

1. *Attractive cover.* The cover includes the title of the project, the name of the project owner, the contractor, date, and a concise title that describes the claim. Use cardstock for the front and back cover and a plastic comb binder or a three-ring binder. Consider using a construction photograph or simple art work from the contract documents as design elements to show your commitment to the project and to create a positive initial impression in the reviewer's mind. Although some claim preparers use a leather cover or other

expensive cover feature on large claims, this may have negative connotations. Consultants should include their company name and logo on the front cover if it lends credibility to the document and their client concurs.

2. *Concise, well-crafted cover letter.* The cover letter should project a cooperative attitude, express a desire for a mutually acceptable resolution, and briefly state the key elements of the claim. Briefly tell how the claim was prepared or include a separate preface explaining how it was prepared.

   The cover letter may also transmit an ultimatum, such as: "If this claim is not resolved within thirty days as described in the contract, it will form the basis of a lawsuit to be filed thirty-one days after its submittal."

3. *Title page.* The title page usually duplicates the cover, but omits the art work.

4. *Table of contents.* Provide section and subsection header descriptions with page numbers. Include a listing of figures, exhibits, and appendixes with page numbers.

5. *Preface.* If a separate preface is included, briefly state not only who prepared the claim, but also the preparer's qualifications, the sources investigated, and how the claim was put together. Preparation by an independent expert lends credibility to a claim and signals that you are serious, but it escalates the dispute.

6. *Description of the parties, definitions of special terms, and abbreviations.* Include a separate section with introductory information either preceding the body of the claim or in section B.1 of the claim. Describe all the parties involved in the claim, especially if the claim will be reviewed by individuals not familiar with the project. Establish the contractor's experience, expertise, financial capability, and thoroughness when bidding the project. Also define any special terms and abbreviations.

### 5.  Executive summary—section A of the claim

The executive summary is the most important element of your claim document. Make it clear and convincing. Have the claim support team review it for grammar, structure, style, and content. If necessary, hire a professional writer to edit it. This is part of polishing your document.

The Executive Summary should be one to four pages in length, with a maximum of ten pages for a large, complex claim. Begin with an overview and summary (one to three paragraphs), and focus on the most important issues. Embed key figures and a table of the damages in the text, or reference one or more graphics for easy access. The executive summary must stand on its own, because the decision maker may read only the summary and rely on staff to confirm that the balance of the claim supports the summary.

Most claims depend on one or two major, strong issues. Address them in the summary to create a clear, strong, favorable opinion in the reader's mind. Present the strongest issues so clearly and forcefully that they are not debatable.

6. **Statement of facts and findings—section B of the claim**

Limit the statement of facts and findings section to 20 or 30 pages. If necessary, provide additional details in the issue analyses section or in an appendix. Organize section B logically with a structure of subsections and sub-subsections, numbered for reference and easy access. Reference statements of fact to source document(s) as noted in Sec. B.12 of Chap. 9.

Remember that the claim reviewer will probably not be as knowledgeable of scheduling, costing, and construction methods as yourself and will certainly not know the dispute and the job as well as you do at this stage. Clearly explain all the issues without being condescending, and make them interesting.

Section B of the claim will normally consist of the following subsections:

1. *Description of the project and the disputed work—determination of the facts.* Assume that the reviewer is not familiar with the project. Create the right atmosphere and focus by describing the project in your terms and in relationship to your claim. Start with the project statistics—name of the owner and designer, size of the project, etc. Provide the bid opening date and bid spread (if favorable), award date, and notice to proceed. Describe the conditions under which the project was built, and summarize what happened along with the problems encountered. Include reduced-size plans or photographs, with labels on areas where disputes arose. Include the pertinent facts needed to prove entitlement, along with references to the source documents.

2. *Description of the original strategy, schedule, and expectations—determination of the facts.* Explain the overall strategy of constructing the project, and summarize the as-planned schedule, focusing on the major phases, key milestones, and critical activities affected by the dispute. The original schedule is a key element of this section; insert a summary timescaled network diagram or bar chart in the text.

3. *Description of progress and problems—determination of the facts.* Summarize the progress of construction and briefly summarize each major claim issue. Provide a summary bar chart of planned versus actual progress, or actual versus would-have-been progress.

4. *Analysis of applicable contract terms and law—the theory of recovery.* You must convince the owner (and their attorney) that you have a legal right to additional payment or a time extension, and would prevail in arbitration or court. Identify the applicable contract provisions and law. Determine liability and at least one theory of recovery for each issue. If you are working with an attorney, this section will be replaced by their legal briefs.

5. *Analysis of linkage from initiating event to resulting costs—causation.* Trace the chain of events and activities from the initiating event, action, omission, or condition for which the owner is responsible to the resulting costs suffered by the contractor.

6. *Schedule analysis—causation.* If scheduling is an issue, summarize the schedule analysis and describe the delays and whether they are compensable, excusable but noncompensable, or nonexcusable. Refer to section D of the claim for detailed analysis of the schedule.

7. *Cost analysis—proof of damages.* Summarize the cost computations and refer to section E of the claim for detailed analysis of the costs.

8. *Overall analysis and conclusions.* This summarizes all the elements of the claim—including the issue analyses, schedule analysis, and cost computations. This section should include a one-page summary of costs.

## 7.   Detailed analyses of issues—section C of the claim

The issue analyses may include the individual claims, the governing contract terms, how the work was planned, how the changes affected the plan, and how the work was actually accomplished. The analyses form the major portion of the claim document, as detailed backup information to the findings. Edit the narrative text to remove cautionary statements, comments, questions, and material that is inappropriate for the claim document.

Each issue analysis section includes sub-subsections for a summary, analysis, conclusions, and chronology.

## 8.   Schedule analysis—section D of the claim

The schedule analysis creates or modifies the as-built schedule and compares it with the as-planned schedule to develop the "would have been, but for..." (WHB) schedule. The WHB schedule is then used to determine the time extension due and the days of compensable delay. The analysis also establishes entitlement for impact and acceleration, if any.

## 9.   Computation and summary of damages—section E of the claim

The cost section details computations of costs and a summary of damages that recaps the damages for each issue and adds markup, interest, bond premium, taxes, and "below the line" amounts due (retainage, etc.).

The summary of damages should occupy a single page and be placed at the end of the cost section. It should also be further summarized as a table in the executive summary. In some cases, the cost computations will be included in the analysis of each issue instead of in a separate section.

## 10.   Placing source documents and other exhibits in the claim

Source documents and prepared exhibits can be placed in one of four different locations in the claim document:

- *Embedded in the text—with the reference.* Embed smaller exhibits in the body of the claim with the text describing it, so the reviewer can examine it while reading. This is effective for figures occupying half a page or less.

It is not effective if the exhibit is too large or too complex. If a portion of a source document is important and understandable out of context, extract and embed it in the text with quotation marks and a source reference.

- *Inserted in the text—opposite the reference.* Insert the most important full-page exhibits in the text on the facing page opposite the description. This works well for crucial exhibits but interrupts the text if more than a few exhibits are inserted.

- *Placed in the appendix—in sequential, numbered order.* The majority of the referenced source documents can be placed in an appendix, numbered, and referenced by an exhibit number in the body of the text. However, this requires the reader to search for the document in the appendix and then flip back and forth to study it while reading the text. The preferred alternative is described below.

- *Separately bound—in sequential, numbered order.* The preferred alternative for full-size exhibits that cannot be embedded in the text is to bind them separately from the claim document. They can then be studied concurrently while the text in the claim document is examined.

Although most project documents will not be referenced in the claim, all must be accessible in the source document files. They may be necessary for rebutting arguments raised by the claim reviewer or for researching information obtained during negotiations.

## 11.  Identifying source documents and other exhibits by exhibit number

Assign a sequential number to exhibits referenced in the text, in the order in which they are presented in the claim. Place the exhibits in an appendix in the same order. If the exhibit is a source document referenced in the narrative text, maintain the date/document code reference (for location in the chronological document file). If in arbitration or litigation, also reference the source documents to the Bates number, if known. Paginate multipage exhibits and refer to the page numbers if needed.

Prepared exhibits (demonstrative evidence) can be identified with letters to differentiate them from the numbered source document exhibits. They may be referenced several times in the claim, and the reviewer can remember the letters better than the numbers used for source document exhibits. Voluminous cost records and large documents that are frequently referenced work well in separately bound appendixes. They are then referenced by the appendix and page number.

Wait until the last minute to number exhibits, as the order in which they are referenced in the text may change. If adding exhibits after the order is set, use the next available number. Consistently place the exhibit numbers in the lower right-hand or upper right-hand corner of the source document to facilitate locating an exhibit.

**If in arbitration or litigation.**  Documents used in depositions, arbitration, or litigation hearings are assigned a sequential exhibit number by the attorney introducing them, plus a letter indicating plaintiff or defendant's exhibit. To avoid confusion between the attorney's exhibits and the claim document exhibits, use a different letter prefix for claim document exhibits (if presenting your claim document as evidence in a hearing). Ideally, the attorneys will agree on a numbering system early, and the same exhibit numbers used for deposition exhibits can be used for hearing exhibits and for your claim document exhibits. This is seldom practical. Consequently, you may need to cross reference your exhibit numbers to your attorney's exhibit numbers when testifying in arbitration or litigation hearings.

## 12.  Appendixes

Place the exhibits in an appendix. Place detailed analyses (when the body of the text exceeds 100 pages) in a second appendix; detailed cost records (e.g., timesheets) in a third appendix; and subcontractor pass-through claims (if those claims are not separately bound) in additional appendixes.

Tab the beginning of each appendix for easy access, or tab groups of documents within each appendix. The exhibits can be tabbed with the section of the claim where first referenced. To speed accessing an exhibit during negotiations or in hearings, sequentially page stamp each appendix. This will allow you to tell the claim reviewer or arbitrator to "look on page 145 of Appendix A to find exhibit 67."

## D.  Preparing Exhibits

Exhibits include copies of source documents and demonstrative evidence prepared to illustrate the facts and your conclusions. Demonstrative evidence is a legal term generally referring to prepared exhibits that communicate information without requiring explanatory testimony. Most exhibits are incorporated into the claim document, usually in an appendix, although some will be enlarged for use in presentations. You may also present physical exhibits (e.g., samples) as well as computer simulations, videos, photographs (either bound in the claim or glued to a large foamboard), slides, and overheads.

## 1.  Highlight key portions of source document exhibits

The most effective means of focusing a reviewer's attention on the important elements of an exhibit is to highlight them with a highlighter pen. This also avoids the confusion of searching through a page for the significant information. Issues that are contradictory or could be confusing can be downplayed by simply not being highlighted. However, avoid appearing to distort the facts.

Yellow highlighting is usually recommended, as it won't copy. If you want a duplicate copy for filing or multiple copies for a hearing, use a highlighting color that leaves a medium to light shadow when copied. This eliminates the need to manually highlight each copy and prevents the possibility of inadvertently creating differences between multiple copies.

## 2. Identifying exhibits

During analysis, source documents are identified by *Type-WhoFrom / To-Date* (e.g., L-GC/ENG-14Aug97). When converting the narrative text to a claim document, add exhibit numbers. As noted, use letters for prepared exhibits to differentiate them from source documents.

## 3. As-planned schedule exhibits

Chapter 10 describes the tools (including graphs and programs) for analyzing schedules. Most of the graphs also can be used as exhibits and are described below, with references to examples in Chap. 10.

**Display the original as-planned schedule.** If an as-planned schedule was prepared, include it as an exhibit, to establish the validity of subsequent exhibits. If it is difficult to read, consider computerizing it and including the computerized schedule as an exhibit.

For presentations, if the original as-planned schedule is not a full-sized drawing, consider having it blown up to approximately the same size as the other full-sized scheduling exhibits.

**Convert bar charts to network diagrams by creating a connected bar chart.** If the original schedule is a bar chart, consider creating a connected bar chart exhibit that shows the relationships between tasks, as explained in Sec. E.3.1 of Chap. 10 (Fig. 10.22). The connected bar chart should have the same general order of tasks as the original bar chart but color and notations can be used for clarity. Normally this is done by computerizing the schedule, adding relationships, and scheduling. However, it can be done manually. The purpose of the connected bar chart is to explain the network logic, in addition to facilitating the transition from the bar chart format to a CPM network format. Support the presentation with an analysis describing how the network logic was determined.

**Create a modified as-planned schedule.** Original as-planned schedules often have obvious errors that would have been resolved during construction. These should be corrected, as explained in Sec. E.3.3 of Chap. 10, and the result can be included as an exhibit. When modifying an original as-planned schedule, record the rationale(s) for the modifications and include them in an appendix to the claim.

**Redisplay the as-planned schedule as a timescale network.** Timescale networks, as explained in Sec. E.3.4 of Chap. 10, are superior for understanding a network. Consider converting the as-planned schedule to a timescale arrow diagram, which makes it much easier to follow the logic and to understand what was planned. Group the tasks in subnetworks, label the network, and annotate it with additional text. See Fig. 10.20.

**Present a summary as-planned schedule.** To explain the plan for construction, summarize the as-planned schedule with a dozen or more activity hammocks and milestones to help explain the overall sequence of construction. If appropriate, add percent complete curves or resource histograms, but avoid making the summary too complex. The result should be simple enough to reduce in size and embed in the text.

## 4. As-built schedule exhibits

**Optionally include schedule updates prepared during construction.** If the original as-planned schedule was updated with progress and changes during construction, the final updates will show actual progress as reported. Consider using one or more updates as exhibits to show status as of specific dates.

**Optionally include short-interval look-ahead schedules prepared during construction.** Some of the short-interval schedules may be used as exhibits, to demonstrate specific facts. This will require, however, that the short-interval schedule tasks be referenced to activities in the main schedule.

**Optionally include schedules to completion prepared during construction.** If a schedule to completion was prepared, it may be included as an exhibit to document how the project was completed.

**Either include an as-built schedule from the final update with adjustments.** The final schedule update can be used as the as-built schedule, after adjustments to supply start and finish dates for the final activities as explained in Sec. E.4 of Chap. 10. The final update must be adjusted with the short-interval look-ahead schedule information, supplemented by activities from the schedule to completion, and corrected for logic errors and minor discrepancies. You should record all corrections and added or deleted activities, and provide the information in an appendix to the claim.

**Or include a detailed as-built schedule.** If accurate updates were not prepared during the course of the project or if more detailed information is needed on construction progress, prepare a detailed as-built schedule as described in Sec. G of Chap. 10. Include the detailed as-built schedule in an appendix to the claim to validate the condensed as-built schedule. For presentations, a clear acetate overlay on the detailed as-built, with felt-tip pen coloring of the activities, can be used to link the detailed as-built to the condensed as-built.

**Create a condensed as-built schedule if a detailed as-built schedule is created.** Condense the detailed as-built schedule with its hundreds or thousands of events to approximately the same level of detail as the as-planned schedule. The resulting condensed as-built schedule uses the same activity descriptions, numbers, and layout where possible but has more tasks than the as-planned schedule. Record the rationale for each condensed activity and include in an appendix to the claim.

**Create and include a summary as-built schedule.** Summarize the as-built schedule to a higher-level summary schedule with only 20 to 40 summary activities, graphical highlighting of key events (milestones), graphical annotations to indicate delay and impact periods, and additional text. The result should have the same layout and the same scale as the condensed as-built schedule, with color to highlight and simplify its logic. It may include superimposed resource histograms and percent completion curves. It can be presented in a modified timescale arrow diagram format, similar to a bar chart, or as a connected bar chart. Some scheduling programs can generate this type of chart automatically through a roll-up feature.

5. **Would-have-been, but for...schedule exhibits**

**Include the would-have-been schedule and create a summary would-have-been schedule.** You may include two would-have-been (WHB) schedules—one at the same level of detail as the as-planned and condensed as-built schedules and the other at a summary level. Their general layout should be the same as the other schedules and should be based on the as-planned schedule but modified by the events chronicled in the as-built schedule. The WHB schedules must represent the most likely course of action, absent the claimed event(s).

**Include an as-planned to as-built comparison schedule.** The as-planned to as-built comparison, described in Sec. E.5.2 of Chap. 10, is a key exhibit for establishing a foundation for the would-have-been schedule. When generating the WHB schedule, document all variances from the as-planned or as-built schedules. Failure to include contractor errors will lead to mistrust and allegations of an attempt to mislead. Include the supporting documentation in an appendix to the claim.

**Create and include a summary as-planned to as-built comparison schedule.** A summary comparison schedule may better convey the important differences between the two schedules than a more detailed comparison.

**Include an as-planned to as-built resource histogram.** Juxtaposing the as-planned versus as-built resource demands will often help explain how the project was impacted.

When comparing two schedules, most software shows the activities of one of the schedules above and larger than the activities of the other schedule. You can use color and hatching to differentiate between the schedules. Hatching is important in case the reviewer is color-blind or if black-and-white copies are made.

**Include an ELIPSE schedule if one was prepared for analysis.** The ELIPSE schedule described in Sec. H of Chap. 10 displays additional information that often explains why work progressed as it did. It can be used to refute counterclaims or alleged contractor errors and to show the effect of multiple RFIs and changes.

**Create and include a banded bar chart if it illustrates your argument.**  A banded bar chart schedule highlights the difference between the as-planned and the as-built schedule. For details, see Sec. 1.2 of Chap. 10.

**Include the would-have-been schedule and the delay computations.**  As explained in Sec. E.5.7, create the WHB schedule, record the delay computations, and include the schedule and computations with the claim.

**Include the as-built to would-have-been schedule comparison.**  Comparing the as-built schedule to the would-have-been schedule will determine the compensable delay, as noted in Sec. E.6 of Chap. 10. It will also show the differences in working conditions, which give rise to impact damages as computed in Chaps. 11 and 12.

## 6.   Other schedule exhibits

**Fragnet schedules.**  Fragnet schedules are portions of a network diagram that show only the relevant portions for easier analysis and understanding. Extract them from the full network, blow them up in size, and display them as separate exhibits to explain specific events and issues. Some software does this automatically.

**Computer simulation.**  Consider preparing computer simulations to show what happened and why.

## 7.   Cumulative curve exhibits for cumulative percent complete, cost, and earned value

Cumulative curves are used to show progress (percent completion, cash flow, and earned value) and to compare planned with actual progress. An abrupt change in the slope of the curve (e.g., a flatter slope, indicating slower progress) coinciding with an event (e.g., an alleged cause of the delay) followed by a sharp rise in the curve when the issue is resolved supports the contention that the event caused the change.

Cumulative curves can overlay a summary schedule to graphically illustrate the pace of construction. The as-built curve can be compared to the as-planned or would-have-been curves to illustrate the differences.

## 8.   Histogram (periodic curve) exhibits for crew size, labor hours, and productivity rates

Histograms (periodic curves) are most often used to show planned and actual resources used (e.g., the number of workers). The planned staffing of a project is ideally a steady increase of crew size followed by a constant crew size and then a slow decline to project completion. This can be compared to actual staffing which (due to impact) may be a smaller initial crew size during good weather, excessive peaks during inclement weather, and wildly fluctuating levels through to completion. This comparison is especially effective if the start and finish dates of the impact are plotted on the diagram and correspond to reduced staffing and then excess staffing to make up for lost time.

Histograms of resource use or rainfall can be overlaid on a summary schedule and combined with a cumulative curve (of progress) to explain what happened. Comparison histograms of as-planned (or would-have-been) crew size versus actual crew size will establish the foundation for impact and inefficiency.

Histograms can be effectively used to show other environmental conditions such as snow or river levels.

## 9. Tabular schedule report exhibits

Tabular data (numbers in tables with some explanation) are usually needed to support testimony and other exhibits and are used extensively in claim documents or written reports. They are seldom used as demonstrative evidence except to lay a foundation for other exhibits, as complete understanding requires careful study and considerable experience. Presenting tabular data is best done in conjunction with graphics.

## 10. Cost exhibits

Most cost exhibits are simple tables or cost columns. Their design can significantly affect how easy they are to understand. Graphics can enhance their usefulness.

Cost data are normally presented on spreadsheets in tabular form, to cover a large amount of detail. However, graphical representations of cost data are far more effective for communicating. For example, variations in productivity rates (cost or labor hours per unit of output) can be plotted on a vertical scale with a line drawing for each reporting period. The learning curve effect (increased efficiency based on experience) becomes apparent, as do abrupt changes in productivity at the beginning and end of the period of impact.

Combine labor productivity data with crew size, RFIs, drawings issued, etc., to show the impact of those variables on productivity. Or use bar charts to display jobsite overhead labor—planned versus actual—for extended overhead claims. Include a cumulative curve of the excess hours or dollars at the bottom to aid in the presentation.

**Multiple small changes.** To show the effect of multiple small changes, create a spreadsheet listing the changes. Group the changes by type of work, work area, or responsible subcontractor, and include:

- A sequential number (if not previously assigned).
- A clear title/description.
- The cost.
- What specification sections and drawing numbers were affected.
- The subcontractor(s) affected.

- When the directive was received or when the need to do the work was recognized.
- The period of delay while waiting for resolution of the problem and delivery of materials.
- The start and finish dates (if the work was done).
- References to RFI numbers, notices of claim, etc.
- Comments.

Plot spreadsheets on large drawing sheets for presentations, and include a reduced-size spreadsheet in the claim document. Also, mark a map of the project with the claim numbers and symbols, indicating where the changes occurred. If suitable, draw the changes with a colored pen or shade the areas affected. This graphical image conveys the extent of the changes and their location on the project. Depending upon the data, the viewer will have an impression of restrictions due to items being on hold, congestion, out-of-sequence construction, etc. Use variable-size circles or other symbols to indicate the size or impact of the change and different colors to indicate either different time frames or the different crews affected.

### 11.  Emphasizing critical cost elements

As noted, to emphasize the magnitude of numbers (i.e., dollar cost), scale the chart so that the peak is near the top of the chart. To deemphasize the cost magnitude, scale the line graphic to fill only one-half or one-third of the chart space.

### 12.  Presenting detailed actual cost data

Detailed actual costs are best presented on a series of nested spreadsheets. Categorize the costs in a logical manner, post to a spreadsheet, and subtotal by category. Each line entry should include the basic facts: the transaction date, any reference numbers, description, amount, and comment, if warranted. Special items should be footnoted. Title each separate spreadsheet and reference it in the summary spreadsheet.

The documents supporting the costs should be organized by category, copied, and then bound in the same order as they are listed in the spreadsheets.

### 13.  Photographs

Photographs show what happened on a project and the impact of conditions resulting in extra costs. Photographs of mud, rain, snow, crowded conditions, etc., generate a feeling for the problems experienced. Organize the photographs chronologically or group them by issue, with the date on the front of each photograph. Label items of interest on the photograph with a caption describing what happened. Show the location of each photograph on a plan of the work area.

Three- by five-inch prints normally give adequate definition for inclusion in a report or for individual review. Four prints can be pasted on $8^{1}/_{2} \times 11$ inch paper with brief captions and copied in color. Aerial photographs (which in some areas can be obtained for around $200 per flight if ordered monthly) and other items needing more definition can be printed at $8 \times 10$ inches and pasted onto $8^{1}/_{2} \times 11$ inch paper for inclusion in a report. Larger blowups or slides may be used for presentations and hearing exhibits.

Digital photography is becoming less expensive and more popular but easily altered. If in arbitration or litigation, the opposing party may challenge them on that basis.

## 14. Videos

Video cameras are easy to use, small enough to carry on a jobsite, and economical. Many project owners and contractors use videos to document the general course of construction and specific problems. A few minutes of videotape illustrating impact and difficult working conditions can be extremely effective in demonstrating the general extent of damages, sometimes out of proportion to the actual impact.

Careful editing of the videotape and a clear narrative can focus attention on important elements. In some cases, freezing selected views, inserting text (supers) and graphics, or simulated events makes the presentation more persuasive. If presenting edited material, make all video field tapes or files available in unedited form.

Videotape can be printed as small stills and the photos mounted as a series on paper or on a presentation board to show overall progress or a specific operation.

## 15. Timelapse photography

Timelapse photography on videotape condenses a full day's progress into a few minutes to provide a quick review of daily progress. It also clarifies work patterns and inefficiencies for nonexperts and can be analyzed with time-and-motion study techniques to determine productivity rates and impacts.

## 16. Pie charts

Pie charts can display costs, resource use, or other numerical information. Pie charts help to focus attention on differences between cost elements and the relative distribution costs or resources.

## 17. Site plans and phasing plans

Use a simplified plan of the project, with boldly lettered descriptions of each work area and the locations at which specific claims occurred or where photographs were taken, to facilitate understanding. Use multiple sheets, large bold text, hatching and halftones, and color to show changes in status over time.

## 18. Drawings

Labels on overall plan or elevation drawings and details help to explain project specifics, especially if they are color-coded. More sophisticated analyses use computer-generated CAD drawings to show impacts and areas affected by change. Isometric drawings showing perspective can often clarify a complex detail. In some cases, a not-to-scale drawing can emphasize the important points better than a scaled drawing.

## 19. Presentation exhibits

The purpose of presentation exhibits is to clearly communicate key concepts and to support the oral presentation. Although they may be blown-up versions of exhibits from the claim document, it is usually better to create them specifically for presentation by simplifying existing exhibits. Or, create short, bulleted lists to support an element of your presentation. Use large lettering and simple graphic elements.

Presentation exhibits should normally be on B or C size sheets (11 × 17 inches or 17 × 22 inches) for table top use to a small group, or E size (34 × 44 inches) for larger spaces. Glue them to foam boards, or attach them with binder clips to a flip chart.

## E. Including Subcontractor and Supplier Claims

When submitting a claim, contractors should include subcontractor claims or specifically exclude them. The former is strongly recommended, as few owners will settle part of an issue and leave the remainder open.

## 1. Timing

Subcontractors are often late in submitting claims. General contractors must not only alert subcontractors to prepare a claim but must track their efforts to ensure coordination of efforts and timely submittal. Late submission by a subcontractor delays resolution of the general contractor's claim.

## 2. Inclusion in general contractor's claim

In many cases, the general contractor will incorporate the subcontractor's claim into the body of the claim document. In other cases, the subcontractor submits an independent claim, and the general contractor refers to it in the text as appropriate, lists it in the cost summary, and adds general contractor markup. Generally, the decision on how to present the subcontractor's claim depends upon its relative and absolute size, the sophistication of the subcontractor, and the subcontractor's ability to prepare a viable claim.

## 3. Supplier claims

Suppliers will sometimes have disputes over the quality of materials furnished, but they seldom make acceleration, delay, or even extra cost claims. General contractors often have claims against suppliers for late material delivery and will backcharge them for the costs.

# 14

# Phase 6—Presenting the Claim and Negotiating an Equitable Adjustment

Once a claim has been prepared, it is presented to the owner and used to negotiate a reasonable settlement. This chapter addresses how to present and negotiate the claim, and contains the following sections:

See Secs. A and B for an introduction and an overview of human behavior and its impact on negotiation. Sections C to E describe the process of preparing for and initiating negotiation. Section F addresses negotiation strategy and tactics. If you are reading this chapter from start to finish, you might want to break your review into related sections (A–B, C–E, and F).

The process of presenting and negotiating a claim is summarized in Fig. 14.1.

## A. General Principles, Style, and Strategy for Negotiators

For greater success in negotiating claims follow the procedures described in Fig. 14.1. In addition, preserve your contract rights and be prepared to mediate and then arbitrate or litigate if negotiations fail.

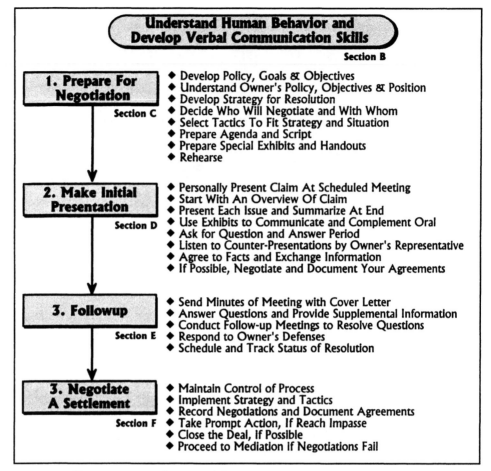

**Figure 14.1.**  How to present and negotiate claims.

## 1. Use a partnering approach and win/win negotiation

This chapter assumes:

1. You are negotiating with someone who is acting in good faith and is empowered to commit an organization.
2. That a partnering approach and win/win negotiation will be used to reach a fair settlement that meets the basic needs of both parties.

Although other negotiation methods (e.g., win/lose) may be effective in some situations, partnering and win/win negotiation usually provide better results.

## 2. Develop your own negotiation style

Style is your approach to negotiation. It should complement your basic philosophy and personality and be used consistently, with slight modifications when necessary.

There are five basic negotiation styles, roughly paralleling the five most common reactions to conflict. Identify your own style and your counterpart's before adopting a strategy and tactics for a specific negotiation. The styles are:

- *Combative or win/lose.* This style pushes for victory, without considering the other party's needs. Although successful against weaker parties or those using the concessional style, it often provokes anger and becomes counter-productive. Combative negotiators tend to anger easily and like to intimidate. Many are closed-minded to alternatives and focus on quick resolution, a weakness exploited by more skillful negotiators.

- *Concessional, superficial, and conflict-avoidant.* These three styles are dysfunctional. Concessional negotiators avoid confrontation, are too ready to share information, and focus on maintaining relationships. They are easy to negotiate with but should not be pushed to where they resent the agreement or their organization repudiates it. Superficial negotiators tend to smooth over differences and avoid dealing with conflicts. Conflict-avoidant negotiators are difficult to bring to the table. Conflict-avoidant and superficial negotiators must be constantly pressured to negotiate and settle.

- *Win/win or problem-solving.* All parties benefit from a win/win style. It builds relationships based on equality and common interests. Win/win negotiators are open-minded to alternatives and creative solutions, allowing many more opportunities for a positive result for both parties.

As noted, a win/win style is the most likely to succeed in the long run. However, you may need to modify the win/win style if your counterpart uses a different style. For guidance on win/win negotiation read *Getting to Yes: Negotiating Agreement without Giving in,* by Roger Fisher, William Ury, and Bruce Patton.[39] It is recommended reading for all negotiators. Their book recommends determining both parties' fundamental interests, determining how the interests overlap, expanding the benefits or creating new benefits, and then sharing those benefits.

## 3.  Consider the relative strengths of the parties

When negotiating, each party has advantages and disadvantages that determine their relative strength and must be considered by the negotiator.

**The contractor's advantages and disadvantages.** Contractors have several advantages over the owner:

- The contractor is the initiator and can characterize the issues in the most favorable light. Contractors can direct the focus, structure, and pace of negotiations, although the owner can thwart those efforts if the process isn't carefully managed.

- Contractors are generally more knowledgeable of the facts and have better cost data.

- The contractor is needed to finish the job as quickly as possible, if work is ongoing.
- Contractors have lien or bond claim rights, which can damage the owner's financial reputation and encourage settlement.

Contractors have two distinct disadvantages:

- The contractor's primary disadvantage is the need to pay subcontractors, suppliers, and labor. This creates tremendous pressure to settle quickly— usually for less. To avoid this problem, promptly submit forward-priced change order requests, push for resolution, and promptly file a claim if the request is rejected.
- The contractor naturally wants to avoid harming business relations and may feel hamstrung in negotiations as a result.

**The owner's advantages and disadvantages.** Owners have three major advantages:

- The owner controls the money and can force a compromise if the contractor needs prompt payment and cannot wait for extended negotiations.
- The contract invariably favors the owner, as it was prepared by the owner or designer.
- The owner normally has a separate contract with the designer and controls the flow of information from the designer to the contractor.

Owners have several disadvantages:

- The owner is reactive rather than proactive.
- If the owner is the government, the representative's negotiation flexibility is limited by numerous regulations.
- The owner's representative needs to satisfy superiors while maintaining a working relationship with the contractor.

**Subcontractor or supplier advantages and disadvantages.** Subcontractors and suppliers have numerous advantages, including:

- Subcontractors have superior knowledge of their work, trade practices, and costs.
- Subcontractors are needed to finish the job and cannot be replaced without delay and extra costs.
- Subcontractors' liens can damage the relationship between the general contractor and the owner and between the owner and the lender.
- The amounts that subcontractors claim are often relatively small compared to the total claim; the owner may settle subcontractors' claims in order to initiate movement toward settlement of the overall claim or to reduce exposure and to focus on the general contractor's claims.

- General contractors seldom have sufficient time, resources, or incentives to carefully examine a subcontractor's proposal. The result is a pass-through claim that is quickly endorsed with little review by the general contractor.

- The owner's claim reviewers often don't scrutinize price proposals from subcontractors as closely as they examine proposals for work by the general contractor's forces. Reviewers have less knowledge of specialty trade work practices and prices, are reluctant to dispute an expert in a field which they don't understand, and usually accept a number that appears reasonable if they trust the subcontractor. (To build this trust and facilitate acceptance of their change order requests and claims, subcontractors need to establish their expertise and reputation for fairness early.)

Subcontractors' and suppliers' disadvantages are:

- They have no direct access to the owner, without the general contractor's approval.

- They are often left out of the communication loop and not involved in scheduling.

- They depend upon the general contractor to submit (and in some cases, negotiate) their claims.

- They are paid last, and payment is often delayed, especially if there is a dispute.

- Their payments are subject to backcharges by the general contractor until a dispute is resolved.

- Retainage may be held until the end of the job, even though their work is done early.

- Subcontractors depend upon the general contractor for future work, which makes them reluctant to pursue claims that could damage business relationships and the prospect of future business.

- Third-tier subcontractors and suppliers are too far down the line on government projects to make a Miller Act bond claim, as only subcontractors, subsubcontractors, and suppliers to the general contractor and subcontractors are covered by the act.

**Objectivity, perspective, and perceptions.**  Be objective in identifying your own and your counterpart's relative strengths and weaknesses to avoid failure or an inadequate settlement. Then use the strategies and tactics in sec. F to counter the disadvantages of a weaker position. Remember that the owner's negotiator has a different perspective and may misperceive the facts and the relative strength of your position. If so, you need to correct misperceptions before negotiating.

## 4.  Present claims as soon as possible

*Contractors should submit a change order request or claim as promptly as possible,* even if the owner has verbally denied responsibility. *Immediate action may pressure the owner to resolve an ongoing problem and stop the impact.* With the right effort and a partnering approach, you also may be able to enlist the owner's and designer's help in resolving problems.

Many contractors don't initiate the disputes process when a change order request is denied but wait until the work is complete before filing a claim. The reasons for this include:

- Unwillingness to antagonize the owner's representative, who may retaliate by tighter inspection and contract interpretation.
- Lack of staff time to prepare the claim.
- Concern that the facts aren't yet fully known and future impacts may occur that won't be reimbursed if the claim is settled now.

However, the disadvantages of waiting outweigh the apparent benefits. The disadvantages include:

- *You lose negotiation leverage, especially on federal contracts, when the project is complete.* During construction, the owner's representative and designer are anxious to keep the contractor working and to finish on time. They don't want you to slacken your efforts, even while they deny time extension requests and impose liquidated damages. After completion, they move on to the next job and won't take the time to discuss issues from old projects. In short, when the job is done, the contractor is no longer needed, the owner has the money, and the contractor has the burden of proof to convince the owner to pay. Disputes are turned over to the owner's legal department or attorneys, and often the individuals who created the problem are no longer involved. On the other hand, if you file a claim while the owner's representative and designer are still on the project, their inability to resolve the dispute tarnishes their reputation.
- Job personnel may quit and be unavailable to work on the claim or to testify years later in court. Memories fade. More often, personnel move on to another job. Good supervisors shouldn't be kept in the office after completing a job, preparing a claim for money already spent, when their time and talent would be better spent building another project. Furthermore, a good superintendent or project manager is seldom a good claim expert. Many superintendents loathe paperwork and are unfamiliar with contract law or claims procedures.
- On negotiated settlements, contractors seldom recover interest from the time the work is done until they submit a change order request or claim. In most cases, contractors also forgo interest from the time a claim is submitted until payment is made. (The federal government pays interest from the date the contracting officer receives the claim.) Every day of delay is a day the contractor finances the work for the owner. If due, interest is paid only at the legal rate of interest, not at the interest rate paid by the contractor.

If interest isn't paid, a 2-year delay at 9 percent bank interest is 19 percent compounded, wiping out any profit and overhead on the extra work. In contrast, if the issue is resolved during construction, only a few progress payments will be missed.

- Legal fees also increase with delays. A few hundred dollars for legal advice on a change order request can grow to thousands for a negotiated claim, and 20 to 50 percent of the recovery in arbitration or litigation.

- Too often when a contractor finally starts preparing a claim, the records turn out to be inadequate. Prompt filing of claims enables contractors to obtain records that would otherwise be lost, or to record notes when memories are still fresh.

## 5. Be reasonable, persistent, and timely

**Be reasonable.** Don't claim too much, but start with a reasonable number that includes some issues that can be compromised. Then aim for a fair and reasonable settlement.

**Be persistent.** Do not expect a major claim to settle without going through the dance of information exchange and negotiation. Be persistent, especially if time is in your favor. For example, government owners often settle claims more easily at the end of a fiscal year when funds are left over or at the beginning when new funds are available. Patience is especially important if people are being asked to change their opinions. Don't press too hard when proposing something that is radically different from what the owner expects. Allow time for your counterpart to absorb new information.

**Be timely.** Timing is important both on the macro level, when choosing the time to submit a claim, and on the micro level, in meetings. For example, at a meeting, you can time the moment you bring up crucial, make-or-break issues or press for a global settlement. Wait until a pattern of cooperation and progress has been established or until everyone is tired of arguing and ready to settle. Or wait until negotiations are nearing completion and everyone is psychologically primed to finalize the agreement to bring up a final issue.

Be prepared for any opportunity to negotiate and capable of either a quick, abbreviated discussion or an extended, ritualized negotiation session. Although most negotiations are conducted in formal meetings, the most important are sometimes conducted on an impromptu basis—over coffee, at a social setting, or during a casual encounter.

## 6. Insist on equity, good faith negotiations, and mutually acceptable solutions

Establish the ground rules for negotiation up front, preferably at the preconstruction conference or partnering workshop.

1. *Use equity as your most powerful negotiating tool. At the partnering workshop, agree that all parties will negotiate changes based on equity, the contract, and contract law. If the project isn't partnered, reach this agreement during negotiations on the first change order.*

2. *Insist on a good faith effort. Do not negotiate with someone who you believe is not acting in good faith.* If you think this is the case:

- Shift resolution to another forum, such as mediation.
- Shift negotiations to another individual or involve another person to moderate the individual who is being unreasonable.
- Refuse to negotiate with that individual and try to force their replacement.
- Raise negotiations to the next level of authority for both organizations, from you and your counterpart in the owner's organization to your respective supervisors.

3. *Use win/win negotiation to reach a mutually acceptable settlement.* Agree with the owner's negotiator to strive for a multually acceptable settlement and use win/win negotiation to achieve it. Win/win negotiation includes focusing on identifying the parties' basic needs, identifying alternatives that expand the benefits for both parties, and using fair negotiation procedures and objective criteria for agreement.

## 7.   Improve your negotiation skills

Negotiation is an acquired skill, which can be improved by study and practice. Read this chapter and continue with independent study. Review books and articles on negotiation, attend seminars, and observe others negotiating. Practice negotiation in your everyday life.

## B.   Human Behavior—Personality Styles, Conflict, and Communication

Negotiation is a "people skill." Your negotiation style reflects your basic personality style. You can be more successful if you improve your communication skills, modify your behavior in response to the situation, and understand how to deal with conflict and interpersonal power.

## 1.   Understand the importance of the human element

How you interact with your counterpart is more important to success in negotiation than in any other aspect of construction claims. For win/win negotiation, have empathy for your counterpart, respect your counterpart's position on the issues, and consider the emotional impact of your actions.

## 2.   Recognize behavioral styles in yourself and others

People are different. Differences in personalities often lead to conflict and prevent effective negotiation. The same approach to negotiation used with different personalities will give widely varied results. It may be helpful to identify your basic behavioral style and your counterpart's to determine the differences and how to deal with those differences when negotiating.

**Basic personality styles.** Experts have been classifying people by their personality styles since the time of the ancient Greeks, and probably before. In the United States, one popular classification system is the Myers-Briggs type indicators, which look at four dimensions of personality: introvert/extrovert, intuition/sensing, thinking/feeling, and perceiving/judging. These dimensions represent continuums, not either/or conditions, as people "tend" toward one end of the spectrum or the other. Determining personality style based on the Myers-Briggs indicators requires testing and expert interpretation.

**Four behavioral styles—analyzers, directors, persuaders, and relators.** A simpler method of identifying people is based on behavioral styles, rather than their underlying personality. It is accomplished by observing their behavior and determining their level of assertiveness and emotionalism, as illustrated by the chart below. It is easier to use but not as accurate as Myers-Briggs.

Do not attempt to determine a person's behavioral style based solely on the following material. It requires training or further study. Even then, it cannot be relied on except as a general guide. Use caution in attempting to identify people by behavioral style and in acting on the result. Most people have acquired behavioral patterns that modify their basic behavior and result in a blend of styles that varies according to the situation. However, under stress most people revert to their basic style(s). No one style is "right" or better for negotiation. Each style has its strengths, and a good negotiator needs to employ elements of each, depending upon the circumstances.

A typical personality/behavioral style chart based on Myers-Briggs follows:

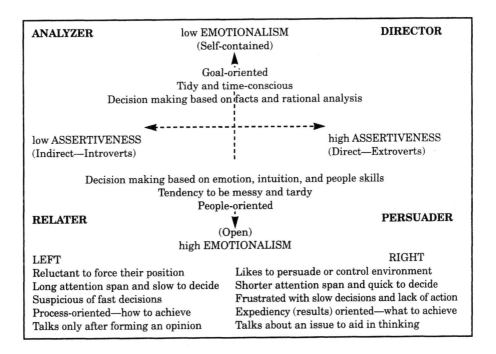

A description of each behavioral style and some suggestions for negotiating with people of each style follow. Remember that most people will display elements of several styles, although one style may tend to be most prominent.

1. *Analyzers—upper-left quadrant.* Analyzers seek accuracy and precision. They like facts and figures to study and they make decisions based on *facts,* not emotion. They are precise and careful when making a decision, and take time to make decisions, since they must first have all of the facts and time to analyze them. They prefer a highly organized process and tend to minimize relationships during negotiations. Many engineers, accountants, and project architects are analyzers.

To negotiate with analyzers, be organized, provide all the facts, and be accurate. Small talk is okay, but focus on technical issues and results. Be patient, avoid being flippant or pushing for a quick decision, and give them a time frame for a response.

2. *Directors—upper-right quadrant.* Directors like to have authority and to control the negotiations. They are time-conscious and businesslike and focus on results instead of relationships. They think deductively (from general information to specifics). They may have their calls screened and receive visitors formally. When you are ushered into their office, you find they are sitting down and formally dressed. Directors tend to talk about business rather than personal interests or their family, and about facts rather than feelings. When negotiating, winning is very important. Determine how you both can win. Many contractors and lead architects are directors.

Get to the point quickly with directors and minimize personal issues. Show them benefits and a logical, goal-oriented approach, but provide options so that they can maintain a sense of control. Expect a fast decision.

3. *Persuaders—lower-right quadrant.* Persuaders value relationships and wish to be liked, to be heard, and understood. They are friendly, open, and comfortable talking about their personal lives. Persuaders avoid being bogged down with technical details or charts and graphs, and prefer to rely on intuition and being able to "read" people. They are fast decision makers and don't waste time on painstaking analysis. Opinions and ideas are very important to persuaders. They may have as much interest in changing your opinion as in winning their objectives. Many salespeople, and some senior executives with marketing responsibility, are persuaders.

When negotiating with persuaders, start with a discussion of personal interests. Then talk about the benefits of settlement and create enthusiasm for your proposed settlement. Present the positive and avoid negatives. Focus on the big picture instead of the details. Expect a fast decision and be aware that they may try to resolve issues by splitting the difference.

4. *Relaters—lower-left quadrant.* Relaters like sincerity and appreciation. They want security and trust and tend to confer with colleagues before making a decision. Relaters are team builders, facilitators, or socializers and promote everyone's involvement and cooperation. They tend to express their emotions if confronted or if the negotiations become adversarial. They are thorough thinkers, and resistant to high-pressure sales. Relaters are as interested in

feelings as in the facts, and want a win/win solution. They build relationships as part of negotiations. They also tend to take on too much and become over-burdened.

Take your time and build a personal relationship and a sense of trust with a relater. Do not pressure them or expect a quick decision. Give them a time frame for decision making. Tell them "We need an answer by _____. Does this work for you?" or "I need your help. Can we have an answer by _____?"

**How to identify the four basic personality styles.** Identify your own style. Then identify the behavior of your counterpart. Use the figure and descriptions above, observe, and listen to what your counterpart(s) say and how they say it. Ask a few discrete questions to help identify their style.

First identify their behavior as being on the right or left of the chart. Are they direct/assertive, or indirect/nonassertive? If they volunteer their names, act as though they've known you for years, and maintain eye contact, they're probably direct. If they don't initiate conversation, seem quiet and reserved, and only briefly make eye contact, they are probably indirect.

Second determine if they're emotional/open or unemotional/self-contained and place them vertically on the chart. If they talk about feelings and relationships first instead of facts, and use body language extensively, they are probably open. If they discuss objectives and process before relationships, use limited body language, and stick to the point, they're probably self-contained.

**Intuitive/factual and confident/insecure styles.** Another dimension not included in the chart, but part of the Myers-Briggs system, is intuitive/factual (sensing). Intuitive people are big-picture/concept-oriented while factual people are detail oriented. Many architects are intuitive, while contractors and engineers tend to be factual.

Also determine whether you are dealing with someone who is confident and self-directed or with someone who is insecure. Self-directed people prefer to reach their own conclusions, based on the facts. Don't force a conclusion on them. Insecure people are more motivated by what they might lose than by possible benefits. They move away from problems rather than toward solutions. Emphasize the negative aspects of failing to reach a settlement. For example, stress the time, expense, and bad image from arbitration or litigation if negotiations fail.

**Multiple styles.** Keep in mind that a person's behavioral style is relative and that most people have multiple styles in varying degrees. Many people tend to have a primary behavioral style that reflects their basic personality and a secondary style (often in an adjacent quadrant) that is largely learned. While they may have skills in an adjacent quadrant, they may be weak in the opposing quadrant. Some people, however, tend to be fairly evenly skilled in three or four quadrants. They also may modify their style for the situation and how they perceive your style. There are few clear-cut personalities, just tendencies.

**Using the information to temporarily modify your behavioral style.**  Once you've identified your counterparts' behavioral styles, modify your *behavior* to complement the behavioral style. Then work on a win/win approach by finding mutually beneficial alternatives. Don't try to imitate another behavioral style—adapt to it as a modest compromise to gain trust and minimize tension.

If you are an analyzer or director and dealing with a persuader or relater, increase your emotional responsiveness by shifting your focus temporarily from the task to relationships. Express your feelings about the issues and accept your counterpart's expression of feelings. Conversely, if you you are a persuader or relater dealing with an analyzer or director, shift from a focus on relationships to the task at hand. Talk about the objectives and minimize emotional content.

If you are a director or persuader dealing with an analyzer or relater, focus on the process as much as the end result. Pay attention to detail and listen instead of talking. Conversely, if you are an analyzer or relater dealing with a director or persuader, deemphasize process and focus on results. Be candid and deal directly with conflict. Don't insist on having all of the details. Pick up your pace.

**Other means of evaluation.**  Every interaction provides opportunities to evaluate behavioral styles and personal values. For example, a preproject partnering workshop included a competitive, team-building game that required cooperation between the teams to maximize benefits for all. All went well until the last session, when the construction superintendent's team switched unexpectedly to a win/lose strategy and "won" the game. As expected, the superintendent used the same win/lose approach during the project.

**More information.**  For more information on personality/behavioral styles, read *People Smarts* by Tony Allessandra and Michael J. O'Connor with Janice Van Dyke (Pfeiffe & Company, 1994)[40] and *Partnering Manual for Design and Construction* by William C. Ronco and Jean S. Ronco (New York: McGraw-Hill, 1996).[41]

## 3. Learn to communicate more effectively

Good communication skills are essential to successful negotiation.

**Methods of communication.**  The three methods of face-to-face communication are:

1. Verbal—the words or content of what is said (or not said).
2. Vocal—the intonation, speed, pitch, inflection, and rhythm of speaking.
3. Visual—the body language of physical movements and body position.

Each method transmits both intentional and unintentional messages that combine to form the total message. Negotiators must closely observe each method and know how to interpret them.

**Listening and paraphrasing.** *Good listening skills are more important than good speaking skills.* Carefully listen to the verbal and vocal messages and observe the visual message. Determine what is being communicated before responding. Restate in your own words the factual and emotional content that you believe your counterpart(s) conveyed and confirm it was what they wanted to communicate. Then you can respond. Paraphrasing is especially important if you or your counterpart are angry. It is reassuring to hear your position and emotions expressed by the other party, even when knowing they don't agree.

**Interpreting what was said.** *What is not said can be more important than what is said.* For example, if you expect your counterpart(s) to vigorously protest a point and they don't, it may mean they will eventually agree with it. If something doesn't sound right, write it down to analyze later, or indicate you didn't hear what was said and either paraphrase what you think was said or ask it to be repeated.

People sometimes use qualifiers that indicate a problem or contradict what was said immediately before. The most common qualifiers are *but* and *however.* People also use introductory phrases which are sometimes the opposite of what they mean. These are unconscious forewarnings. They include:

We can work out the details later...

As you know...

In my humble opinion...

Frankly...

Honestly...

I'll do my best...

Whatever you say...

We have a problem...

You may know more about this than me...

I don't mean to pry, but...

**Asking questions to learn and to communicate.** Questions do more than elicit information. Questions can communicate your position, change the direction of discussions, attack, or defend your position. They can also delay, avoid responding to the other party's questions, and encourage the other party to examine their position. Questions can be leading, with the response suggested; provocative; or conciliatory, depending upon your objectives. Questions can be asked in response to a question: to clarify, confuse, or counterattack.

If you want to force an issue, direct questions to one individual and require a specific response. Or ask general questions directed to no specific individual, but insist on an answer.

The exact phrasing of a question is crucial. Ask questions that either require a specific response such as yes or no (i.e., are closed-ended) or are open-ended (e.g., "How do you feel about that?"). Open-ended questions begin with a verb that requires the other party to:

- Define or clarify their statement or position.
- Illustrate or give an example.
- Classify or compare two things.
- Verify a position or statement.
- Justify a position.

If you are not getting the desired response from your questions and you want more detail or clarification:

- Repeat the question, but with slightly different wording or intonation—"Yes, I understand, but what does it *really* mean?"
- Ask about your counterpart's feelings regarding the issue—"Do you feel your offer is reasonable?"
- Say that you don't understand an answer or ask why the answer is such-and-such—"I'm not sure I understand that. Can you tell me why you are limited to only $450,000?"
- Use the key question words who, what, when, where, and why: "Who decided that was the limit? What was their reason? When can we talk with them? Where are they? Why don't we call them now?"

The setting influences the amount and value of the information obtained. People tend to volunteer more information when comfortable with their environment, at a neutral location, and in an informal setting rather than in a negotiation session.

**To convince—repeat the message and use multiple channels of communication.** Repeating a message helps people remember it, especially if it first identifies a need and then provides the solution. Messages delivered both orally and visually are remembered better than those presented in only one medium. Visual aids are therefore extremely helpful.

**Understanding and communicating with body language.** Research indicates that the majority of information communicated is nonverbal and that body language plays a major part. It can supplement or contradict the verbal message. Body language includes any movement, gesture, or body position that contains a message. Body language can be as obvious as a yawn (of boredom or fatigue) or as subtle as dilation of the pupils of the eyes (indicating receptivity or pleasure). Be aware of body language, learn to interpret others' body language, and use your body language to communicate. Note, however, that interpreting body language is difficult, inexact, and requires considerable experience.

To improve your negotiating position, project a positive attitude with your clothing, voice, choice of words, posture, facial expression, and basic attitude. Never communicate weakness or indecisiveness. Communicate directness and trustworthiness by looking directly at people, without pleading or evasiveness. Use your hands to emphasize, explain, and complement your verbal and non-verbal messages. If appropriate, stand up and go to a flipchart to illustrate a point. In addition:

- Don't fidget, doodle, or fiddle with a pencil.
- Sit with a relaxed but attentive posture.
- If possible, keep your voice pitched low but firm, forcing listeners to concentrate.
- Project confidence even if not feeling confident.
- Don't talk too much. You won't learn anything new from what you are saying, but the other party is learning about you. Do not dominate the conversation.
- Be pleasant, but avoid insincerity.
- Smile only when you feel like smiling or to communicate satisfaction. Do not smile out of nervousness.
- Respect your counterpart's physical space. The space people need depends upon their origins (rural versus urban, U.S. versus Japanese, etc.) and relative physical size.

Some examples of body language indicators and the attitude they *sometimes* convey are:

| **Common Interpretations of Body Language** | | | |
|---|---|---|---|
| *Face* | *Attitude* | *Face* | *Attitude* |
| Frown | angry or concerned | Smile | cooperative, sly,... |
| Yawn | bored, avoid, tired | Pupils of eye dilate | receptive, pleased, tired |
| *Arms, hands, legs* | *Attitude* | *Arms, hands, legs* | *Attitude* |
| Firm posture | confident, rigidity | Hand under chin | uninterested |
| Tilt head slightly | attentive | Point w/index finger | threatening |
| Straight head | inattentive | Fidgeting | nervous, bored |
| Shrug shoulders | indifferent | Hand to back of neck | annoyed, irritated |
| Touch nose, eyes closed | concentrating | Scratching back of hand | frustrated or itching |
| Clench fist, cross arms | angry | Clench fist, stamp foot | defiant |
| Hand closed and to side of face | bored | Edge of chair, unbutton coat | cooperative |
| Reaching out | sympathy | Throw object on table | upset, cutting you off |

**Common Interpretations of Body Language** *(Continued)*

| | | | |
|---|---|---|---|
| Hand to chest (men) | open/sincere | Tug ear | wanting more information |
| Hand to chest (women) | shock/protective | Scratch top of head | uncertain, uncomfortable |
| Hands on back of head | superiority | Tips of fingers touching | confident |
| Legs crossed, leaning back | superiority | Tapping fingers on table | impatient |
| Hand open, on cheek, forward | thinking | Clean eyeglasses or fiddle | needing time to think |
| Stroke chin, knuckles under chin | interested | Put eyeglasses on table | disengaged, eyestrain |

| *Eyes* | *Attitude* | *Eyes* | *Attitude* |
|---|---|---|---|
| Downcast eyes | sympathy or self-pity | More blinking | exaggerating, discomfort |
| Less blinking, look in eye | confidence | Look down over glasses | disapprove, disbelieve |

| *Voice* | *Attitude* | *Voice* | *Attitude* |
|---|---|---|---|
| Soft voice | sympathetic | Low, clearing throat | nervous |

Nonverbal communication is extremely complex, since several different indicators are often combined and may relay multiple messages, some of which are conflicting. Also, there are gender differences, regional differences in the United States, and differences between people from different countries or cultures. In addition, body position and movements may be just a habit or reaction to an itch.

**Summarize your discussions.** When finishing a conversation or a particular topic of discussion, summarize the discussion and state the points of agreement and disagreement. This confirms the common perspective and the consensus toward your position.

**Use technology to improve communication.** New technology and wider application of existing communication tools promise to improve communication but also present risks. For example, communicating by fax instead of letter and broadcast faxing common messages to multiple organizations (e.g., for requests for proposals to subcontractors) reduces time and costs. If indiscriminately used, however, the resulting information overload will cause important messages to be ignored, resulting in more problems.

E-mail also expands communication and can provide a record of communication. However, it eliminates verbal, vocal, and body language communication and should not be relied upon to the exclusion of face-to-face communication or telephone conversations.

Lotus Notes or similar group software can improve open, collaborative communication in a partnering environment.[41] Using a project web page can accomplish a similar result. If implemented widely, and people and organizations take the risk and invest the needed energy, these techniques may help revolutionize communication and cooperation. They eliminate hiding information, which forces the parties to be forthright and avoid gamesmanship.

## 4. Identify and resolve conflicts

Conflict occurs when two or more parties have different needs that are or appear to be incompatible. Conflict is a natural part of human interaction and inherent in contract disputes. Resolving conflict through one-on-one negotiation can be either satisfying or a dreaded undertaking.

**Address conflict in order to resolve it.** Acknowledging conflict is more productive than ignoring it. It allows the parties to resolve their differences and to establish a productive relationship without hidden agendas or resentments.

**Typical responses to conflict.** The most common reactions to conflict usually reflect the individual's personality/behavior style. They are:

1. *Withdrawing* and not addressing the issue by refusing to meet or discuss the claim, thereby avoiding resolution. Analyzers tend to respond by withdrawing. Allow them some time and then propose a process to resolve the dispute and move forward. Minimize emotional responses, press for continuing negotiations, and impose deadlines if possible.

2. *Smoothing over* the conflict on the surface, but not addressing the real issues. Persuaders tend to smooth over conflict. The result is delayed resolution and (if the contractor accepts the superficial resolution) continued work or impact without a change order. Insist on addressing the issues and negotiating.

3. *Capitulating,* which makes negotiations easy. However, the individual's organization may not accept the agreement and may resent what they feel is an unfair settlement, creating problems with other issues or later negotiations. Relaters sometimes respond to conflicts by capitulating.

4. *Forcing* a position on others, or win/lose. This eliminates compromise. Directors tend to respond to conflict by pressuring the other party. If negotiating with a person using this approach, provide a positive environment, work toward goals that can be shared, and persuade them to consider alternatives. Stand up to the other individual(s) to gain respect, but make it clear you're trying to achieve mutually beneficial results.

5. *Compromising* and reaching resolution, or win/win. Both parties win something and build on common goals while accepting differences. Any of the personality types can use this method.

**To deal with conflict, first reduce anger.** It is difficult to resolve conflict if someone becomes angry. First control your own temper. Then defuse the other party's anger by the following actions:

1. Ask for a brief time out. State that a short break is needed (for everyone to get tempers under control) before proceeding with negotiations.
2. If necessary, allow the other party to vent their anger. Then try to reframe it in a more positive tone without denying their anger.
3. Change the subject—either to a nonconfrontational issue or to a positive subject.
4. Resume negotiating the issue that caused anger only after diffusing the anger so that it doesn't interfere with negotiations.
5. Listen more attentively. Determine the underlying reason for the other party's anger. Modify your behavioral style to adapt to the individual's behavioral style, in order to reduce tension and to make your counterpart feel more comfortable.
6. Paraphrase and validate feelings to ensure clear communication and avoid further anger.

**Additional information.** See *Partnering Manual for Design and Construction*[41] for additional information on conflict avoidance and resolution.

## 5.  Understand power and how to deal with it

Power over another individual, or the belief that one individual has power over another, influences negotiations immensely. To successfully negotiate, be aware of power, the sources of power, and how to deal with power.

Title power comes with a position and the trappings of the position. Situational power is slightly different, as it reflects an individual's current assignment instead of position in an organization (e.g., the owner's negotiator has situational power, whereas the owner's vice-president has title power). Reward and coercive power are opposites sides of the same coin. The owner's negotiator has the ability to reward by agreeing to a change order request, or to punish by refusing to seriously discuss an issue.

Consistency power results from being reliable to the point that others come to trust you. For example, your ability to deliver on your promises gives you consistency power, especially if the other party sometimes does not deliver. Charismatic power comes from radiating a certain confidence, power, and empathy. Most of us have some charisma and can enhance it by focusing more on relationships.

Expertise power comes from possessing greater skill and experience. Expertise in CPM scheduling, cost estimating, and construction methods gives the contractor's negotiators expertise power over the owner's negotiator. Informational power is similar but results from having information that the other party lacks and sharing or withholding it, depending upon the circumstances.

Effective contractor negotiators focus on the use of the power of consistency, charisma, expertise, and information. The combination often balances the owner negotiator's power, which is usually situational and is based on the power to reward or punish. The resulting balance of power helps both parties negotiate a win/win solution.

## C. Preparing for Negotiation—Issues, Personalities, and Position

Thorough preparation is essential to success. Establish negotiation objectives, strategies, and tactics, and in some cases rehearse your presentation.

### 1. Understand the issues and determine the facts

It is assumed that you have identified all the issues and determined the facts as described in Chaps. 6 through 13 before starting negotiations. This is essential for successful negotiation, but sometimes overlooked.

### 2. Confirm company policy and set objectives

The contractor's negotiator must understand company policy on change orders and claims. Ideally, the policy favors a partnering approach that emphasizes win/win negotiations. If the policy is unclear, clarify it with company management.

Then tentatively outline your objectives. Include upper and lower monetary settlement figures, time extension needs, and the resolution of any ancillary issues, such as a change in inspection procedures. Include maintenance of a good working relationship with the owner's representative and the designer as one objective. Obtain authorization of a settlement number, or at least agree on a settlement range. And determine a process of conferring prior to commitment. Your objectives may change again during negotiation, as more facts become known and the possibilities become better defined. It helps, therefore, to classify your objectives as (1) nonnegotiable (seldom strictly true), (2) important, and (3) expendable.

### 3. Determine the owner's policy, needs, objectives, and positions on the issues

Understand the owner's policies, procedures, needs, objectives, and positions on the various issues to be resolved. Determine the owner's understanding of the dispute, negotiating strategy, probable expectations, alternatives, and attitudes. Analyze any constraints on bargaining procedures or settlements. Who has the strongest technical expertise? Who knows the costs and financial status? Which party is supported by their management? Who really grasps the facts? Who is better at negotiating?

The FARs (Federal Acquisition Regulations) are the main constraint on negotiating federal contract changes. Other constraints may include political and public pressure or the owner's need for quick completion of the work. Identify owner positions or basic interests that are (1) mutually advantageous to both the owner and the contractor, (2) advantageous to the owner and present few disadvantages to the contractor, or (3) advantageous to the contractor and present few disadvantages to the owner. Also try to determine the owner's agenda(s). Be aware that hidden agendas are often more important than the open agenda and may be quite different or even in conflict with the open agenda. In addition, the members of the owner's negotiation team may have differing agendas.

Research the owner's negotiator. Identify the negotiator's limitations and capabilities, personality type, attitudes, biases, preferences, and probable agendas. Determine the negotiator's style of negotiation, personal needs and biases affecting the negotiations, and how to meet those needs.

Start your research while preparing the claim. Talk to the owner's representative and ask what the policies and procedures are. Gain a clear understanding of the owner's positions on the issues. If working with a public agency, file a Freedom of Information Act request for needed documents. Ask questions during the initial presentation and during follow-up. Ask other contractors about the owner and individual negotiator, and talk to other individuals in the owner's organization.

Analyze the differences between your position and the owner's. Determine the extent of the differences and the relative importance of each position. This helps you to avoid giving away something that you value that has little perceived value to the other side.

Evaluate relative strengths and weaknesses. What will happen if you don't negotiate? In other words, what are the Best Alternatives To a Negotiated Agreement for you and the owner—your BATNAs?

## 4. Develop a strategy for each negotiation

Strategy is your overall plan for a specific negotiation. It provides the framework for negotiation and is specific for each claim, depending on the issues in dispute and the relative strengths, and the negotiation styles of the parties. For success, clearly define and carefully follow your strategy. It may be as simple as "keep insisting on full compensation until they agree" or as complex as the multielement general strategy described below.

Develop a strategy that leads to the best possible resolution, within the framework of your policies, objectives, and negotiating style. Also prepare alternative strategies, in case parts of your plan must be abandoned. An alternate theory of entitlement should be developed for major issues if the primary recovery theory has contractual or legal shortcomings. Likewise prepare a strategy for responding to counterclaims and expected owner defenses. For instance, know the facts needed to disprove a concurrent delay allegation when negotiating a delay claim.

A good strategy includes flexibility on the settlement amount. Otherwise, true negotiation is nearly impossible. The owner's negotiator needs to achieve some benefit from negotiation, which requires that you compromise on some issues. Do this by presenting one or more nonessential issues with disputed entitlement or with aggressive damage estimates, that you can afford to give up in negotiations.

*Truthfulness is mandatory* for win/win negotiation. Do not lie, deliberately mislead, or grossly exaggerate. At the same time do not assume your counterpart shares your ethical standards.

**Recommended general strategy for most negotiations.**  The following multielement strategy will be successful for most negotiations and can be modified for specific circumstances. It is both a general strategy and a standard procedure.

1. *Agree on settlement criteria and procedures. First agree that both parties will act fairly, consider equity, and negotiate in good faith.* Then agree on procedures and the major steps to settlement.

2. *Understand each other's position.* Reaching agreement is difficult if both parties don't understand each other's position. Therefore, clearly communicate your claim, ensure that the crucial facts are known, and understand the owner's position and objectives. More importantly, identify both parties' fundamental interests. Only then can you identify win/win solutions and create new mutually beneficial solutions.

3 *Attempt to establish independent issues.* Before proceeding, break the negotiations into separate issues that can be negotiated and settled independently of each other, and agree whether your agreements on each issue are final or only tentative depending on a satisfactory final agreement on all issues. The more independent issues you can identify, the higher the probability that at least some of them can be resolved successfully. Successfully resolving easier issues early creates momentum that helps resolve the more difficult ones. In addition, separating the issues prevents the owner's negotiator from reopening issues you thought were settled. If the owner's negotiator won't agree to separate agreements on each issue, ensure that your agreement is also tentative. Otherwise you may be forced to negotiate again, starting at your supposedly "final" position.

4. *Establish a reasonable range.* Before initiating serious point-by-point negotiation on an issue, establish a reasonable initial position for both parties. Initial positions should be defensible but contain room for compromise. If the owner's position isn't reasonable, insist it be revised to a good faith counterproposal. After a reasonable range has been established for each issue, you may wish to push for a global settlement.

5. *Negotiate individual issues.* Then negotiate each issue point by point. The recommended sequence of negotiation is:
   - Start with one or two easy issues, preferably ones with a very strong position where the owner is clearly wrong. The owner's negotiator will feel forced to concede, in order to show good faith. This establishes momentum and puts you in the best light. However, don't resolve all the easy issues immediately, as they may be needed later to "jump start" stalled negotiations.

- Then negotiate an issue important to the owner but unimportant to you or even an issue that you intend to concede.
- Only then tackle an issue that is important to you but will not automatically be resolved in your favor. Invoke the earlier progress and your concessions on earlier issues to help reach your objectives.
- Proceed with the next issue, depending upon progress and the specific tactics being used.
- If an impasse is reached on one issue, move on to another. Be certain that your agreement to set aside an issue isn't misinterpreted as agreement to the owner's current proposal.

Select a sequence that appears like to work; stick with it if it does work; or change if it is not working. If negotiators have roughly equal strength of position and know each other well, they may decide to address the tough issues first.

6. *At some point, reach for a global settlement.* Hopefully, your positions will be close enough together that you can agree on a final number for all remaining issues. Otherwise an impasse on global settlement is possible.

**Owner strategies on revealing an initial position.**   One owner-negotiator strategy is to conceal their position and to negotiate the contractor down until reaching the owner's objective or the lowest possible point the contractor will accept. This is the only viable approach if the owner's negotiator hasn't made an independent analysis because of either lack of time or insufficient information. The owner's negotiator will ask a series of questions or go step-by-step through each item of the claim—demanding justification, seeking inconsistencies, and pursuing weaknesses until you concede or compromise. The negotiator may abruptly switch to an earlier topic or an unrelated topic, to throw you off balance or to pursue newly revealed information.

Hiding their initial position is intended to reduce your proposal amount before starting serious negotiation. To counter, don't concede points, except for errors and obviously overaggressive assertions. Be prepared, however, to provide sufficient supporting detail so that the owner's negotiator can evaluate each point based on historical cost figures or reliable estimates.

Another owner-negotiator strategy is to reveal their initial position as a counterproposal. If it is grossly low, demand justification or that they concede it isn't valid and revise it with something defensible.

A third owner-negotiator strategy is to start with their final offer, which provides little room for compromise. To counter, ask them to justify each point and to negotiate those that are not reasonable.

**Make flexibility a key element of your strategy.**   Negotiating is like jazz. To be successful, you must improvise and be willing to:

- Adjust your means to your ends, and your ends to your means. In other words, use an approach and tactics that will achieve your objectives, and adjust your objectives to what is achievable.
- Follow the line of least resistance, if it leads to your objectives.

- Preserve a line of retreat from any position.

- If one approach fails, try another unless you are trying to win through persistence.

**Agree on principles.** If having difficulty agreeing on specifics, negotiate basic principles and then move back to specifics.

## 5.   Decide who will attend meetings and find out which owner personnel will attend

Decide whether an individual or team will negotiate for your side, and who will be involved.

**Consider using someone other than project personnel to present and negotiate claims.** Change order requests are normally presented by project team members—often the project manager. This works best, as team members are familiar with the project, and introducing a new party at this stage tends to escalate the dispute. However, if the change order request has been rejected and a claim is being submitted, the situation changes, and a different person may obtain a better result.

Job relations are better preserved if a different person negotiates claims rather than someone who is dependent upon the good will and cooperation of inspectors and designers. Escalating the level of attention (not the intensity) by bringing in a more senior company representative (or a consultant) to present and negotiate the claim can help achieve resolution. This is especially true if the project team members are too busy getting the work done or lack the negotiation experience for resolving disputes without antagonizing the owner's representative. Jobsite personnel can then blame a claim or unpopular contractor position on management or a consultant, while maintaining good working relations with the owner's representative.

**Claim negotiation team.** If more than one person is involved in negotiations, designate a team leader. Assign responsibilities to each team member, and define how the leader will control responses to questions by the owner's representative(s). Prepare a budget and a schedule for preparation and negotiation, but don't let it dictate the final negotiated settlement.

If the claim preparer is a project team member or if the claim is small, the project team member will normally negotiate. If a consultant prepares the claim, the project manager or construction executive may negotiate, with the consultant supporting. A large negotiation team includes a lead negotiator, the claim preparer to present and defend the claim, a project team member who knows the project, key subcontractors, and a construction executive to represent the company and make final decisions. On large claims or claims that will probably be arbitrated or litigated, an attorney may be on the team or even be the lead negotiator.

Some contractors assign senior managers to their negotiation team, to obtain a psychological advantage over staff-level owner negotiators. This is expensive but may help achieve a better settlement.

**Consider involving superintendents, subcontractors, and suppliers.** General contractor claims should normally include subcontractor pass-through claims. Even though it makes the total amount larger and more difficult to sell, it lends authenticity. Subcontractors with significant claims that support your position should be asked to attend.

Involving field personnel in the presentation can be very convincing. An explanation of the problem by a superintendent or foreman may convince at a gut level and can be more effective than a more sophisticated presentation with charts and graphs.

## 6.  Select the most effective tactics for your negotiation style, strategy, and situation

Tactical decisions include which issues to present orally and which to leave for the claim reviewer to study in the claim document. Tactics include the order of presentation, level of detail, exhibits to be used, who will address each issue, etc. Your team must have consensus on all the significant issues and tactics and agree on how to deal with internal conflicts.

## 7.  Prepare an agenda and script

Being organized and thorough will improve your presentation, give you confidence, and communicate to the owner that you are serious. A written outline of the planned presentation ensures organization and thoroughness. Send it to the owner's representative who will probably agree to it, ceding you control. Controlling the agenda is an important part of negotiation and allows you to time the negotiation to best meet your objectives.

Expand the outline to a rough script for important issues, filling in the details and covering each point as planned. Refer to your script but do not read it or slavishly follow it, which would create a forced, wooden delivery.

## 8.  Prepare special exhibits and handouts

Although most exhibits are prepared during the claim preparation effort, some may be identified during preparations for negotiation. For details, see Sec. D.19 of Chap. 13. You also may wish to hand out an agenda and reduced-size copies of key exhibits.

## 9.  Identify and prepare a response to rebuttals and counterclaims

Good claims analysis includes identifying and evaluating potential owner rebuttals and counterclaims. Some of these issues may be referred to in the narrative text, but you will probably avoid recording the most sensitive issues if discovery is likely. However, you should identify likely objections and plan a response for each. This is a key step in preparing for negotiations and a primary purpose of early discussions with the owner's negotiator.

## 10. Rehearse

Experienced negotiators rehearse their presentations. Rehearsing reduces the use of notes and ensures that essential issues are not overlooked. The result is a more relaxed, convincing presentation.

Rehearsing is most important for team negotiations. Team members must integrate their presentations to avoid conflicting messages. Have someone act as the owner's negotiator, to respond during the rehearsal and to critique at the end. Rehearsing with team members often brings out internal conflicts which must be resolved before meeting with the owner's negotiation team. Rehearsing is an excellent preparation process for dealing with owner questions and objections. Videotaping the rehearsal for team review also helps.

## 11. Adopt a successful attitude

For success, *enjoy what you're doing*. Claim resolution should be a positive part of your job and not a dreaded task. When you are skilled at negotiating, you will approach it as a satisfying challenge leading to measurable results.

## D. The Initial Presentation

Most contractors initiate a formal claim either by plunking down a copy of the claim document on the owner representative's desk or by mailing it to the owner's project manager. There is no oral introduction, explanation, or discussion. This may be satisfactory for most change order requests and small claims but is a serious mistake for a major claim or an issue that is likely to be contested. It fails to set the right tone, level of importance, and urgency. It also degrades the partnering approach to conflict resolution.

Initiate the negotiation process with an oral presentation and discussion to establish the facts and present your position. You do not want the owner to form a position yet, unless it is positive, but to keep an open mind. The initial meeting is most productive if the owner's position does not become fixed. You are communicating, not arguing.

An effective oral presentation facilitates settlement of a weak or poorly prepared claim. A poor presentation, or none, can negate weeks of painstaking work.

## 1. Schedule the presentation meeting

First schedule a meeting with the right individual(s) from the owner's organization. Otherwise little will be accomplished. Identify those whom you prefer to attend from the owner's side and insist that either a decision maker or an influential adviser attend. The owner's representative should also attend. If they intend to have an attorney, claim consultant, or a senior manager present, bring an equivalent person from your organization. Keep in mind that a large group (more than four or five from each side) can be counterproductive unless all interests need to be represented.

Shortly before the meeting, call and confirm attendance of critically important individuals. It is more difficult for them to "forget" if reminded shortly before the meeting.

Agree on the location and purpose of the meeting. Inform the owner's representative of your objectives and agenda. Agree to be open-minded and willing to discuss the issues in depth, and ask them to do the same.

The length of the meeting depends upon the circumstances and the willingness of the owner's staff to spend the time. If aware of the importance of the meeting and the advantages of attending the presentation, they will be more likely to devote sufficient time.

**2.   Formally present the claim at an official scheduled meeting**

The initial presentation meeting has different objectives and different procedures than subsequent negotiation meetings.

**Be on time.**   Be prompt, arriving a few minutes early in order to be set up but not so early as to appear overeager.

**Establish an appropriate atmosphere or tone at the meeting.**   Some negotiators adopt a negative tone to lower their opponent's expectations and put them on the defensive. Others will adopt a more positive tone to engender cooperation and a sense of common interests. Adopt the tone that is consistent with the effect you want to have: businesslike, informal, frustrated, firm, etc. The most productive tone, however, reflects mutual cooperation. Be positive, not defensive or apologetic.

Achieve the desired tone with your clothing, facial expression, tone of voice, choice of words, and the procedures that you follow. Appropriate humor, for example, helps set a friendly tone. A friendly, cooperative tone puts the owner's negotiators at ease, enabling them to focus your presentation. When the owner's team starts to arrive, engage them in casual conversation and discreetly determine their style and attitudes.

**Remember to sell.**   You are there to "sell" your proposal. It's an unusual sales job, as it carries an implicit threat (of pursuing litigation if your claim isn't accepted). This requires not just a presentation but a performance, with everyone focused on the objective.

**Start the meeting with introductions, identification of roles, and validation of basic assumptions.**   Start with a friendly greeting. Ask all parties to identify themselves by name, interest in the negotiations, and their role. Next state the purpose and agree on the agenda, procedures, and schedule for the presentation.

The initial fact-finding portion of the negotiation meeting is not the time to argue. It is for presenting the claim, exchanging information, and agreeing on the basic issues to be negotiated.

**Start the presentation with an overview of the claim.** After introductions, start your presentation with an overview of the claim (e.g., an executive summary) by the leader of your presentation team. Address the most important issues. Discuss business relationships as well as factual and contractual issues. Keep the overview brief, clear, and positive in tone. Use summary exhibits to provide visual effects and communicate the most important elements of the claim.

**Present the significant issues.** Then present each important issue in detail, complete with a summary of the analysis, conclusions, and exhibits for the issue. The most knowledgeable person on each issue should make the presentation. Ask that substantive questions, requiring a detailed response or discussion, be held for the question and answer period. Accept clarifying questions either during the presentation or at the end of the presentation on each separate issue. Request that questions be directed to the presentation team leader, who may assign and intercept questions to others.

During the presentations, provide detailed information only as necessary to build a foundation, establish credibility, and demonstrate thoroughness. Briefly describe detailed exhibits, explain how they were prepared (to reassure the owner's negotiator of their completeness and accuracy), describe specific contents, and explain how each exhibit is used for analysis. Then focus on the summary exhibits.

Vary the pace to meet the needs of the reviewers. They may prefer details or just a brief presentation. Be ready to address any side issues, but then move back to the agenda.

**Use exhibits to communicate and complement the oral presentation.** Exhibits are very important. Visual aids communicate essential facts more efficiently and effectively than exclusively auditory presentations but are best used in conjunction with oral presentations. Well-prepared graphics demonstrate a high level of effort, the importance of the claim, and your readiness to pursue the dispute through to litigation if necessary.

**Summarize at the end.** The closing summary by the team leader briefly repeats the key points of the presentation and ends with a formal request for a prompt and fair resolution. Present a word picture, one or two graphs, and a summary cost exhibit to create a graphical image that remains etched in their minds.

**Schedule a question and answer period.** After the formal presentation, encourage the owner's representatives to ask questions and discuss specific issues. Suggest covering each issue in the order presented. If necessary, ask them if they have questions about each issue covered. Focus on clarifying and providing additional information. Do not let discussions deteriorate into an argument. If you are unable to answer a question or provide requested information, note the request and promise to respond as quickly as possible.

**Listen to counterpresentations by the owner's representative or designer.**   In some cases, a member of the owner's team will present counterarguments. Although this may provide valuable insights to the owner's initial reactions, discourage rebuttals that have been heard before or that don't specifically respond to your claim. Avoid an argument at all costs. Steer the counterpresentation to the reactions of the owner's decision maker. Discover what the decision maker readily accepts, what additional information is needed, and how to satisfy the owner's needs. Postpone in-depth discussions to the next meeting, and limit your rebuttal of counterpresentations (at the initial meeting). Even outrageously wrong counterpresentations may be left unrefuted, other than by a dismissive remark or gesture, as refutation may lend credence to them.

**Confirm the owner's positions on the issues.**   Paraphrase the counterpresentation and ask if it truly represents what was said. Then try to determine what was behind the statements. Was it an emotional response, was there an underlying but unstated issue that needs to be addressed, do you or the owner lack crucial information, do you need to seriously consider the argument and prepare a rebuttal, or do you need to concede the point? Either respond immediately or state when you will respond. Take detailed notes of the counterpresentation and of all discussions. Confirm the owner's tentative positions on the issues in dispute and ask for an open mind until after the claim is reviewed.

**Agree to exchange information and stipulate to the facts.**   Agree to exchange documents to facilitate the owner's review of your claim and your evaluation of rebuttals. Identify key facts and contractual interpretations that can be agreed to or can be addressed incrementally as the negotiations progress.

**Document and sign any agreements.**   If you do negotiate and resolve some issues, document the points agreed to while at the presentation meeting. As a minimum prepare a brief handwritten summary of the points agreed to and those still not resolved and have both parties initial it. Preferably write up a complete agreement and have both parties sign. Even if not signed or initialed by both parties (which is preferred), a written record is an effective tool for retaining the agreements that were reached.

**Agree on the next step.**   End the meeting with a positive commitment from both sides for further action. Get the claim reviewer's commitment to review the claim with an open mind, to request further information, if needed, and to promptly respond by a certain date. Ask for a second meeting after their initial review but *before* they make a decision.

Schedule a second meeting at this time, outline the agenda, and get commitments from key owner representatives to attend. This prevents the owner from postponing review and response.

**Wrap up and summarize.** At this time, thank everyone for attending, summarize the results, and restate the next step.

### 3. Prepare minutes of the meeting and a memorandum of understanding

Immediately after the initial presentation, meet with your team and confirm what was agreed to at the meeting, the apparent positions of the owner's representatives, and the best strategy and tactics to follow in further negotiations. Write up your notes in a memorandum to file. Prepare minutes of the meeting for sending to the owner. Prepare a memorandum of understanding for any agreements not written up at the initial presentation, sign it, and send it to the owner for signatures.

### 4. Focus on entitlement, if disputed

If entitlement is strongly disputed, focus on proving your contract and legal rights before continuing with proof of damages.

## E. Follow Up

The follow-up helps you to reach a fair settlement and requires a reasonable effort to ensure that it is done well. The following suggestions apply to a large claim but can be scaled down for a smaller claim.

### 1. Send a letter and minutes of the presentation meeting

Write or fax the owner's representatives a brief letter thanking them for meeting and summarize the points discussed and any agreements reached. Note that requested information has been attached or will be forwarded by a separate transmittal. Restate the agreed actions by both parties and express your willingness to meet again to continue negotiations.

### 2. Answer questions and provide supplemental information

Review your notes, and the notes of the rest of your presentation team, for questions needing a response. Prepare a response to the owner's questions and send it as soon as possible. Call again later to confirm that it was received (and to encourage a prompt review), and again later to discuss the information provided and answer any further questions.

### 3. Meet to resolve questions of fact

For some large claims, staff personnel will meet to resolve questions of fact and to negotiate details. This simplifies the negotiation process and saves time for the lead negotiator, who will handle the final negotiations.

#### 4. Respond to owner's defenses

Respond to all owner defenses and counterclaims, presented either at the meeting or in subsequent communications. Be positive, but firm, when doing so.

#### 5. Track the status of disputes, change order requests, and claims

When there are more than just a few disputes, keep a log of actions and the current status of all disputes, change order requests, and claims. Review the log periodically, provide a copy to the owner (the nonconfidential portions), and press for continued action and resolution. Insist on some progress, even if it is necessary to escalate negotiations to the next level of both organizations' managements.

### F. Negotiations—Style, Strategy, Tactics, and Procedures

The recommended negotiation strategy is based on a win/win approach, supported by a partnering effort by both parties. There are numerous other methods of negotiation, some of which *may* be more appropriate for a specific situation. Remember, however, that even with a win/win approach, negotiation is competitive.

#### 1. Select and set up the meeting location

The physical environment affects communication and can help you reach a favorable settlement. Try to conduct negotiations in your office, which gives you an advantage. If this is not possible, try to meet at the jobsite or some other neutral location. If you do meet at the owner's location, which is most common, attempt to establish favorable working conditions. One advantage of meeting at the owner's location is that you can leave anytime. It's rather awkward to walk out of negotiations in your own office.

Personal comfort and convenience help to facilitate resolution. Be sure the meeting space is large enough for all parties, supporting staff, and their materials. It should be well lit, well ventilated, and at a comfortable temperature. Other desirable features include a white board or flip chart, availability of reproduction equipment, a separate caucus room, access to telephones, and refreshments. A little generosity, like providing lunch, helps promote positive attitudes, saves time, and provides an opportunity for building personal relationships and informal negotiations.

If the meeting isn't at your location, request the features listed above and offer to provide lunch. Avoid being put at a disadvantage. For example, if the light from windows is too intense, sit with your back to the windows with the other party facing the glare. If you're facing the light, draw the shades or change sides. To promote cooperation, scatter your team around the table or on two sides of the table instead of all being across from the other team. Or place your team on one side with the other team opposite, and sit at one end to establish yourself as the facilitator.

## 2.  Confirm the negotiator's authority to settle

Confirm that the owner's claim negotiator has authority to settle and determine the limits of that authority. The negotiator for a public agency should have either authorization for a specific level of settlement or general endorsement by the governing body or supervisor. Don't be trapped in what you thought was negotiation, but which ends up with the owner's negotiator saying that *your offer* will be taken to the real decision maker. If that happens, retract your agreement by asserting mutual misunderstanding and state that your position will change if the mutual "agreement" is not accepted in whole.

If the negotiator lacks authority to make commitments, treat the meeting as an information exchange session. Explain your position, learn the owner's position, and attempt to win the negotiator over to your side, but do not reduce your claim (except for obvious overstatements). Alternatively, negotiate a portion of the claim at each meeting, with the owner's negotiator obtaining approval of that portion before proceeding to the next.

If there are multiple reviewers, ask all to attend negotiation sessions, or at least the first session to establish basic positions, negotiation criteria, and procedures. Otherwise those attending must justify any agreements to those not attending and you may not reach consensus on the eventual settlement.

## 3.  Try to organize (manage) the negotiations

The contractor's negotiator should manage the initial presentation and should attempt to manage the entire process of follow-up and negotiations. Start with taking the lead in discussions. Suggest the issues to discuss, their order, when to move from one issue to another, when to table an issue that has become too controversial or is at an impasse, etc. If done diplomatically and fairly, this may not be noticed by the other party. If noticed, it may be appreciated for the efficiency it provides. Effective control of the agenda, tone, and pace of meetings usually results in faster, more favorable settlements.

Timing is an important element of control. Make every effort to start negotiations as soon as possible and continue negotiations until the issue is resolved. Press the owner to commit to a schedule, agenda, and meeting dates.

If the owner's negotiator attempts to control the negotiations, work collaboratively by jointly establishing agendas and agreeing on procedures. If conflicts arise, remember that successful negotiations are more important than who manages the process.

## 4.  Insist on an equitable adjustment

At the first negotiation session, adopt *equitable adjustment* as your mantra and repeat it as frequently as possible. Ask your counterpart to agree that if you are due an equitable adjustment, the owner will not ask you to settle for anything less. Point out your earlier agreement to consider equity in negotiations, to negotiate in good faith, and to strive for a mutually acceptable settlement using win/win negotiations. Also point out the contract provision that refers to equitable adjustment. (Most contracts include such a statement or something similar.) Few owner negotiators will refuse to agree to such a request. Once they have agreed, you can use it throughout the negotiations. Bring it up, for example, when arguing against strict interpretation of an onerous contract clause such as no damages for delay.

## 5.  Apply tactics as needed  and be ready to respond to your counterpart's tactics

Tactics are the "tools" used to implement a strategy. Some tactics will be preplanned and used repeatedly during negotiations, while others might be needed unexpectedly. Some might not be used for a specific negotiation and others might be used repeatedly. Although planning to use certain tactics when preparing for negotiations, be ready to adopt whatever tactic is needed.

To be successful, tactics must be compatible with your personality, style, and strategy. Methodical personalities often use technical/factual-based tactics, while dynamic personalities prefer rapid-fire, shoot-from-the-hip tactics.

Your use of tactics must also take into account the owner's negotiator (your counterpart), the specific circumstances, and the status of negotiations. You must be ready to respond to your counterpart's tactics, some of which may be win/lose.

Different tactics are employed at different stages in negotiations—the opening, the middle, and the end. Others may be used at any time, in response to your counterpart's tactics.

**Opening moves.**    Certain tactics are used to initiate negotiations.

1. *Agree on a process that will lead to a favorable settlement.* First agree on procedures (such as step-by-step or global resolution), the criteria for settlement, and an agenda. Make certain that the agreed-on process supports a mutually agreeable solution and that it will respond to your negotiation style and strategy.

2. *Create a positive environment.* To ensure win/win negotiations, create a positive environment based on cooperation, openness, and positive relationships:
   - Be calm and relaxed, but don't appear casual or flippant.
   - Exude confidence. Assume that your proposal is reasonable and will be accepted.
   - Avoid appearing too smart or overly sophisticated; it will make your counterpart competitive or reluctant to accept a settlement.
   - Be respectful of your counterpart. Don't use first names unless you have permission or are certain it won't be resented.
   - *Smile and appear to be enjoying yourself,* especially when challenged, pressured, or attacked. Appear impervious to pressure. It may throw any attacker(s) off track and will probably cause them to change tactics.

3. *Present your proposal.* Start negotiations and the beginning of each session by restating your proposal and the goals of the session, to ensure that your counterpart understands the facts and your position.
   - Start with a summary of what you plan to discuss and your objectives.
   - Cover each issue thoroughly but without excessive detail. Use a logical progression (e.g., chronological, by trade) to describe each point.
   - Invite questions either during your presentation or after you've finished.
   - Answer questions as completely as possible, and provide additional information later if necessary.
   - Appeal to fairness and equity, especially if your position under the contract or contract law is weak.
   - Ensure that your counterpart is aware that you are prepared to proceed to arbitration or litigation if required.
   - When finished with the presentation, recap the agenda and request a response.

4 *Sell the benefits of agreement and the risks of impasse.* When making your initial "pitch" for resolution, emphasize the benefits of agreement and the risks of reaching an impasse. Some personality styles are motivated by benefits, others by avoiding risks.

5. *Understand your counterpart's position and interests.* After presenting your proposal, ask your counterpart to clarify the owner's position. Then explore the owner's basic interests, which may or may not be fully considered in their position.

6. *Identify additional benefits for both parties.* Before negotiating based on your respective positions, determine if there are additional benefits that could be created as part of the agreement. Identify options with mutual benefit or options that cost you little but greatly benefit the owner (and vice versa). To identify these options, look at assumptions both parties have taken for granted, identify assumptions that may be incorrect, determine whether there are exceptions to the general rules, and ask if there is a better approach.

7. *Negotiate based on interests as well as relative position.*

**The middle game.** The middle part of negotiations is the longest, but not necessarily the most difficult. It begins when you move from statements of opposing positions toward a middle ground. It requires a wide range of tactics to meet a constantly changing dynamic.

1. *Ask questions. Use questions as a tactic,* as noted in Section B.3. Ask questions, listen to answers, and watch body language.

   Prepare your questions before meeting. If a question occurs to you during a meeting, write it down for use at an appropriate time. Don't ask a question immediately when it occurs to you unless it is appropriate, and don't let your thoughts about the question keep you from listening to what the other party is saying. However, you may ask a question to block your counterpart's statements or the statements of one or your own team members. This is especially useful if you don't want them to continue a comment or question and you don't want to appear controlling.

Generally avoid questions that can be answered with a yes/no unless a yes or no is what you are seeking. Avoid two-part questions. Clearly word your questions and frame them so that you get the type of response needed. Use a tone of voice and body language that reinforces your message. Your tone may be forthright, demanding, firm, questioning, etc. Only those who are attacking and who are prepared for the consequences will use sarcasm or ridicule.

Ask questions either as part of an informative ongoing discussion, or as a blunt *demand* for information. Ask some questions that you already know the answers to, in order to test your counterpart's knowledge and honesty. Do not directly challenge your counterpart's honesty unless you wish to terminate all effective negotiation.

Allow time for an answer, unless you must immediately correct a misstatement or defend your position. Listen carefully to the wording of the answer, the tone of voice, and what isn't said. If an answer is not given or is not adequate, try your question again with different wording. If it is still not answered, write it down and raise it again later.

2. *Respond to owner proposals and statements.* Always respond, either verbally or with body language, to your counterpart's proposals. For example, if you don't flinch slightly when a street vendor says the sandals are 20 dollars, he may then add "for each shoe." If you still don't flinch, he may continue that the pair you like is more, and so on. But if you flinch, he may throw in something extra and at least will stop asking for more. Don't overdo your reactions. Respond to incorrect statements without starting an argument.

3. *Respond to objections.* As noted, you should prepare responses to potential objections and countercliams before negotiations. When objections are raised, answer them immediately, firmly, politely, and with quiet confidence. Demonstrate that you have considered all reasonable objections. Otherwise, the owner's negotiator will explore inadequately refuted objections and start looking for more.

Don't panic when an objection is raised. First, make sure that you understand it. You don't want to respond with the wrong rebuttal; that may open up a new issue. Asking for clarification also gives you time to develop a response. In some cases, however, you may not want to clarify the issue but to respond to the objection in as limited a manner as possible in order to avoid more objections.

Don't agree with a partially valid objection. Instead of saying "yes, but..." state something like "that may appear reasonable, but..."

Sometimes you can avoid a full response or deflect an objection by answering it with another question. But don't try to be smart. That would be counterproductive.

4. *Admit errors. If you make a mistake, admit it, correct it, and move on.* Don't make excuses or try to cover it up.

5. *Be silent, if appropriate.* Silence is a powerful tactic that may cause your counterpart to start babbling. Use silence and a disapproving look when insulted, to elicit an apology. Use silence when asked a challenging question if you need time to prepare a response. Or if the answer to your question is obviously insufficient, use silence and an expectant expression and body language to elicit more information.

When your counterpart uses silence as a tool, wait them out using body language to communicate your satisfaction with the lapse in discussions or to indicate that you are expecting further comments from them. Or use the silence to change the discussion to an issue you wish to pursue.

6. *Propose your own alternatives.* Propose alternatives that will give your counterpart some choices but will tend to favor the option you prefer. Downplay or don't include options you find undesirable.

7. *Launch a trial balloon.* To explore your counterpart's reaction to an idea without committing yourself, you can say, "I haven't explored this in detail, but...." If they respond positively, follow up. If they shoot it down, move on to another idea. If your counterpart launches a trial balloon consider it to be a tentative offer.

8. *Do not disclose information unnecessarily.* Do not reveal information or make concessions without a reason or without reciprocation by the other party.

9. *Negotiate before doing the work to allow more options.* It is much easier to negotiate forward-priced change orders. You can clarify, adjust, or limit the scope of work (e.g., exclude certain conditions such as rain during paving, which would then become an additional change if it occurred). You can also adjust the schedule, quality, terms and conditions of payment, etc. Few of these options are available when negotiating after-the-fact change orders. However, there are some things you can negotiate, besides price, on after-the-fact changes:

   - An additional time extension, which doesn't "cost" the owner out-of-pocket.
   - Relaxation of specific quality issues that don't materially affect function of the finished work.
   - A limited agreement for direct costs only, with impact and delay costs to be resolved later.

   When negotiating the price for completed work, you need not prove that your cost was competitive with a specialist in that area. You only need to show that it was reasonable for the conditions encountered and the experience level of your forces, and that you didn't make any gross errors. Since the owner benefits when your forces perform better than the industry average, the owner also should bear the costs for less-than-optimum performance, short of mismanagement.

10. *Develop and maintain good personal relations to facilitate resolution. Do not* falsely create friends or take advantage of friendships. Do make every reasonable effort to develop friendly relations with your counterpart, which will tend to improve communication and reduce conflict. It is hard to demonize someone you like and respect.

    There are many ways to initiate and cement friendly relations—the best being to act naturally and to like and respect the other person. Have coffee with your counterpart and discuss nonwork issues. Asking for a small favor unrelated to the negotiations can build good relations, especially if it reflects positively on the other person. For example, requesting a copy of a paper your counterpart has written or asking for an opinion on a technical issue will gain information and also show you respect his or her knowledge and judgment.

    Avoid damaging personal relations by separating business and personal interactions. Do not argue over contentious negotiation issues during breaks, at lunch, or when not specifically negotiating.

    When negotiating, discuss the issues, not personalities. Minimize any appearance of criticizing individuals, focusing instead on their actions and the impact of those actions.

11. *Use informal meetings to continue (or to facilitate) negotiations.* Many issues can be resolved in informal meetings, over coffee or outside the office, as it is easier to get at the heart of the issues.

12. *Win the game, not points.* Prove only what is necessary to substantiate entitlement and damages; don't try to prove the owner or designer is wrong unless this is both true and certain to benefit you. Make it easy for the owner's representative to approve your claim.

13. *Maintain a sense of humor.* Humor is a good tactic, in the right tone, at the right time. Humor helps establish personal relationships, encourages the other party to relax, defuses anger, and can make a difficult point without evoking an angry response. Use humor to counter an attack and gently or strongly criticize the other party's position. But avoid flippant "humorous" comments. They can be detrimental.

14. *Rely on established procedures or a recognized authority.* People tend to accept existing or previously accepted procedures from other negotiations. Therefore identify and follow procedures that have been used previously and that lead to your objectives.

    Use the endorsement of an authority on the subject to support your position. For example, refer to a quote in a reference book that supports your position. Or, identify someone who would support your position as an authority figure and then quote that person or ask them for comments on the issue.

15. *Use written material and printed forms.* The written word is highly regarded by most people. Therefore, back up important statements with a

written document. A handwritten note is given more credence and remembered longer than a verbal statement. Printed material is accorded more weight than handwritten material. Therefore use preprinted forms as recommended in Chaps. 4 and 5.

If you are a specialty subcontractor, create standard prices for typical work order tasks or use catalog pricing for extra work. Submit preprinted labor and equipment rates from an official price list with your first change order request. If requested, resubmit the rates with computed fringe benefits and taxes. Then print the agreed rates and provide a copy to the owner's representative to ensure that you don't have to negotiate them again. General contractors can use *R. S. Means* or other commonly accepted estimating manuals for official price lists.

16. *Bring in an expert.* If applicable, obtain a brief report from an expert. It should be on their letterhead or even neatly bound. Or have an expert join you for part of a session. Alert your counterpart that you are bringing an expert to "help you" with a technical issue. Do not bring an attorney to a meeting without informing your counterpart.

When bringing experts to a meeting, downplay their involvement to prevent overreaction. Call them in partway through the session, introduce them, establish their expertise, and ask them to present their findings. They should leave after presenting their material and answering questions. If the owner brings an expert to the meeting, avoid disputes between the experts, as it will get you nowhere.

17. *Explore dissension within owner's team.* If the owner's technical person or field representative appears dissatisfied with the owner's position, explore the differences and use them to help reach agreement. If appropriate, have the counterpart on your team discuss the technical issues in an attempt to determine the source of the dissent.

18. *Present both sides of an argument.* When presenting your points, it is often better to list the alternatives, both pro and con, than to list only your arguments. When doing this, present your own viewpoint last, as people remember best what is said last. They also better remember what is said first than anything said in the middle of negotiations.

19. *Do not argue.* Arguing will get you nowhere. If your counterpart expresses a position you don't agree with, acknowledge the position but don't agree. "I hear what you say; you are disappointed with our proposal; however,...." If your counterpart becomes emotional about an issue, validate the feelings and qualify your validation by adding "and...." Do not say "but..." since that negates the validation. Then, move on to the facts.

20. *Do not allow personal attacks.* Do not allow anyone to verbally attack you on a personal level. Demand an apology or walk out and insist on an end to personal attacks.

21. *Maintain self-control and keep your perspective.* Do not allow yourself to become emotionally entangled. Show passion and interest, but don't let

your emotions dictate your actions. Do not show anger, except for a reason, and then keep your expressed anger controlled.

Do not get rattled. One of the most effective counterattacks against a strong attack by a dominating personality is to be calm and cool and simply ignore the attempt to intimidate. Continue your efforts as if the tirade didn't exist. That frustrates an attacker, sometimes to the point where that person loses focus and can be brought into a win/win resolution.

22. *Preserve a line of retreat.* Avoid a frontal attack on a counterpart's position, except from overwhelming strength (e.g., when the facts, equity, and the contract all support your position). Keep a line of retreat and leave your counterpart a line of retreat. This will prevent anyone from losing face and the repercussions that result, and it will ensure flexibility for both parties.

23. *Bypass nonnegotiable issues or an impasse.* If stuck on a crucial issue, especially an emotional issue, set it aside and discuss other topics. Build momentum and return to the issue later. If still at an impasse when you return to the issue, bring in a third party as an informal mediator or escalate the issue to the next level of management for both parties.

24. *Take a break.* If negotiations are going badly because of anger or an impasse or because you are losing on a weak position, take a recess and resume later. Tempers will cool or the interruption will throw the owner's negotiator off pace, and you can regroup.

25. *Don't negotiate against yourself.* If the owner's representative rejects your offer, don't lower it (unless it was obviously too high) until a counteroffer is made. Otherwise you are negotiating against yourself. If your counterpart says your proposal is too high and doesn't make a counteroffer, ask just how much better it needs to be and consider the response to be a counteroffer.

26. *Keep a number of issues open until the final settlement.* Avoid allowing negotiations to get down to one major issue (usually price) without having ancillary issues that can be included in a compromise. Otherwise you may reach an impasse and be unable to move forward.

27. *Hold back your strong points.* Don't use your strongest arguments until they are needed to obtain a significant concession.

28. *Establish an allowance for undefined scope.* If a portion of the work depends upon a future owner decision (e.g., fixture selection), agree on an allowance with the owner to pay the actual costs when the work is complete.

29. *Perform undefined work on force account.* If a portion of the work is insufficiently defined to accurately estimate and the owner will not accept a reasonable contingency, do the work on force account.

30. *Use unit price agreements when the quantity of work is uncertain.* If the quantity of work is uncertain, propose a unit price agreement. Be certain to include all fixed costs (for mobilization, demobilization, equipment purchase, etc.) and spread those costs over a somewhat smaller quantity of

work than will probably occur. I learned this lesson on my first construction claim when our superintendent proposed a unit price for guniting (based on material use) that I felt was barely break-even, and spread mobilization costs over the estimated quantity. Neither I nor the government's negotiator realized just how much waste there is in guniting. The agreement created resentment when the actual quantity was an order of magnitude larger than estimated and our profit became excessive.

31. *Remember that timing is everything.* Timing can be essential to successful negotiations. For example, if your counterpart quits at 5:00 P.M., don't press for resolution at 10:00 A.M. or wait until 4:45 P.M. Make your proposal at 4:00 P.M. before it's time to go home but there is still time to settle the issue.

    When negotiating forward-priced change orders, include deadlines with all proposals. If these deadlines are not met, revise and resubmit your proposal with revised values. Otherwise the owner's negotiator will not respect your deadlines and you will end up negotiating prices under one condition and doing the work under another.

    *In some cases, you may not want to reveal your deadline:* unless it is a mutual deadline, or unless you have enough relative power to force a resolution by that deadline. If you have to go to their city to negotiate, appear to settle in for the long run. Rent an apartment (with a cancellation option) instead of a hotel room if appropriate. Make yourself comfortable and let them know you are. Bring along your spouse or arrange for your spouse to visit, etc. However, first ensure that they will negotiate continuously and not force you to travel back and forth.

    If timing is in your favor, don't rush into an agreement. If timing is in the owner's favor, be aggressive and keep pushing for resolution. Don't let the pace slow down but always have a proposal, alternative, or argument on the table.

32. *Be persistent and patient.* If you keep negotiating, you will eventually achieve your objectives (if they are reasonable). Do not become frustrated. Patience can be a powerful tool as your counterpart realizes that you are determined to win and that you will spend the time and effort required. The owner's representative will be more likely to settle after realizing you can't be easily dissuaded.

33. *Avoid negotiating by telephone.* Avoid negotiating by telephone, unless you are resisting the other party's proposal. It is easier to say no over the phone, but harder to convince someone when you cannot use body language and exhibits, or read their body language. Another drawback is that you cannot shake hands when you're reached agreement.

    If you must negotiate by telephone, don't allow it to begin with an unscheduled call. You would be at a disadvantage, without your files organized and not psychologically prepared. Tell the party you can't discuss the issue without having your files or being able to make notes and agree to call them back at a specified time. When negotiating by phone, use a checklist to ensure that you've covered all the issues and keep extensive notes. When finished, write up the key points of discussion and agreement and send a copy to your counterpart to avoid misunderstanding. If you misstate something or erroneously agree to an issue, call back immediately and correct your error. It is a lot easier to rescind an offer made in telephone negotiations than in face-to-face negotiations.

34. *Learn to negotiate with the U.S. government.* There are special procedures and laws governing the negotiation of change orders with federal agencies. For details, see the *Negotiating Construction Contract Modification (NCCM) Guide*[23], published by the U.S. Army Corps of Engineers for government negotiators.

35. *Experiment.* Try unfamiliar tactics to see if they work for you. If they don't work, try them again later under different circumstances, with a different negotiator, or when you are more experienced.

**The ending game.**   Completing negotiations successfully is the most difficult part of negotiations.

1. *Remember that timing is crucial.* Eighty percent of the concessions in negotiations are made in the last 20 percent of the time available to negotiate. People are usually more flexible during this time and more willing to compromise. They have invested a lot of effort in the negotiations and are anxious to achieve success and finish. Some negotiators use this time to spring last-minute surprises to force you into accepting something you wouldn't otherwise tolerate. To avoid that, try to identify all the issues early and address each one. Remember that your counterpart is also under pressure.

2. *Do not move too quickly to the bottom line.* Successful negotiation requires some give and take before trying to reach a bottom line. This is especially true when negotiating with people from other cultures.

3. *Don't immediately accept an initial offer.* If the other party makes an initial offer that satisfies your needs, don't accept it without careful thought. You can probably get more. If you immediately accept the offer, the owner may think that a better deal could have been negotiated, regret the offer and resent you in the belief that you have won an advantage. This will interfere with future negotiations. The owner also may suspect that there is something wrong with the agreement or may seek concessions on peripheral issues. Your negotiations must make the other party feel that they "won" something.

   Another reason for not accepting an initial offer is that you may unwittingly agree to a bad settlement. Before accepting an offer that's too good to be true, recheck your numbers to ensure that you haven't made a mistake.

   Don't appear greedy. Some negotiators start with their bottom line and they may become offended if it is rejected. If the proposal meets your needs and you're afraid a counterproposal may cause the owner to reconsider, ask for some time to consider or reluctantly concede, giving some reason for not asking for more.

4. *Caucus before making a decision.* If negotiating as a team, consult with your team members before accepting an offer or conceding a major point. This gives your concession more value, ensures that you don't overlook an important point, and allows you to raise the ante if necessary.

5. *Compromise only for a reason.* Freely concede errors and overly aggressive claim assertions. Do not, however, compromise points of substance without getting something in return. That will prevent your position being whittled down. When you do compromise, don't undervalue your concession, or your counterpart also will undervalue it.

6. *Make smaller-and-smaller concessions.* When making price concessions, give a reasonable initial concession, and then make smaller and smaller concessions as negotiations progress. This leads to an obvious end, as it tells your counterpart that you are approaching your final price. The opposite technique, making bigger offers in an attempt to settle, only encourages further negotiations.

7. *Clarify vague or assumed terms and conditions before settling.* Never assume anything; confirm all of your basic assumptions. Clarify any vague terminology used by your counterpart and any terms and conditions that you have assumed or considered to be details. These can kill a settlement, or can be used by one party to back out of a deal if they decide it isn't to their advantage. Quantify as much as possible. Don't rely on *reasonable* or other vague terminology that may mean something entirely different to the owner.

   Attempt to define all conditions that are assumed to be part of the agreement. For example, timely approval of the agreement can be specified to avoid owner delay.

8. *Present a signed agreement.* If the owner isn't a sophisticated buyer of construction and the designer isn't administering the contract, you can present a signed change order form with your latest offer for their signature. Use the standard change order request form displayed in Fig. 5.2, but with the optional change order agreement section of Fig. 5.3, or use the standard AIA change order form. If the price isn't too far from the owner's latest offer, the owner's representative may sign it just to avoid negotiations. This allows you to include minor details in the agreement that might not otherwise get included. Any future negotiations are then based on your proposed terms and conditions, which gives you an advantage.

**Countering win/lose tactics.**  Not everyone will use win/win tactics, and many of those who generally seek win/win solutions will sometimes use tactics that aren't very positive. You need to be aware of these tactics and know how to parry them.

1. *Bluffing.* Don't be surprised if your counterpart tries to bluff. Be ready to call a bluff without being confrontational. If you bluff, be prepared in case your bluff is called.

2. *Misinformation.* Some negotiators will leak misinformation that leads their counterpart to overconfidence, incorrect strategies, or a focus on "straw issues" instead of the important points. Some will go as far as disseminating deliberately misleading information. For example, they might leave their notes behind or allow their opponent to "accidentally" view them, in an effort to make the other party think they have found confidential information. The other party may try to take advantage of this apparent mistake and instead be taken advantage of themselves. Don't let yourself be drawn into such a trap.

3. *Deliberate errors.* Some negotiators need to put their "fingerprint" on their counterpart's proposal, or are just extremely meticulous. Their counterparts will sometimes make a few deliberate but unimportant errors in their proposal to give the negotiator something to do and thereby avoid hours of nit-picking. These errors can be in the mathematics, spelling, grammar, or content (as in a "straw" issue that is certain to be rejected). Once the error is found and corrected, both parties are ready to go on to the next issue.

    Deliberate errors can also deceive, or even defraud. An example of fraud would be deliberately changing the wording of a tentative agreement to favor a position. The change, which they blame on an honest error, may not be noticed when you sign the document. To prevent this from happening, carefully review all draft agreements. If a change is initially missed but discovered later, insist on correction.

4. *Playing dumb.* Some negotiators play dumb to elicit information. This is sometimes used when negotiating with someone from a different cultural or language group. An extreme form of playing dumb is deliberately creating confusion and misunderstanding the facts. This may be used by a party with a very weak position, to conceal the facts or to misdirect their opponent. On the other hand, be careful when explaining issues to avoid revealing too much.

5. *Claiming false constraints.* Some negotiators impose nonessential deadlines, to force an early settlement. Or they make escalating demands for more concessions, instead of compromising on their existing position. Call their bluff.

6. *Straw issues.* negotiators sometimes claim a number of "straw" issues, to conceal their hidden agenda. Or they ask for things that they eventually concede either to demonstrate a willingness to compromise or in exchange for more significant concessions from their opponent.

    The best defense against straw issues is to know your counterpart's agenda, both open and hidden. Carefully consider offers to trade concessions to ensure that you don't make a major compromise to obtain a minor concession.

7. *Blaming an absent party.* Negotiators may blame an absent party who has the information needed for a response, or who has set a policy that precludes accepting your proposal. Ignore the limitation, or state your confidence in your counterpart's ability to deal with the problem, and press on.

8. *Referring agreements for approval—a variation of the absent party tactic.* Some negotiators will claim that they must refer agreements to another individual or group for final approval after you think you have agreed to a final number. To avoid this, confirm their level of authority before entering negotiations. Or state that any changes by a higher authority would cause you to retract your commitments and reopen negotiations on all issues. To reinforce your firmness, note the specific points that you would renegotiate. Also appeal to their ego; say "don't they always agree with your recommendations?" If you can't avoid referral for final approval, get their commitment to strongly recommend the settlement and draw up and sign a written agreement that is "subject to final approval." That makes it psychologically more difficult for them to change an agreement. And keep in mind that you also can claim the need to consult with a higher authority.

9. *Whittling.* After apparent agreement, some negotiators will try to whittle you down by asking for small but significant concessions on a number of issues that you assumed were resolved. Whittling is often successful because most people become more confident after making a decision, and want to convince themselves that they negotiated a good deal. Paradoxically, this makes them open to giving a little more. For example, some retailers exploit this by negotiating a competitive price on the basic item and then proposing high-margin options.

   To avoid whittling, ensure that all of the ancillary items are included in the basic agreement. To stop whittling, ask for "extras" of your own or shame them with the comment that they really can't be serious since they already did so well in negotiations. Contractually, you don't have an agreement until the other party agrees. Their whittling becomes a counteroffer. If you have made an offer that isn't accepted, and feel that you are being whittled down from a reasonable offer, withdraw your offer and revert to either an earlier or a completely revised offer. You should offer a reason for the change.

   Always be willing to walk away from negotiations (for the moment), as you seldom need to get an issue resolved immediately.

10. *Nonnegotiable demands.* Some negotiators will make a take-it-or-leave-it offer or a nonnegotiable ultimatum in an effort to obtain their objectives, even when they really would accept less. This tactic may be counterproductive, except when the party making it enjoys a much stronger position. However, a nonnegotiable demand can sometimes break a deadlock; it may strengthen the negotiating team's resolve over other issues; and it may weaken the counterpart's position on certain issues.

   If a nonnegotiable demand is used against you, first defuse any emotional heat. Next discuss the demand in order to confirm that you truly understand their position and that they understand yours. Then ask to set it aside for the moment (promising to return to it later) and attempt to make progress on another issue. There are a number of tactics that you can later use to try to break through the demand. For example, you can make acquiescence to their demand contingent on their conceding a crucial point to you. Or you can whittle at their demand and try to strip out some of the elements. Persistence pays. Most important, however, is continued progress and successful negotiations on other issues. The more the other party has invested in negotiations, the less likely they are to walk away without resolution.

Another way to handle unreasonable demands is to simply tell your counterpart(s) that the demands are unreasonable and that there is no point in pursuing them further. If they know that is true, they may drop the demands at that point.

11. *Walkouts.* Walking out or threatening to walk out, after expressing dissatisfaction with negotiations, may be an honest expression of frustration. Or it may be a ploy to force concessions or to escape a losing argument.

   If the owner's negotiator hasn't yet walked, offer a face-saving opportunity to return to the table. Emphasize the progress you have made and the need to continue. Be positive. If you are in the wrong, apologize and attempt to restart negotiations. Change the subject; use humor to defuse the situation. Offer an alternative, even if it only rewords the proposal you just made. If the other party has already walked out, take your time; they are likely to offer to return to the table. If you walked out, regroup and return within a reasonable time.

12. *Good cop/bad cop performances.* Some owner negotiation teams will attempt to use this classic approach with one negotiator being unreasonable and the other "on your side" and "helping" negotiate the unreasonable team member down. The result will be what they planned when they started. If this happens, identify the tactic and shame them into a more reasonable approach. Tell them that they must settle on a single strategy before negotiating, or ignore the "bad cop" and negotiate only with the "good cop."

## 6.   Record discussions and document agreements

Take the time to make brief notes of your proposals, the owner's counterproposals, and any agreements. Ask your negotiation team members to take notes. Record nonverbal communication, the general atmosphere, and your impression of how the owner's team felt about the issues. Your notes may be invaluable in settling ancillary issues that arise later, such as whether certain points were included in the agreement. You can study your notes between sessions to help you decide how to proceed. In addition, maintain a current log of both parties' positions so that you don't agree to a compromise based on the wrong numbers.

Your records of negotiations should include:

- The date, times, and the names of those attending each negotiation.

- Topics discussed, the highlights, and as much detail as possible of the discussions.

- Important quotes by the other party that reveal their position on the issues.

- A summary of offers and counteroffers.

- Items for which there was a general consensus and specific agreements.
- Points of disagreement and each side's position.
- Calculations and other work papers.
- Copies of the agenda, visual aids, handouts, and meeting minutes.
- Copies of any signed agreements.

Use a laptop computer to take notes. It is much faster than handwriting; more can be recorded; a fast typist can record a lot more information, and the output is always readable.

When agreement is reached on an issue, write it up immediately and have both parties sign or initial it. Although preferable, it isn't necessary to define every detail. But as soon as possible, have a detailed agreement drawn up and signed.

## 7. Know how and when to "close the deal"

Closing is an art. Bringing the negotiations to an agreement requires that you close the deal. Too many negotiations just wander along, when they could be closed much earlier.

**Winning through persistence.** If no progress has been made and everyone is ready to quit, keep negotiating until something has been agreed. Insist on some progress on some issue. Then try to keep negotiating until all issues are settled. This approach is used by many mediators, who have found that physical exhaustion or hunger can dull the appetite for *justice, winning,* or even *"principle."*

**Don't drive too hard a bargain.** Don't get greedy. It will make future negotiations more difficult, and you may lose what agreement you've reached and have to start over.

**Splitting the difference.** The most frequent method of closing the final gap between offer and counteroffer is to split the difference. If you have carefully positioned yourself and insisted on reasonable compromises during negotiations, this may be a fair way to close negotiations. If not, you can decline and offer some alternative, but you will have to justify it. For example, you can simply state that you can't afford to give up that much. Or you can develop some quantifiable formula that lends credence to an unequal split. For example, if some of your costs are clearly out-of-pocket direct costs (e.g., a materials invoice), you can claim the full amount on those costs and split the difference on the remaining costs. Pleading inability to commit absent subcontractors can also help skew the split in your favor.

If a 50-50 split is unreasonable, enticing the other party to propose splitting the difference provides an opportunity to obtain a better settlement. Comment on how much effort you've both put into the negotiations, what a shame it would be to not settle now, and how close your positions are. If they offer to split the difference, the have just moved halfway toward your position. You can take that as their new position while you come down only a small amount. They can then settle based on your new relative positions.

**Tossing a coin.**    If their counterproposal is acceptable as a minimum, you can offer to toss a coin to decide whose figure will be used. This is preferable to an impasse, and it theoretically gives you half of the difference.

**Asking for a favor.**    If unable to reach an agreement and the difference isn't too large, explain that you just can't come down and ask them to come up, as a favor to you that you will *eventually* reciprocate. If you've built up a good relationship, this may suffice.

**Turn it over to the principals.**    If unable to reach resolution with all negotiation team members present, turn over final resolution to a principal (i.e., a senior manager) from each side. This removes project team members and the owner's representatives, who have become emotionally involved, and it reduces resolution to a business decision. It also puts pressure on the principals to perform. The result is almost always a quick resolution, which will be satisfactory if your principal is well prepared. As an example, I once spent 6 months of off-and-on negotiations over a $1.3 million acceleration claim against the U.S. Army Corps of Engineers. After having thoroughly examined the issues, my client (the owner of the construction company) and his counterpart from the Corps met and settled for $1 million in fifteen minutes.

Your principal, if well briefed and a good negotiator, will often do far better than you can do, especially if the owner's principal isn't as skilled or well prepared. This can be one of your most successful tactics.

**Help justify the settlement.**    After essentially reaching agreement, help your counterpart justify it to his or her supervisor.

## 8.   How to proceed if the claim is rejected or negotiations excessively delayed

If the owner doesn't meet within a reasonable period of time, won't continue with negotiations, fails to address the issues in negotiation, or refuses reasonable compromise—take one or more of the following actions:

- Change your bargaining position.
- Focus on the terms and conditions: schedule, quality, payment terms, etc.
- Eliminate contingencies for identifiable problems and agree that if problems do occur, they will be considered changes.
- Bring in new negotiators.
- Involve a third party (a mediator or project neutral) to help resolve one or more major issues.

- Request mediation. If refused or unsuccessful, demand arbitration or file for litigation, depending upon the contract language.

If the owner rejects your claim:

- Review the contract-specified procedures for appeals, etc.
- Gather additional information and prepare further submittals, if warranted.
- Seek assistance from an attorney or independent claims consultant.
- Appeal the rejection as specified in the contract, or bring in your next level of management to appeal directly to the next level of the owner's management.
- Request mediation and file for arbitration or litigation depending upon the contract language. That may break a logjam and apply some pressure to settle.
- File a lien, if warranted.

## 9.  Mediate

*If you are unable to resolve a dispute through negotiation, mediate before filing for arbitration or litigation.* Mediation is far more effective and much less expensive than arbitration or litigation. In some cases, however, it may be necessary to demand arbitration or file for litigation before the owner will get serious.

# Bibliography

1. *Survey of profitability of construction contractors: Municipal and utility contractors.* 1970, 1984. Washington, D.C.: Associated General Contractors of America.
2. *Construction industry annual financial survey.* 1995. Princeton, NJ: Construction Financial Management Association (CFMA).
3. Carringer, Michael. *Unfavorable contract terms.* Hensel Phelps Construction, SE District. [unpublished manuscript]
4. McManamy, Rob. 1994. Industry pounds away at disputes. *Engineering News Record,* 11 July 11: 24–27.
5. Vidogah, William, and Issaka Ndekugri. 1997. Improving management of claims: Contractors' perspective. *Journal of Management in Engineering,* Sept./Oct.: 37–44.
6. Schriener, J. 1991. Partnering paying off on projects. *Engineering News Record,* 14 Oct.: 26–27.
7. Pinnell, S., and Jeffrey S. Busch. 1994. *Dispute management programs: Partnering, claims management and dispute resolution.* Paper presented at the 25th Annual Project Management Institute Seminar/Symposium, 17–19 October, in Vancouver, British Columbia, Canada.
8. Godfrey, Kneeland, A., Jr. 1996. *Partnering in design and construction.* New York: McGraw-Hill.
9. Civitello, Andrew M., Jr. 1987. *Contractor's guide to change orders.* Englewood Cliffs, NJ: Prentice-Hall.
10. Denning, James. 1993. More than an underground success. *Civil Engineering Magazine,* December: 42–45.
11. Matyas, Robert M., A. A. Mathews, Robert J. Smith, and P. E. Sperry. 1996. *Construction dispute review board manual.* New York: McGraw-Hill.
12. Tittes, Pamela Raye. 1997. Managing "killer" clauses in construction contracts. *CFMA Building Profits Magazine,* Jan./Feb.: 24–26.
13. Association of Soil and Foundation Engineers. 1985. *A guide to forensic engineering and service as an expert witness.* Silver Springs, MD.
14. Martell, R., and D. Poretti. 1988. Using experts to "prove up" your construction case. *The Construction Lawyer,* 8(2): 11–19.
15. Schwartzkopf, William, John J. McNamara, and Julian F. Hoffar. *Calculating construction damages.* New York: Wiley Law Publications.
16. Pinnell, Steven S. 1992. Construction scheduling disputes: Proving entitlement. *The Construction Lawyer,* 12(1): 18–30.
17. Wickwire, Jon M., Thomas J. Driscoll, and Stephen B. Hurlbut. 1991. *Construction scheduling: Preparation, liability, and claims.* New York: Wiley Law Publications.
18. Toomey, Daniel E., and Mark R. Berry. 1995. The scheduling expert: A primer on preparing direct and cross. *The Construction Lawyer,* 15(2): 63–75.
19. Leary, Christopher P., and Barry B. Bramble. 1988. *Project delay: Schedule analysis modes and techniques.* Paper presented at the Project Management Seminar/Symposium, 17–21 September, in San Francisco: 63–68.
20. Driscoll, Thomas J. 1989. *Time impact analysis: A key for successful proof of delay.* Paper presented at the Fourth Annual Construction Litigation Superconference, 8 December, in San Francisco.
21. Hoshino, Kenji P. 1995. *Collapsed as-built cpm analysis.* Construction Management Resources. An unpublished paper.
22. Hewitt, Roger. 1988. *Winning contract disputes: Strategic planning for major litigation.* Sacramento, CA: Ernst & Young.
23. *Negotiating Construction Contract Modification (NCCM) Guide.* 1996. Huntsville, AL: U.S. Army Corps of Engineers, Huntsville Division, Directorate CE Training Management. Control #368.

24. Hughes, Frank, and Debera Masahos. 1997. Statutes permitting recovery of attorney's fees in construction cases. *The Construction Lawyer,* 17(4).
25. Parker, Henry W., and Clarkson H. Oglesby. 1972. *Methods improvement for construction managers.* New York: McGraw-Hill.
26. Borcherding, John D., and Luis F. Alarcon. 1991. Quantitative effects on construction productivity. *The Construction Lawyer,* 11(1): 1, 36–47.
27. Oglesby, Clarkson H., Henry W. Parker, and Gregory A. Howell. 1989. *Productivity improvement in construction.* New York: McGraw-Hill.
28. Halligan, David W., Laura A. Demsetz, James D. Brown, and Clark B. Pace. 1994. Action-response model and loss of productivity in construction. *Journal of Construction Engineering and Management,* March 120(1): 47–64.
29. *Modification impact evaluation guide,* ER 415-1-3 1979. Washington, D.C.: Department of the Army, Office of the Chief of Engineers.
30. Schwartzkopf, William. 1995. *Calculating lost labor productivity in construction claims.* New York: Wiley Law Publications.
31. Thomas, H. Randolph. 1992. Effects of scheduled overtime on labor productivity. *Journal of Construction Engineering and Management,* 118(1): 60–76.
32. American Association of Cost Engineers, 1973. Effect of scheduled overtime on construction projects. *Report Bulletin,* 15(5), *155-160*
33. Leonard, Charles. 1988. *The effects of change orders on productivity.* Thesis presented to Concordia University, Montreal, Quebec, Canada, in partial fulfillment of the requirements for the degree of Master of Engineering (Building).
34. Work efficiency decreases at abnormal temperatures. 1972. *CONSTRUCTOR Magazine,* May.
35. Koehn, Enno, and Gerald Brown. 1985. Climatic effects on construction. *Journal of Construction, Engineering and Management* June 3(2): 000.
36. Howell, Gregory, Alexander Laufer, and Glenn Ballard. 1993. Interaction between subcycles: One key to improved methods. *Journal of Construction Engineering and Management.* 119(4).
37. Strunk, William, Jr., and E. B. White. 1959. *Elements of style.* Toronto, Ontario, Canada: The Macmillan Company.
38. Tufte, Edward R. 1983. *The visual display of quantitative information.* Cheshire, Conn.: Graphics Press.
39. Fisher, Roger, William Ury, and Bruce Patton. 1991. *Getting to yes: Negotiating agreement without giving in.* New York: Penguin Books.
40. Allesandra, Tony, Michael J. O'Connor, and Janice Van Dyke. 1994. *People smarts: Bending the golden rule to give others what they want.* San Diego, CA: Pfeiffer & Company.
41. Ronco, William C., and Jean S. Ronco, 1996. *Partnering manual for design and construction.* New York: McGraw-Hill.
42. Dunham, Clarence W., and Robert D. Young. 1971. *Contracts, specifications, and law for engineers.* New York: McGraw-Hill.
43. Nash, Ralph C., Jr. 1989. *Government contract changes.* Washington D.C.: Federal Publications.
44. Sweet, Justin. 1997. *Sweet on contract law.* Chicago: American Bar Association.
45. Pinnell, Steven S. 1980. Critical path scheduling: An overview and a practical alternative. *Civil Engineering.* 50(7): 66-70
46. Pinnell, Steven S. 1981. Easier, better method of construction scheduling. *Pacific Builder & Eng. I lineer.* Feb.: 17-18.

# Index

A page number appearing in **boldface** type indicates that an illustration appears on that page.

## ABOUT THE AUTHOR

Steven S. Pinnell is an engineer and construction manager with over 25 years' experience in the management of engineering and construction projects. He is a principal of Pinnell/Busch, Inc., a firm based in Portland, Oregon, with a nationwide and international practice in project management services for the design and construction industry. Mr. Pinnell has lectured widely and authored numerous papers on project management, contract methods, contract negotiations, and construction dispute management.